Methods in Molecular Biology

Series Editor
John M. Walker
School of Life and Medical Sciences
University of Hertfordshire
Hatfield, Hertfordshire, AL10 9AB, UK

For further volumes:
http://www.springer.com/series/7651

Bacterial Cell Wall Homeostasis

Methods and Protocols

Edited by

Hee-Jeon Hong

Department of Biological and Medical Sciences, Faculty of Helath and Life Sciences,
Oxford Brookes University, Oxford, UK;
Department of Biochemistry, University of Cambridge, Cambridge, UK

Editor
Hee-Jeon Hong
Department of Biological and Medical Sciences
Faculty of Health and Life Sciences
Oxford Brookes University
Oxford, UK

Department of Biochemistry
University of Cambridge
Cambridge, UK

ISSN 1064-3745 ISSN 1940-6029 (electronic)
Methods in Molecular Biology
ISBN 978-1-4939-3674-8 ISBN 978-1-4939-3676-2 (eBook)
DOI 10.1007/978-1-4939-3676-2

Library of Congress Control Number: 2016939971

Printed on acid-free paper

This Humana Press imprint is published by Springer Nature
The registered company is Springer Science+Business Media LLC New York

Preface

We live in a bacterial world. Bacteria have inhabited this planet for a few billion years longer than humans, and they have made themselves at home in every corner of it. From oceans to deserts, from frozen glaciers to hydrothermal vents, few environments are apparently too hostile for them to occupy and exploit. This of course also includes our own body tissues, and those of the animals and plants we rely on for our nutrition. They are even, unless especially treated, in the water that we drink. We have never known a world without them, but for most of our history we were completely unaware of their existence. Only 350 years ago—after nearly 200,000 years of evolution in their presence—did human technology advance to produce an optical microscope and enable us to begin to appreciate the extent of their proliferation, and the impact they have upon our lives.

The bacterial cell wall is central to their successful lifestyle. It not only gives their cells shape and rigidity but also provides a physical barrier through which every transaction between the cell contents and the external environment must take place. This is a crucial role in a single-celled organism. The habitat of bacteria can be diverse, unpredictable, and often hostile, and a major function of the cell wall is to enable cells to persist through, or adapt and respond to, changes in this environment. The integrity of the cell wall is consequently essential for bacterial survival, and efficient homeostatic mechanisms have evolved for monitoring and maintaining it. Not surprisingly, finding chemicals which are capable of overwhelming these mechanisms to inflict lethal damage to the cell wall has also developed as an important method for killing the bacteria which cause us harm. Tremendous effort has therefore been expended on researching and understanding bacterial cell wall homeostasis, and this volume is intended to bring together the most widely used and important protocols currently being employed in this field. In Chapter 1 we see how modern microscopy techniques, and other biophysical methods, are being used to characterize the subcellular structure of the bacterial cell wall and to visualize some of the machinery responsible for its construction and maintenance. Chapter 2 considers the high-throughput approaches which can be used both to identify all the genes and proteins that participate in the correct functioning of an organism's cell wall and to characterize the genome-wide changes in gene expression occurring in response to cell wall stressors. Protocols for assaying individual gene products for specific cell wall functions or identifying chemicals with inhibitory activity against the cell wall are detailed in Chapters 3 and 4, while later chapters cover methods for analyzing the nonprotein components of the cell wall, and the increasing use of computational approaches for predicting and modeling cell wall-related functions and processes. It is our hope that this volume serves to emphasize the diversity of the research taking place into bacterial cell wall homeostasis, and how the integration of information from across multiple disciplines is going to be essential if a holistic understanding of this important process is to be obtained.

I express my gratitude to all the authors contributing to this volume and offer my special thanks to the series Editor, Professor John Walker, for the wonderful opportunity of editing this book.

Cambridge and Oxford, UK *Hee-Jeon Hong*

Contents

Contributors

LAURA ALVAREZ • *Laboratory for Molecular Infection Medicine Sweden, Department of Molecular Biology, Umeå Centre for Microbial Research, Umeå University, Umeå, Sweden*
EMMA S. ARGUIJO-HERNÁNDEZ • *Departamento de Genética y Biología Molecular, Centro de Investigación y de Estudios Avanzados del IPN (CINVESTAV-IPN), C.P., México, DF, Mexico*
XIAO-HUI BAI • *Hefei National Laboratory for Physical Sciences at the Microscale and School of Life Sciences, University of Science and Technology of China, Hefei, Anhui, People's Republic of China; College of Life and Environment Science, Huangshan University, Huangshan, Anhui, People's Republic of China*
ELISA BINDA • *Department of Biotechnology and Life Sciences, University of Insubria, Via Dunant, Varese, Italy; "The Protein Factory" Research Center, Politecnico of Milano, ICRM CNR Milano University of Insubria, Via Dunant, Varese, Italy*
BENJAMIN P. BRATTON • *Lewis-Sigler Institute for Integrative Genomics, Princeton University, Princeton, NJ, USA; Department of Molecular Biology, Princeton University, Princeton, NJ, USA*
MARCO A. CARBALLO-ONTIVEROS • *Departamento de Genética y Biología Molecular, Centro de Investigación y de Estudios Avanzados del IPN (CINVESTAV-IPN), C.P., México, DF, Mexico*
LÙCIA CARRANO • *Fondazione Istituto Insubrico Ricerca per la Vita (F.I.I.R.V.), Gerenzano, Italy*
FELIPE CAVA • *Laboratory for Molecular Infection Medicine Sweden, Department of Molecular Biology, Umeå Centre for Microbial Research, Umeå University, Umeå, Sweden*
YUXING CHEN • *Hefei National Laboratory for Physical Sciences at the Microscale and School of Life Sciences, University of Science and Technology of China, Hefei, Anhui, People's Republic of China*
STUART J. CORDWELL • *School of Molecular Bioscience, The University of Sydney, Sydney, NSW, Australia; Discipline of Pathology, School of Medical Sciences, The University of Sydney, Sydney, NSW, Australia; Charles Perkins Centre, The University of Sydney, Sydney, NSW, Australia*
GONÇALO COVAS • *Laboratory of Bacterial Cell Surfaces and Pathogenesis, Instituto de Tecnologia Química e Biológica António Xavier, Universidade Nova de Lisboa, Oeiras, Portugal*
VANINA DENGLER • *Department of Molecular and Cellular Biology, Harvard University, Cambridge, MA, USA*
JOSEPH P. DILLARD • *Department of Medical Microbiology and Immunology, University of Wisconsin-Madison, Madison, WI, USA*
ANDREW DUONG • *Department of Biology, McMaster University, Hamilton, ON, Canada; Department of Biochemistry and Biomedical Sciences, McMaster University, Hamilton, ON, Canada*

ALEXANDER J.F. EGAN • *Institute for Cell and Molecular Biosciences, The Centre for Bacterial Cell Biology, Newcastle University, Newcastle upon Tyne, UK*
MARIE A. ELLIOT • *Department of Biology, McMaster University, Hamilton, ON, Canada; Michael G. DeGroote Institute for Infectious Disease Research, McMaster University, Hamilton, ON, Canada*
SÉRGIO R. FILIPE • *Laboratory of Bacterial Cell Surfaces and Pathogenesis, Instituto de Tecnologia Química e Biológica António Xavier, Universidade Nova de Lisboa, Oeiras, Portugal; UCIBIO@REQUIMTE, Departamento de Ciências da Vida/ Faculdade de Ciências e Tecnologia, Universidade Nova de Lisboa, Caparica, Portugal*
SIMON J. FOSTER • *Krebs Institute, University of Sheffield, Sheffield, UK*
TATSUYA FUKUSHIMA • *Division of Gene Research, Department of Life Sciences, Research Center for Human and Environmental Sciences, Shinshu University, Ueda, Nagano, Japan*
JAMES C. GUMBART • *School of Physics, Georgia Institute of Technology, Atlanta, GA, USA*
GABRIELA HENRIQUES • *Laboratory of Bacterial Cell Biology, Instituto de Tecnologia Química e Biológica António Xavier, Universidade Nova de Lisboa, Oeiras, Portugal*
SARA B. HERNANDEZ • *Laboratory for Molecular Infection Medicine Sweden, Department of Molecular Biology, Umeå Centre for Microbial Research, Umeå University, Umeå, Sweden*
ANDY HESKETH • *Department of Biochemistry, Cambridge Systems Biology Centre, University of Cambridge, Cambridge, UK; Cambridge Systems Biology Centre, University of Cambridge, Cambridge, UK*
JAMIE K. HOBBS • *Krebs Institute, University of Sheffield, Sheffield, UK*
HEE-JEON HONG • *Department of Biological and Medical Sciences, Faculty of Health and Life Sciences, Oxford Brookes University, Oxford, UK; Department of Biochemistry, University of Cambridge, Cambridge, UK*
GRANT J. JENSEN • *California Institute of Technology and Howard Hughes Medical Institute, Pasadena, CA, USA*
YONG-LIANG JIANG • *Hefei National Laboratory for Physical Sciences at the Microscale and School of Life Sciences, University of Science and Technology of China, Hefei, Anhui, People's Republic of China*
LUIS KAMEYAMA • *Departamento de Genética y Biología Molecular, Centro de Investigación y de Estudios Avanzados del IPN (CINVESTAV-IPN), C.P., México, DF, Mexico*
KALINKA KOTEVA • *Department of Biochemistry and Biomedical Sciences, McMaster University, Hamilton, ON, Canada; Michael G. DeGroote Institute for Infectious Disease Research, McMaster University, Hamilton, ON, Canada*
JONATHAN D. LENZ • *Department of Medical Microbiology and Immunology, University of Wisconsin-Madison, Madison, WI, USA*
QIONG LI • *Hefei National Laboratory for Physical Sciences at the Microscale and School of Life Sciences, University of Science and Technology of China, Hefei, Anhui, People's Republic of China*
GIORGIA LETIZIA MARCONE • *Department of Biotechnology and Life Sciences, University of Insubria, Via Dunant, Varese, Italy; "The Protein Factory" Research Center, Politecnico of Milano, ICRM CNR Milano University of Insubria, Via Dunant, Varese, Italy*
FLAVIA MARINELLI • *Department of Biotechnology and Life Sciences, University of Insubria, Via Dunant, Varese, Italy; "The Protein Factory" Research Center, Politecnico of Milano, ICRM CNR Milano University of Insubria, Via Dunant, Varese, Italy*

EVA MARTÍNEZ-PEÑAFIEL • *Departamento de Genética y Biología Molecular, Centro de Investigación y de Estudios Avanzados del IPN (CINVESTAV-IPN), C.P., México, DF, Mexico*
ALESSANDRA M. MARTORANA • *Dipartimento di Biotecnologie e Bioscienze, Università di Milano-Bicocca, Milan, Italy*
PIERLUIGI MAURI • *Istituto di Tecnologie Biomediche, Consiglio Nazionale delle Ricerche, Milan, Italy*
NADINE MCCALLUM • *Marie Bashir Institute for Infectious Diseases and Biosecurity, University of Sydney, Westmead, NSW, Australia*
SARA MOTTA • *Istituto di Tecnologie Biomediche, Consiglio Nazionale delle Ricerche, Milan, Italy*
JEFFREY P. NGUYEN • *Lewis-Sigler Institute for Integrative Genomics, Princeton University, Princeton, NJ, USA; Department of Physics, Princeton University, Princeton, NJ, USA*
LAM T. NGUYEN • *California Institute of Technology and Howard Hughes Medical Institute, Pasadena, CA, USA*
MIGUEL A. DE PEDRO • *Centro de Biología Molecular "Severo Ochoa", Universidad Autónoma de Madrid-Consejo Superior de Investigaciones Científicas, Madrid, Spain*
MARIANA G. PINHO • *Laboratory of Bacterial Cell Biology, Instituto de Tecnologia Química e Biológica António Xavier, Universidade Nova de Lisboa, Oeiras, Portugal*
ALESSANDRA POLISSI • *Dipartimento di Biotecnologie e Bioscienze, Università di Milano-Bicocca, Milan, Italy*
MADELEINE RAMSTEDT • *Department of Chemistry, Umeå University, Umeå, Sweden*
MELISSA ELIZABETH REARDON-ROBINSON • *Department of Microbiology and Molecular Genetics, University of Texas Health Science Center, Houston, TX, USA*
RUTH REYES-CORTÉS • *Departamento de Genética y Biología Molecular, Centro de Investigación y de Estudios Avanzados del IPN (CINVESTAV-IPN) C.P., México, DF, Mexico*
RYAN E. SCHAUB • *Department of Medical Microbiology and Immunology, University of Wisconsin-Madison, Madison, WI, USA*
JUNICHI SEKIGUCHI • *Department of Applied Biology, Graduate School of Science and Technology, Shinshu University, Ueda, Nagano, Japan*
DANIELLE L. SEXTON • *Department of Biology, McMaster University, Hamilton, ON, Canada; Department of Biochemistry and Biomedical Sciences, McMaster University, Hamilton, ON, Canada*
JOSHUA W. SHAEVITZ • *Lewis-Sigler Institute for Integrative Genomics, Princeton University, Princeton, NJ, USA; Department of Physics, Princeton University, Princeton, NJ, USA; Department of Molecular Biology, Princeton University, Princeton, NJ, USA*
ANDREY SHCHUKAREV • *Department of Chemistry, Umeå University, Umeå, Sweden*
NESTOR SOLIS • *School of Molecular Bioscience, The University of Sydney, Sydney, NSW, Australia; Department of Oral Biological and Medical Sciences, Centre for Blood Research, University of British Columbia, Vancouver, BC, Canada*
PAOLA SPERANDEO • *Dipartimento di Biotecnologie e Bioscienze, Università di Milano-Bicocca, Milan, Italy*
HUNG TON-THAT • *Department of Microbiology and Molecular Genetics, University of Texas Health Science Center, Houston, TX, USA*
ROBERT D. TURNER • *Krebs Institute, University of Sheffield, Sheffield, UK*

FILIPA VAZ • *Laboratory of Bacterial Cell Surfaces and Pathogenesis, Instituto de Tecnologia Química e Biológica António Xavier, Universidade Nova de Lisboa, Oeiras, Portugal*
WALDEMAR VOLLMER • *Institute for Cell and Molecular Biosciences, The Centre for Bacterial Cell Biology, Newcastle University, Newcastle upon Tyne, UK*
CHENGGANG WU • *Department of Microbiology and Molecular Genetics, University of Texas Health Science Center, Houston, TX, USA*
JING-REN ZHANG • *Center for Infectious Disease Research, School of Medicine, Tsinghua University, Beijing, People's Republic of China*
CONG-ZHAO ZHOU • *Hefei National Laboratory for Physical Sciences at the Microscale and School of Life Sciences, University of Science and Technology of China, Hefei, Anhui, People's Republic of China*

Part I

Analyzing Physical Properties of the Bacterial Cell Wall

Chapter 1

Atomic Force Microscopy Analysis of Bacterial Cell Wall Peptidoglycan Architecture

Robert D. Turner, Jamie K. Hobbs, and Simon J. Foster

Abstract

Atomic force microscopy (AFM) has been used extensively to characterize the surface structure and mechanical properties of bacterial cells. Extraction of the cell wall peptidoglycan sacculus enables AFM analysis exclusively of peptidoglycan architecture and mechanical properties, unobscured by other cell wall components. This has led to discoveries of new architectural features within the cell wall, and new insights into the level of long range order in peptidoglycan (Turner et al. Mol Microbiol 91:862–874, 2014). Such information has great relevance to the development of models of bacterial growth and division, where peptidoglycan structure is frequently invoked as a means of guiding the activities of the proteins that execute these processes.

Key words Atomic force microscopy, Peptidoglycan, Bacteria, Cell wall

1 Introduction

AFM is a scanning probe microscopy technique in which a sharp tip with a point diameter of about 10 nm is moved back and forth over a sample, building up an image from the height profile of each sequentially scanned line. AFM is not diffraction limited and therefore yields very high resolution. Combined with an intrinsically good signal-to-noise ratio, AFM is an excellent technique for analysis of disordered surfaces on length scales from a few nanometres to a few micrometres. The capability to measure force additionally enables nanoscale mapping of mechanical properties. AFM has been applied to analyze both the cell walls of living bacteria and extracted peptidoglycan sacculi which are either left intact or broken open to reveal the inner surface [1, 2]. Here, we restrict ourselves to the techniques required for imaging of sacculi.

Sacculus imaging has now been successfully applied to *Escherichia coli*, *Pseudomonas aeruginosa*, *Campylobacter jejuni*, *Caulobacter crescentus*, *Bacillus subtilis*, *Staphylococcus aureus*, *Lactococcus lactis*, *Streptococcus pneumoniae*, and *Enterococcus faecalis* [3–7]. This

Hee-Jeon Hong (ed.), *Bacterial Cell Wall Homeostasis: Methods and Protocols*, Methods in Molecular Biology, vol. 1440,
DOI 10.1007/978-1-4939-3676-2_1, © Springer Science+Business Media New York 2016

protocol should therefore work for these and similar species. The protocol varies mainly between Gram-positive and Gram-negative species, but there are some species-specific variations. For a peptidoglycan sacculus purification protocol applicable to a very broad range of species, but which has not been tested by AFM, see [8].

The sample preparation has two phases. The first is extraction of sacculi from the organism of interest and the second is mounting the sample for AFM imaging. The AFM imaging should be carried out in an intermittent contact mode (e.g., tapping mode) in ambient conditions. Operation of the AFM itself is beyond the scope of this protocol and should be carried out in accordance with the instrument manual.

2 Materials

Follow all local safety and waste disposal procedures when following this protocol. Purification of Gram-positive peptidoglycan involves the use of hydrofluoric acid which can be particularly hazardous. It has been assumed that normal microbiology lab equipment is available to the reader, but more specialised items have been listed here.

2.1 Purification of Peptidoglycan Sacculi

1. Distilled water.
2. HPLC-grade water.
3. Phosphate-buffered saline (PBS): Make from tablets (Gibco 18912-0140) following the manufacturer's instructions. Autoclave.
4. Sodium dodecyl sulfate (SDS) solution (5 % w/v): Add 100 mL distilled water to 5 g SDS in a suitable container and mix. Store at room temperature.

2.1.1 Purification of Peptidoglycan Sacculi from Gram-Positive Bacteria

1. For breaking *B. subtilis*: French Press.
2. For breaking *S. aureus*: FastPrep (MP Biomedicals FastPrep-24), FastPrep tubes (MP Biomedicals Lysing Matrix B 6911-100).
3. Tris buffer (50 mM, pH 7): Add 3 g Tris base to 500 mL distilled water. Adjust to pH 7 using 1 M hydrochloric acid. Autoclave.
4. Pronase stock solution (20 mg/mL):

   (a) Make a Tris buffer (1 M, pH 7.5): Add 60.6 g Tris base to 500 mL distilled water. Adjust to pH 7.5 using 1 M hydrochloric acid.

   (b) Make a sodium chloride solution (4 M): Add 116.9 g sodium chloride to 500 mL distilled water.

   (c) Combine 50 mL distilled water, 0.5 mL Tris buffer (1 M, pH 7.5), 0.125 mL sodium chloride solution (4 M) and 1 g protease (Sigma P6911).

   (d) Incubate the solution for 1 h at 37 °C.

(e) Filter sterilize the solution using a 0.22 μm filter.

(f) Split into 1 mL aliquots and store at −20 °C until required.

5. Hydrofluoric acid (48 % v/v).
6. pH indicator strips.

2.1.2 Purification of Peptidoglycan Sacculi from Gram-Negative Bacteria

1. SDS solution (10% w/v): Add 100 mL distilled water to 10 g SDS in a suitable container and mix. Store at room temperature.
2. Sodium phosphate buffer (50 mM, pH 7.3):

(a) Add 1.2 g sodium dihydrogen phosphate to 200 mL distilled water and mix.

(b) Add 1.4 g disodium hydrogen phosphate to 200 mL distilled water and mix.

(c) While mixing, gradually add sodium dihydrogen phosphate solution to disodium hydrogen phosphate solution until the pH is 7.3.

(d) Autoclave.

3. Ultracentrifuge capable of reaching 400,000 × *g*.
4. Ultracentrifuge tubes.
5. α-Chymotrypsin stock solution (1 mg/mL): Suspend 10 mg α-chymotrypsin in 10 mL in sodium phosphate buffer (50 mM, pH 7). Store 1 mL aliquots at −20 °C until required.
6. Tip sonicator (if broken sacculi are required) (MSE Soniprep 150 with 3 mm diameter probe).

2.2 Mounting of Peptidoglycan Sacculi for AFM Imaging

1. Steel pucks.
2. Mica discs.
3. Epoxy glue.
4. Tip sonicator (Gram-positive bacteria) (MSE Soniprep 150 with 3 mm diameter probe).
5. Scotch tape.

3 Methods

3.1 Purification of Peptidoglycan Sacculi from Gram-Positive Bacteria

1. Grow one liter of liquid bacterial culture to an OD_{600} of approximately 0.5.
2. Chill on ice.
3. Centrifuge at 15,950 × *g* for 10 min.
4. Resuspend in 1 mL PBS. This will result in more than 1 mL of bacterial suspension.
5. Split the suspension equally between four 1.5 mL Eppendorf tubes.

6. Boil in a water bath for 10 min (kills bacteria).
7. Allow bacterial suspension to cool to room temperature.
8. Follow "Breaking *S. aureus*" or "Breaking *B. subtilis*" sub-protocols, if desired.
9. Centrifuge for 3 min at 20,000 × *g* and discard supernatant.
10. Resuspend pellets in 1 mL 5 % (w/v) SDS (*see* **Note 1**).
11. Boil for 25 min.
12. Centrifuge for 3 min at 20,000 × *g* and discard supernatant.
13. Resuspend pellets in 1 mL 5 % (w/v) SDS.
14. Boil for 15 min (*see* **Note 2**).
15. Centrifuge for 3 min at 20,000 × *g* and discard supernatant. Repeat this five times, resuspending the pellet in distilled water each time.
16. Resuspend pellets in 0.9 mL Tris–HCl (50 mM, pH 7) and add 0.1 mL pronase stock solution. Incubate the sample at 60 °C for 90 min.
17. Centrifuge for 3 min at 20,000 × *g* and discard supernatant. Resuspend in distilled water, then centrifuge again using the same conditions, and discard supernatant.
18. Carefully observing local safety and waste disposal procedures, resuspend the pellets in 250 μL hydrofluoric acid (HF).
19. Incubate for 48 h at 4 °C. This removes teichoic acids.
20. Centrifuge for 3 min at 20,000 × *g* and safely discard supernatant. Repeat this, resuspending the pellet in distilled water each time, until the pH of the supernatant is 5 as measured using an indicator strip.
21. Resuspend in a minimal quantity of HPLC-grade water (about 100 μL).
22. Store at −20 °C until further use.

3.2 Purification of Peptidoglycan Sacculi from Gram-Negative Bacteria

1. Grow 1 L of liquid bacterial culture to an OD_{600} of approximately 0.5.
2. Chill on ice.
3. Centrifuge at 15,950 × *g* for 10 min.
4. Resuspend in 1 mL PBS (or distilled water for *C. crescentus*). This will result in more than 1 mL of bacterial suspension.
5. Follow "Breaking Gram-negative bacteria" sub-protocol, if desired.
6. Heat 3 mL 5 % (w/v) SDS to 100 °C in a 50 mL Falcon tube using a dry heat block.

7. Add bacterial suspension to this dropwise.
8. Leave to boil for 30 min.
9. Transfer bacterial suspension to an ultracentrifuge tube.
10. Collect pellet by ultracentrifugation at 400,000 × *g* for 15 min at room temperature.
11. Resuspend pellet in distilled water using a fine-tipped plastic Pasteur pipette (*see* **Note 1**) and repeat ultracentrifugation four times.
12. Resuspend in 3.6 mL sodium phosphate buffer (50 mM, pH 7.3). Add 0.4 mL α-chymotrypsin stock solution. Incubate overnight at 37 °C, with agitation on an orbital shaker.
13. Mix with 4 mL 10 % (w/v) SDS in a 50 mL Falcon tube.
14. Boil for 30 min using a dry heat block.
15. Collect pellet by ultracentrifugation at 400,000 × *g* for 15 min at room temperature.
16. Resuspend pellet in distilled water and repeat ultracentrifugation twice.
17. Resuspend in a minimal quantity of HPLC-grade water (about 100 μL).
18. Transfer to Eppendorf tubes.
19. Boil tubes for 10 min in a water bath.
20. Store at 4 °C until further use.

3.3 Breaking *S. aureus*

Samples should be kept as close to 4 °C as possible throughout these steps.

1. Transfer from 1.5 mL Eppendorfs to FastPrep tubes.
2. FastPrep treat the samples 6× at speed 6 for 30 s with a ~1-min pause between each run. Check for breakage by optical microscopy using a 100× oil immersion objective (*see* **Note 3**). If the cells have not broken, keep repeating the FastPrep cycles until >95 % breakage is observed by optical microscopy.
3. Spin down for 30 s at 1000 rpm, to separate beads (pellet) from the supernatant (cell extracts).
4. Transfer supernatant to Eppendorfs.

3.4 Breaking *B. subtilis*

Samples should be kept as close to 4 °C as possible throughout these steps.

1. Pool samples in a 50 mL Falcon tube.
2. Run through French Press at 500 psi.
3. Check for breakage by optical microscopy using a 100× oil immersion objective. If the cells are not broken, keep repeating

French Press step until >95 % breakage is observed by optical microscopy.

4. Transfer sample to Eppendorfs.

3.5 Breaking Gram-Negative Bacteria

Samples should be kept as close to 4 °C as possible throughout these steps.

1. Transfer sample to a 20 mL plastic universal tube.
2. Top up to approximately 3 mL with PBS.
3. Dip sonicator probe about 1.5 cm into the sample and switch on the sonicator for 30 s at an amplitude of 5 μm.
4. Check sample by optical microscopy using a 100× oil immersion objective. If the cells are not broken, keep repeating the sonication step until >95 % breakage is observed by optical microscopy.

3.6 Preparation of Steel-Mica Stubs for AFM

1. Use a minimal amount of epoxy to stick mica discs to steel stubs and leave to cure. These can be stored indefinitely.
2. Immediately before use, apply scotch tape sticky side down upon the mica surface and pull away to remove the top layer of mica (*see* **Note 4**).

3.7 Preparation of Sacculus Suspensions (Gram-Positive)

1. Dilute peptidoglycan stock in HPLC-grade water, e.g., 5 μL peptidoglycan stock to 400 μL HPLC-grade water (*see* **Note 5**).
2. Use a tip sonicator to disperse the sacculus suspension (e.g., dip the sonicator probe about 0.5 cm into the sample and activate the sonicator for 3 × 30 s bursts at 5 μm amplitude for *S. aureus* or 1 × 20 s burst at 5 μm amplitude for *B. subtilis*).

3.8 Preparation of Sacculus Suspensions (Gram-Negative)

1. Dilute peptidoglycan stock in HPLC-grade water, e.g., 1 μL peptidoglycan stock to 50 μL HPLC-grade water (*see* **Note 5**).
2. Mix briefly with a vortex mixer.

3.9 Mounting of Peptidoglycan Sacculi for AFM Imaging

1. Pipette 2 μL of sacculus suspension onto a freshly cleaved mica stub (*see* **Note 6**).
2. Direct a gentle flow of nitrogen gas onto the stub until the water has evaporated (*see* **Note 6**).
3. Using two pipettes, load 50 μL of HPLC-grade water into one leaving the other with an empty tip. Pipette the water onto the mica (*see* **Note 6**), then aspirate with the empty pipette, and discard the aspirated solution. Repeat this process for a total of three times.
4. Direct a vigorous flow of nitrogen onto the stub to "blow off" the remaining solution (*see* **Note 7**). Ensure that the sample is completely dry.

4 Notes

1. Resuspending in SDS and washing SDS out results in a lot of foam. Take care that this does not overspill the Eppendorf.
2. The sample can be stored overnight at this point.
3. Take great care not to transfer the glass FastPrep beads onto the microscope slide as this will make it difficult to apply a cover slip. If beads do end up on the slide, use the cover slip to push them out of the way.
4. Ensure that a complete layer has been removed as parts of the surface that had previously been exposed are contaminated. Stubs can be reused until there is no mica left.
5. You may need to try several different dilution factors to find an appropriate working concentration.
6. Do not let the drop overspill the mica as it will wash contaminants from the steel stub into your sample. If the drop does overspill, dry off the mica with a tissue and cleave it again before starting over.
7. At this point try and blow the liquid clean off the mica.

References

1. Turner RD, Vollmer W, Foster SJ (2014) Different walls for rods and balls: the diversity of peptidoglycan. Mol Microbiol 91:862–874
2. Dufrêne YF (2014) Atomic force microscopy in microbiology: new structural and functional insights into the microbial cell surface. MBio 5:e01363–14
3. Turner RD, Hurd AF, Cadby A et al (2013) Cell wall elongation mode in Gram-negative bacteria is determined by peptidoglycan architecture. Nat Commun 4:1496
4. Yao X, Jericho M, Pink D, Beveridge T (1999) Thickness and elasticity of gram-negative murein sacculi measured by atomic force microscopy. J Bacteriol 181:6865–6875
5. Hayhurst EJ, Kailas L, Hobbs JK, Foster SJ (2008) Cell wall peptidoglycan architecture in *Bacillus subtilis*. Proc Natl Acad Sci U S A 105: 14603–14608
6. Wheeler R, Mesnage S, Boneca IG et al (2011) Super-resolution microscopy reveals cell wall dynamics and peptidoglycan architecture in ovococcal bacteria. Mol Microbiol 82:1096–1109
7. Turner RD, Ratcliffe EC, Wheeler R et al (2010) Peptidoglycan architecture can specify division planes in *Staphylococcus aureus*. Nat Commun 1:26
8. Wheeler R, Veyrier F, Werts C, Boneca IG (2014) Peptidoglycan and nod receptor. Glycoscience: biology and medicine. Springer, Japan, pp 1–10

Chapter 2

Ultra-Sensitive, High-Resolution Liquid Chromatography Methods for the High-Throughput Quantitative Analysis of Bacterial Cell Wall Chemistry and Structure

Laura Alvarez, Sara B. Hernandez, Miguel A. de Pedro, and Felipe Cava

Abstract

High-performance liquid chromatography (HPLC) analysis has been critical for determining the structural and chemical complexity of the cell wall. However this method is very time consuming in terms of sample preparation and chromatographic separation. Here we describe (1) optimized methods for peptidoglycan isolation from both Gram-negative and Gram-positive bacteria that dramatically reduce the sample preparation time, and (2) the application of the fast and highly efficient ultra-performance liquid chromatography (UPLC) technology to muropeptide separation and quantification. The advances in both analytical instrumentation and stationary-phase chemistry have allowed for evolved protocols which cut run time from hours (2–3 h) to minutes (10–20 min), and sample demands by at least one order of magnitude. Furthermore, development of methods based on organic solvents permits in-line mass spectrometry (MS) of the UPLC-resolved muropeptides. Application of these technologies to high-throughput analysis will expedite the better understanding of the cell wall biology.

Key words UPLC, HPLC, Reverse-phase liquid chromatography, Cell wall, Peptidoglycan, Muropeptide

1 Introduction

The interest on the biology of bacterial cell wall peptidoglycan (PG) is rising fast since the realization that, in addition to its structural role, this unique macromolecule plays critical roles in the interactions between bacteria and their environment, including other living organisms. The bacterial cell wall is a covalently closed polymeric macromolecule which is subjected to a number of modifications related with the physiological state of the cells and the environmental conditions. The canonical monomeric subunit consists of the disaccharide pentapeptide GlcNAc-(β1-4)-MurNAc-L-Ala-D-Glu-(γ)-(di-amino acid)-D-Ala-D-Ala, where meso-diaminopimelic acid and L-lysine are the more frequent di-amino acids [1]. Monomers are converted into linear polymers by means of

Hee-Jeon Hong (ed.), *Bacterial Cell Wall Homeostasis: Methods and Protocols*, Methods in Molecular Biology, vol. 1440,
DOI 10.1007/978-1-4939-3676-2_2, © Springer Science+Business Media New York 2016

MurNAc-(β1-4)-GlcNAc glycosidic bonds, and then linear polymers are covalently linked by means of peptide bridges between the peptide moieties. The final result is a net-like macromolecule which encloses the cell body. Further metabolic activities result in a series of modifications in the chemical nature of PG subunits, and on the relative proportions of the different subunits [1]. A detailed knowledge of the subunit composition under particular conditions is often relevant to understand to what extent PG variations influence bacterial adaptation to environmental challenges, resistance to antibacterial agents, immune-modulatory activity, and toxin release and signaling [2, 3].

The strategy to obtain compositional and structural information of the bacterial PG is based on the insolubility of this molecule in boiling SDS, and on the availability of specific enzymes (muramidases or lysozymes) able to split the MurNAc-(β1-4)-GlcNAc glycosidic bonds which hold the structure together. The first property permits a straightforward way to obtain fractions highly enriched in PG and the second provides a reliable way to disassemble the PG into individual subunits. A far more formidable task is to devise sensitive and reliable methods appropriate for the resolution, identification, and quantification of the different subunits, over 40 in *Escherichia coli*. The high-performance liquid chromatography (HPLC) method, originally devised by Glauner et al. in 1988 [4], was a breakthrough which revealed a completely unforeseen complexity in PG structure, and provided a reliable and sensitive analytical tool. This method has been in use for more than 25 years essentially unchanged, in spite of the dramatic improvements in instrumentation, and HPLC column materials. However, this method suffered from three critical limitations: the requirement for inorganic buffers, which complicated identification of subunits by mass spectrometry (MS), and prevented the use of in-line MS-spectrometers; the very low sample turnover (3 days for sample preparation and ca. 3 h of HPLC run time per sample); and a requirement for relatively large sample amounts (>200 μg PG/sample). During the last years, we have been working out new methods to circumvent these limitations trying to make the best of the new instrumentation, in particular the introduction of UPLC technology and the superb properties of new materials for reverse-phase chromatography. UPLC technology allows the use of new stationary phases with a very small particle size (in the range of 2 μm) that withstand very high pressures, increasing resolution, speed, and sensitivity. This together with improved detectors with high sampling rates and low-volume sample injectors has led to the use of smaller sample volumes (1–10 μL) and shorter run times (5–20 min), which are essential requisites for high-throughput analysis [5].

Here we present protocols which cut down sample preparation and run times dramatically, are MS-compatible, and require about one-tenth the amount of sample. We describe sacculi isolation from bacterial cultures, both Gram-negative and Gram-positive peptidoglycan purification, muramidase digestion, and sample preparation for LC chromatography. We describe two LC methods for the UPLC: inorganic method, which uses phosphate buffer as mobile phase, and organic method, which uses organic solvents as mobile phase and is MS compatible. The methods described can be easily adapted for the more frequently available UPLC machines by anyone with a basic knowledge of UPLC techniques. We finally provide some general instructions on data processing.

2 Materials

Prepare all solutions with fresh MilliQ water. Use only ultrapure water from a distillation or deionization unit with a resistance of 8 MΩ/cm at 25 °C and analytical grade reagents. pH is critical in most solutions used (*see* **Note 1**). Solutions are stored at room temperature unless otherwise indicated.

As several reagents are toxic or harmful, part of the work must be carried out in a fume hood, e.g., use of HF for Gram-positive bacteria muropeptide isolation or preparation of mobile phases. Consult the product safety information and material safety data sheets of chemicals and dispose of the products conveniently.

2.1 Sacculi Preparation

1. Phosphate buffer saline (PBS 1×): Dissolve 8 g NaCl, 0.2 g KCl, 1.44 g Na_2HPO_4, and 0.24 g KH_2PO_4 in 800 mL MilliQ water. Adjust to pH 7.4 using concentrated HCl. Adjust volume to 1 L with MilliQ water. Sterilize by autoclaving.
2. Lysis buffer: SDS 5 % (w/v) in MilliQ water.
3. 12–50 mL tubes (*see* **Note 2**).
4. Hot plate stirrer and magnets (*see* **Note 3**).

2.2 Gram-Negative Bacteria Peptidoglycan Purification

1. Benchtop ultracentrifuge, TLA-100.3 rotor, and ultracentrifuge tubes (3 mL polycarbonate tubes).
2. MilliQ water.
3. 2 mL Eppendorf tubes.
4. 1 mg/mL Pronase E stock: Dissolve 10 mg Pronase E (EC 3.4.24.31) in 10 mL Tris–HCl 10 mM pH 7.5 NaCl 0.06 % (w/v). Prepare 1 mL aliquots and store at −20 °C. Pronase E must be activated prior to digestion by incubating for 30 min–1 h at 60 °C (*see* **Note 4**).
5. SDS 10 % (w/v) solution in MilliQ water.

2.3 Gram-Positive Bacteria Peptidoglycan Purification

1. Benchtop ultracentrifuge, TLA-100.3 rotor, and ultracentrifuge tubes (3 mL polypropylene tubes) (*see* **Note 5**).
2. MilliQ water.
3. 1.5 and 2 mL Eppendorf tubes.

4. 10–15 mL tubes (*see* **Note 6**).
5. Glass beads (diameter 0.1 mm) and vortex.
6. Lyophilizer (or SpeedVac concentrator).
7. SDS 10% (w/v) solution in MilliQ water.
8. 100 mM Tris–HCl pH 7.5: Dissolve 12.11 g Tris in 900 mL MilliQ water. Adjust to pH 7.5 using concentrated HCl. Adjust volume to 1 L with MilliQ water.
9. 10 mg/mL α-Amylase stock: Dissolve 100 mg α-amylase (Sigma-Aldrich) in 10 mL MilliQ water. Prepare 500 μL aliquots and store at −20 °C.
10. 1 M $MgSO_4$: Dissolve 12 g $MgSO_4$ in 80 mL MilliQ water. Adjust volume to 100 mL with MilliQ water.
11. Nuclease mix: 100 μg/mL DNase I (EC 3.1.21.1) and 500 μg/mL RNase A (EC 3.1.27.5). Mix enzymes in 1 mL MilliQ water, prepare 200 μL aliquots, and store at −20 °C.
12. 2 mg/mL Trypsin stock: Dissolve 20 mg trypsin (EC 3.4.21.4) in 10 mL MilliQ water. Mix and prepare 1 mL aliquots. Store at −20 °C.
13. 50 mM $CaCl_2$: Dissolve 555 mg $CaCl_2$ in 80 mL MilliQ water. Adjust volume to 100 mL with MilliQ water.
14. 8 M LiCl: Dissolve 33.9 g LiCl in 80 mL MilliQ water. Adjust volume to 100 mL with MilliQ water.
15. 100 mM EDTA pH 7.0: Dissolve 2.9 g EDTA in 80 mL MilliQ water. Stir vigorously on a magnetic stirrer. Adjust pH to 7.0 with concentrated NaOH and adjust volume to 100 mL with MilliQ water.
16. Acetone.
17. Hydrofluoric acid 48% (v/v) (HF).

2.4 Muramidase Digestion

1. 1.5 mL Eppendorf tubes.
2. Digestion buffer: 50 mM Phosphate buffer pH 4.9. Dissolve 682.6 mg NaH_2PO_4 and 14.3 mg $Na_2HPO_4 \cdot 7H_2O$ in 80 mL MilliQ water and stir until completely dissolved. Adjust pH to 4.9 with diluted orthophosphoric acid (*see* **Note 7**). Adjust volume to 100 mL with MilliQ water.
3. 1 mg/mL Muramidase stock: Dissolve 10 mg Cellosyl (EC 3.2.1.17) in 10 mL digestion buffer (phosphate buffer 50 mM pH 4.9). Vortex thoroughly until completely dissolved and prepare 1 mL aliquots. Store at −20 °C.

2.5 Sample Reduction and Filtration

1. 1.5 mL Eppendorf tubes and long glass tubes.
2. Borate buffer: Borate buffer 0.5 M pH 9.0: Dissolve 3.1 g boric acid in 80 mL MilliQ water and stir until completely

dissolved. Adjust pH to 9.0 with concentrated NaOH. Adjust volume to 100 mL with MilliQ water.

3. Freshly prepared 2 M $NaBH_4$ solution: Dissolve 76 mg $NaBH_4$ in 1 mL MilliQ water (*see* **Note 8**).
4. Orthophosphoric acid 25 % (v/v).
5. pH-indicator strips: pH range 5.0–10.0 and pH range 0.0–6.0.
6. 96-Well filter plate (regenerated cellulose filter, 0.2 μm pore size), multititer 96-well plates (350 μL, V-bottom (*see* **Note 9**), and pierceable adhesive seal (*see* **Note 10**).
7. Vacuum manifold and pump.

2.6 UPLC Separation

1. Acquity UPLC system (Waters).
2. Analytical columns: Kinetex C18 UPLC Column 1.7 μm particle size, 100 Å pore size, 150 × 2.1 mm (Phenomenex).
3. Precolumn filters or guard columns: Security guard ultra cartridges C18 for 2.1 mm ID columns and holder (Phenomenex).
4. Inorganic buffer A: 50 mM Phosphate buffer, pH 4.35. Prepare a 10× buffer A solution by dissolving 40 g NaOH in 1800 mL MilliQ water. Adjust pH to 4.35 with orthophosphoric acid (*see* **Note 11**). Adjust volume to 2 L with MilliQ water and add 2 mL sodium azide 2 % (w/v). Keep stock solution at room temperature. To prepare the working solution, mix 200 mL 10× buffer A with 1800 mL MilliQ water and adjust pH to 4.35 (if necessary, carefully adjust pH with orthophosphoric acid 25 % (v/v)). Filter buffer with a filter membrane (nylon or nitrocellulose, 0.45 μm pore size) and a vacuum pump.
5. Inorganic buffer B: 50 mM Phosphate buffer, pH 4.95, 15 % methanol (v/v). Prepare a 10× buffer B solution by dissolving 60 g NaOH in 1800 mL MilliQ water. Adjust pH to 4.95 with orthophosphoric acid (*see* **Note 12**). Adjust volume to 2 L with MilliQ water. Keep stock solution at room temperature. To prepare the working solution, mix 170 mL 10× buffer B with 1530 mL MilliQ water and adjust pH to 4.95 (if necessary, carefully adjust pH with orthophosphoric acid 25 % (v/v)). Add 300 mL HPLC-grade methanol and mix. Filter buffer as described above.
6. Organic buffer A: Formic acid 0.1 % (v/v). Dilute 2 mL HPLC-grade formic acid in 2 L MilliQ water and mix.
7. Organic buffer B: Formic acid 0.1 % (v/v), acetonitrile 40 % (v/v). Dilute 2 mL HPLC-grade formic acid and 800 mL acetonitrile in 2 L MilliQ water and mix.

3 Methods

3.1 Sacculi Preparation

1. Grow cultures to the desired optical density in the appropriate culture medium for the bacteria (*see* **Note 13**).
2. Harvest the cells at 3000 g for 15 min and resuspend the pellet in 1.5 mL PBS 1× or its own media (*see* **Note 14**).
3. Transfer the sample to tubes and place a stirring magnet (*see* **Notes 2** and **3**). Add 1.5 mL boiling lysis solution and place tubes on a beaker with boiling water on a magnetic hot stirrer plate (*see* **Note 15**). Let the samples boil for 30 min–3 h (*see* **Note 16**). Finally, switch off the hot plate and let the lysate stir overnight (*see* **Note 17**).

3.2 Gram-Negative Bacteria Peptidoglycan Isolation

1. Wash sacculi by spinning down the samples for 10 min at 20 °C and 150,000 × *g* using 3 mL polycarbonate ultracentrifuge tubes (*see* **Note 18**). Fill the tubes with 3 mL lysate. Make sure that the rotor is properly balanced. After centrifugation, all soluble compounds will remain in the supernatant. Carefully discard it with a vacuum pump without removing the pellet (*see* **Note 19**). Resuspend the pellet in 900 μL MilliQ water and check for the presence of SDS (*see* **Note 20**). If needed, add 2 mL MilliQ water, mix, and centrifuge again. Repeat this wash step until SDS is completely removed (*see* **Note 21**).
2. Activate Pronase E 1 mg/mL by incubation for 30 min–1 h in a 56–60 °C water bath (*see* **Note 22**).
3. Transfer the resuspended pellet from Subheading 3.2, **step 1** (~900 μL), to 2 mL Eppendorf tubes (*see* **Note 23**). Add 100 μL activated Pronase E 1 mg/mL. Incubate samples at 56–60 °C during 1 h (*see* **Note 24**). Stop the reaction by adding 110 μL SDS 10% (w/v) and boil for 5 min in water (*see* **Note 25**).
4. Let the sample cool down before transferring to the ultracentrifuge tubes. Wash Pronase E-digested sacculi by adding 2 mL MilliQ water. Mix and ultracentrifuge as described above. Carefully discard the supernatant and, after total removal of SDS, resuspend the pellet in 100 μL MilliQ water or digestion buffer (*see* **Note 26**).

3.3 Gram-Positive Bacteria Peptidoglycan Isolation

1. Once the samples are cooled down to room temperature, concentrate the sacculi by ultracentrifugation during 10 min at 20 °C and 150,000 × *g* using 3 mL polypropylene ultracentrifuge tubes (*see* **Note 18**). Fill the tubes with 3 mL lysate. Make sure that the rotor is properly balanced. After centrifugation, soluble compounds will remain in the supernatant. Carefully discard it with a vacuum pump without removing the pellet (*see* **Note 27**). Resuspend the pellet in 500 μL

MilliQ water and check for the presence of SDS (*see* **Note 20**). If needed, add 2 mL MilliQ water, mix, and centrifuge again. Repeat this wash step until SDS is completely removed (*see* **Note 28**).

2. Transfer the resuspended pellet from Subheading 3.3, **step 1** (~500 μL), to 2 mL Eppendorf tubes. Add 200 mg glass beads and break the cells vigorously vortexing at 4 °C during 15 min (*see* **Note 29**).
3. Leave the tubes stand for 1 min on the bench or make a short spin at 2000 × *g* to allow the glass beads and unbroken cells to precipitate to the bottom of the tube (*see* **Note 30**) and carefully pipette the supernatant into an ultracentrifuge tube (*see* **Note 31**).
4. For maximum sample recovery, add 500 μL MilliQ water to the glass beads and unbroken cell pellet and repeat **steps 2** and **3**. Mix the recovered sample in the ultracentrifuge tube used before (final volume of sample ~1 mL).
5. To concentrate the sacculi, add 2 mL MilliQ water, mix thoroughly, and ultracentrifuge during 10 min at 20 °C and 150,000 × *g*.
6. Resuspend the pellet in 1 mL 100 mM Tris–HCl, and transfer it to a 15 mL tube (*see* **Note 6**). Add 10 μL 10 mg/mL α-amylase and incubate for 2 h at 37 °C with vigorous shaking (*see* **Note 32**).
7. For nucleic acid degradation, add 1 mL 100 mM Tris–HCl, 40 μL 1 M $MgSO_4$, and 2 μL nuclease mix. Incubate the samples for 2 h at 37 °C with shaking.
8. Treat the sample with trypsin by adding 100 μL 2 mg/mL stock solution and 50 μL 50 mM $CaCl_2$. Incubate for 16 h at 37 °C with magnetic stirring (*see* **Note 33**). Inactivate the digestion by adding 200 μL SDS 10 % (w/v) and boil the samples for 10 min.
9. Transfer the samples to ultracentrifuge tubes and centrifuge during 10 min at 20 °C and 150,000 × *g*. Remove the supernatant and wash the insoluble material as described in Subheading 3.3, **step 1**, until total removal of the SDS.
10. Resuspend the SDS-free pellet in 1 mL 8 M LiCl and incubate for 10 min at 37 °C (*see* **Notes 34** and **35**). Add 2 mL MilliQ water and ultracentrifuge the sample during 10 min at 20 °C and 150,000 × *g*. Carefully remove the supernatant.
11. Resuspend the pellet in 1 mL 100 mM EDTA and incubate for 10 min at 37 °C (*see* **Notes 34** and **36**). To remove EDTA, add 2 mL water and ultracentrifuge the samples during 10 min at 20 °C and 150,000 × *g*.

12. Remove the supernatant and wash the pellet with 1 mL acetone (*see* **Note 37**). Add 2 mL MilliQ water and ultracentrifuge during 10 min at 20 °C and 150,000 × *g*. Carefully remove the supernatant (*see* **Note 38**). To completely remove acetone, resuspend the pellet in 1 mL MilliQ water, increase volume up to 3 mL with MilliQ water, and ultracentrifuge again.
13. Resuspend the pellet in 500 μL–1 mL MilliQ water, transfer the sample to Eppendorf tubes, and dry them using a lyophilizer (*see* **Note 39**).
14. To remove the teichoic acids, resuspend the pellet in 1 mL chilled 49 % HF and transfer the sample to a 10–15 mL plastic tube. Stir for 48 h at 4 °C with a magnetic stirrer (*see* **Note 40**).
15. Transfer the samples to ultracentrifuge tubes, add 2 mL MilliQ water, and mix. Centrifuge for 10 min at 20 °C and 150,000 × *g* (*see* **Note 41**). Discard the supernatant in the proper waste container.
16. Resuspend the pellet in 1 mL MilliQ water, increase volume up to 3 mL with MilliQ water, and ultracentrifuge for 10 min at 20 °C and 150,000 × *g*. Repeat this step once and finally resuspend the pellet in 100 μL MilliQ water or digestion buffer (*see* **Note 26**).

3.4 Muramidase Digestion

1. Transfer resuspended sacculi (~100 μL from Subheadings 3.2, **step 4**, or 3.3, **step 16**) to 1.5 mL Eppendorf tubes.
2. Add 2 μL muramidase 1 mg/mL and let the reaction work for 2–16 h at 37 °C (*see* **Note 42**).
3. Muropeptides are now in the soluble fraction. Boil the samples for 5 min to stop the reaction (*see* **Note 43**). Centrifuge for 15 min at room temperature and 20,000 × *g* in a benchtop centrifuge and transfer the muropeptide-containing supernatant to 1.5 mL Eppendorf or to long glass tubes (*see* **Note 44**).

3.5 Sample Reduction and Filtration

1. Add borate buffer to the sample to adjust pH to 8.5–9.0. For a 100 μL reaction, 15–20 μL borate buffer is typically used. Check pH using indicator strips (*see* **Note 45**). The pH now is more alkaline to make the reduction step not very fast (*see* **Note 46**).
2. Add 10 μL freshly prepared $NaBH_4$ 2 M and let the sample reduce at room temperature for 20–30 min (*see* **Note 47**).
3. Adjust sample pH to 2.0–4.0 with orthophosphoric acid 25 % (v/v) (*see* **Note 48**).
4. Transfer reduced samples to a 96-well filter plate (0.2 μm pore size, regenerated cellulose). Using a vacuum manifold recover the filtered samples in a 96-well multititer plate (*see* **Notes 49** and **50**).

3.6 UPLC Inorganic Separation

1. Set the column temperature to 35 °C.
2. Prepare mobile phases and refill bottles A and B with inorganic buffer A (phosphate buffer 50 mM, pH 4.35) and inorganic buffer B (phosphate buffer 50 mM, pH 4.95, methanol 15 % (v/v)), respectively. Purge pumps and tubes according to the UPLC system instructions (*see* **Note 51**).
3. Equilibrate the column with inorganic buffer A, flow 0.25 mL/min, until pressure is stabilized (*see* **Notes 52** and **53**).
4. Using the system auto-sampler, inject 10 μL sample (*see* **Note 54**).
5. Perform the LC run using the gradient described in Fig. 1 and measure absorbance at 204 nm (*see* **Notes 55–60**).

3.7 UPLC Organic Separation

1. Set the column temperature to 45 °C.
2. Prepare mobile phases and refill bottles A and B with organic buffer A (formic acid 0.1 % (v/v)) and organic buffer B (formic acid 0.1 % (v/v), acetonitrile 40 % (v/v)). Purge pumps and tubes according to the UPLC system instructions (*see* **Note 51**).
3. Equilibrate the column with organic buffer A, flow 0.175 mL/min, until pressure is stabilized (*see* **Notes 53** and **60**).
4. Using the system auto-sampler, inject 10 μL sample (*see* **Note 54**).
5. Perform the LC run using the gradient described in Fig. 2 and measure absorbance at 204 nm (*see* **Note 55–60**).

Time (min)	% A	% B	Comment
0	100	0	
1	100	0	
20	0	100	To wash the column
22	0	100	
23	100	0	To re-equilibrate the column
28	100	0	

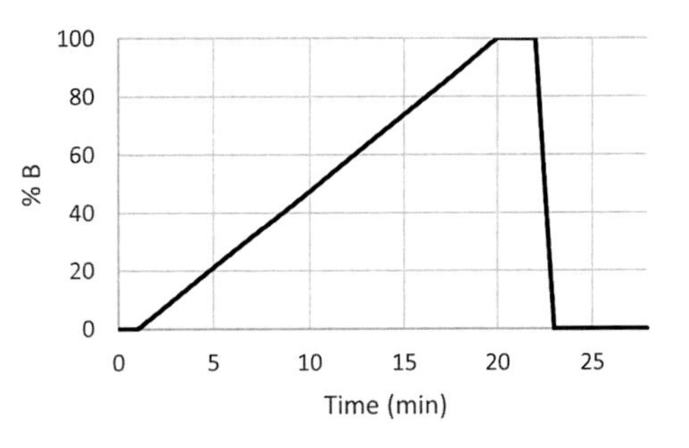

Fig. 1 UPLC inorganic gradient

Time (min)	% A	% B	Comment
0	95	5	
0.1	93	7	
3	82	18	
11	50	50	To wash the column
12	50	50	
12.1	95	5	To re-equilibrate the column
15	95	5	

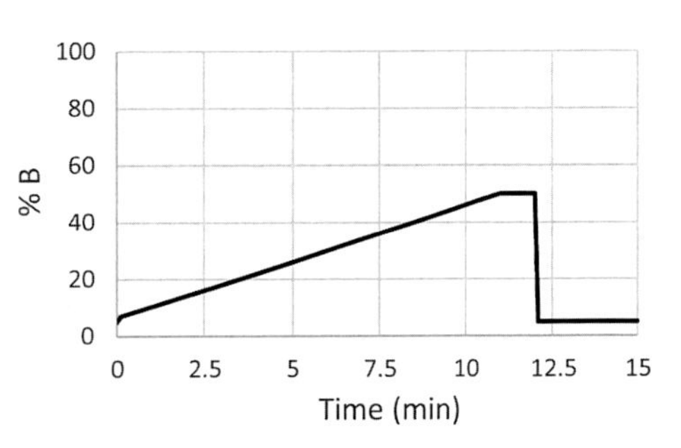

Fig. 2 UPLC organic gradient

3.8 UPLC Data Processing

1. Extract the raw data and represent the chromatogram by plotting absorbance at 204 nm (arbitrary units) against retention time (min) (*see* **Note 60**).
2. Define chromatographic processing regions removing unuseful data (*see* **Note 61**).
3. Use appropriate software for peak alignment (*see* **Note 62**).
4. Calculate the area of each peak by integration using the appropriate software (e.g., UPLC manufacturer's software, MATLAB) and determine the relative abundances for each peak (*see* **Note 63**).
5. Determine the identity of each peak by comparison to a known reference chromatogram, e.g., Fig. 3a, b (*see* **Note 64**).
8. Represent the results as a muropeptide table that typically contains retention time and relative abundance for all detected muropeptides (Fig. 3c).

4 Notes

1. When measuring the pH, always fix the pH electrode in a vertical position and gently stir the solution. The pH of most solutions is temperature dependent; therefore adjust at a temperature as close as possible to the temperature the buffers are going to be used.
2. Make sure that the tubes are suitable for boiling. We usually use conical centrifuge tubes (e.g., 15 mL Falcon tubes or similar).
3. Magnets should be small enough to fit in 12–50 mL tubes. We usually use 12 × 6 mm stirring bars.
4. Pronase E is a mixture of at least three caseinolytic activities and one aminopeptidase activity used for removal of peptidoglycan-bound proteins such as Braun's lipoprotein. The mixture retains activity in SDS 1% (w/v). The product can be completely inactivated by heating above 80 °C for 15–20 min.

   Other proteases as chymotrypsin (EC 3.4.21.1) and trypsin (EC 3.4.21.4) can also be used.
5. It is absolutely necessary to use polypropylene tubes or other acetone and HF-resistant material for Gram-positive peptidoglycan extraction. Otherwise, tubes will degrade and sample will be lost.
6. We usually use 15 mL conical centrifuge tubes.
7. Finely adjust pH using orthophosphoric acid 25% (v/v). Concentration of Na is critical in the inorganic system; therefore if pH goes below the indicated values by more than a few hundredths of a unit, it should not be readjusted with NaOH.

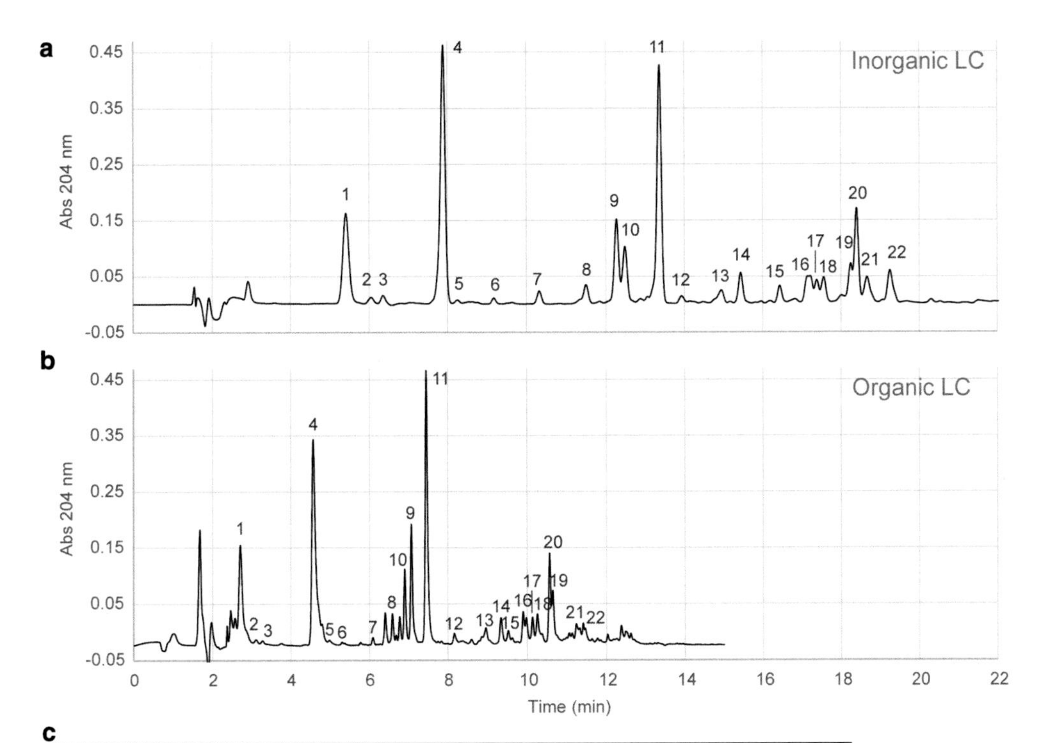

c

Peak	Murop.	Inorganic LC		Organic LC	
		RT (min)	%	RT (min)	%
1	M3	5.42	10.17	2.70	11.32
2	n.d.	6.06	0.58	3.10	0.24
3	M4G	6.36	0.63	3.27	0.30
4	M4G	7.89	23.63	4.56	25.20
5	M2	8.58	0.55	4.97	0.54
6	M5	9.18	0.38	5.30	0.15
7	M3-Lpp	10.33	1.00	6.08	0.50
8	D33	11.50	1.91	6.58	1.77
9*	D34	12.28	7.11	7.06	7.76
10*	D43	12.49	5.13	6.89	4.61
11	D44	13.37	19.94	7.44	20.67
12	D45	13.93	0.65	8.16	0.87
13	n.d.	14.94	1.44	8.96	1.72
14	T444	15.44	2.42	9.34	2.08
15	n.d.	16.43	1.30	9.52	0.93
16	D34N/D43N	17.17	3.45	9.93	3.77
17	n.d.	17.37	1.68	10.13	1.76
18	n.d.	17.55	2.09	10.25	2.36
19*	D44N	18.25	2.77	10.64	3.82
20*	D44N	18.40	7.24	10.56	5.42
21	D45N	18.67	2.74	11.27	2.24
22	T444N	19.25	3.20	11.43	1.97

Fig. 3 *Vibrio cholerae* peptidoglycan profiles. Representative chromatograms from a *Vibrio cholerae* sample analyzed using the inorganic (**a**) and the organic (**b**) LC methods. (**c**) Relative abundance of muropeptides. *RT* retention time (min). %: relative abundance. *n.d.* not determined. *: muropeptides with a shift in the retention time dependent on the LC method

8. Sodium borohydride ($NaHB_4$) is a reducing agent highly soluble in water and lower alcohols. It reacts with these solvents to produce H_2 in a quite violent reaction. Do not close any tube lids.
9. V-bottom multititer 96-well plates ensure maximum recovery of the sample upon injection.
10. Alternatively, samples can be individually filtered using 1 mL syringes and 4 mm syringe filters (4 mm regenerated cellulose filters, 0.2 μm pore size) and maximum recovery vials with caps.
11. We usually add 55 mL orthophosphoric acid 85 % (v/v) and then finely adjust the pH mL by mL with orthophosphoric acid 25 % (v/v).
12. We usually add 50 mL orthophosphoric acid 85 % (v/v) and then finely adjust the pH mL by mL with orthophosphoric acid 25 % (v/v).
13. For optimal sample preparation, a minimum of 10^{10} bacteria cells are required. 10^9 bacteria produce a small pellet after ultracentrifugation that is easily lost during the washing steps. We usually grow 10 mL cultures up to stationary phase. Larger volumes require scaling up reagents and are more time consuming, but provide better results.
14. Resuspension volume is not critical, although it is preferred to use the minimal volume possible to reduce the number of washes by ultracentrifugation.
15. When boiling SDS is added, cells will lyse immediately and the lysate will become transparent, proteins will solubilize while the sacculi remain intact. For larger sample volumes, optimal lysis is achieved by pouring the sample suspension drop by drop into an equal volume of boiling lysis solution, in tubes with stirring bars, inside a beaker of boiling water.
16. Samples need stirring during the boiling process to shear the DNA, which otherwise will interfere with the formation of a compact pellet upon centrifugation. For Gram-positive bacteria boiling with SDS can be shortened to 30 min since nuclease treatment will be performed later. Also, cell amount will affect the efficiency of the lysis, so longer boiling times are required for higher cell concentrations. Remember to add water to the beaker from time to time due to the high evaporation rate.
17. For Gram-positive bacteria, overnight stirring is not required.
18. Centrifuging at low temperatures will cause SDS precipitation.
19. The resulting pellet is transparent and can be difficult to localize, so it is advisable to mark one side of the tube and place the tube with this side up in the rotor. After centrifugation, the pellet will be located at the bottom in the opposite side of the tube.

20. Resuspend the pellet and agitate the bottom of the tube. Bubble formation is indicative of detergent presence. High sacculi concentration can also produce bubbles, but they usually disappear immediately after the agitation.
21. Usually, for 10^{10} cells only two washes are required.
22. To save time, this step can be performed during the previous washing steps. Activated Pronase E is stable at 4 °C during 24 h.
23. Pronase digestion and inactivation can be performed in the ultracentrifuge tubes. However, this will shorten the lifetime of these tubes and hence disposable tubes are preferred.
24. Pronase E will remove peptidoglycan-bound proteins such as Braun's lipoprotein. Some bacteria do not require Pronase E treatment, but it is recommended for removal of other peptidoglycan-associated proteins.
25. Boiling in SDS 1 % (w/v) will denature Pronase E and it will be removed from the supernatant during the next washing steps.
26. Muramidase digestion is equally effective in both solutions. This final volume is important because after the digestion the muropeptides will remain in the soluble fraction. If peak intensity after LC is too low, this volume can be decreased. However, scaling down the volumes can cause difficulties during the reduction and filtering steps.
27. Gram-positive bacteria sacculi pellets are not transparent like in Gram-negative bacteria. However, provided that the polypropylene tubes are translucent it is recommended to mark one side of the tube for helping localizing the pellet (*see* **Note 19**).
28. Usually, for Gram-positive bacteria at least three washes are required.
29. Cells can also be broken by sonication with glass beads as described in [6] or with the help of a FastPrep instrument as described in [7].
30. Long centrifugations lead to loss of sample due to an excess of cell precipitation. In our experience, letting the tubes stand for a few minutes gives the best results.
31. Avoid pipetting the precipitated pellet in the ultracentrifuge tube.
32. α-Amylase hydrolyzes α bonds of large, α-linked polysaccharides. This treatment is used to remove glucose polymers from the sample. Its use depends on the composition of the peptidoglycan of each bacterial species and can be skipped sometimes.
33. Trypsin is a protease that degrades peptidoglycan-bound proteins.

34. Transferring the sample to new tubes is not required.
35. LiCl is used for extraction of peptidoglycan-associated proteins [8–10].
36. EDTA is a chelating agent that will remove the LiCl from the sample, which could interfere with the muramidase digestion.
37. Acetone removes phospholipids from the cell wall fraction and thus facilitates protein extraction [11]. Due to its corrosive nature, do not add more than 1 mL to avoid spilling and rotor damaging.
38. After the acetone wash, pellets are usually not very compact and they easily detach from the tube walls. Hence, removal of acetone is better performed using the pipette instead of decanting the tube.
39. Resuspension volume is not critical and will depend on the amount of pellet. Smaller volumes will be dried faster. Aliquoting the sample in several tubes reduces drying time. Alternatively, samples can be dried using a SpeedVac concentrator.
40. When working with HF, follow the local safety and laboratory regulations, use the flow hood, and make sure that the materials are suitable for the use of this acid. HF is precooled on ice before use. Incubating the samples for less than 48 h gives bad or poor PG isolation.
41. If volumes are scaled up, do not centrifuge more than 1.5 mL HF to avoid spilling and rotor damaging. Rinse the rotor, lid, and O-ring with distilled water after use to completely remove HF and prevent corrosion.
42. After 2 h of reaction, 95 % of the sample has been digested. We usually leave the reaction overnight to ensure total digestion.
43. It is critical not to add detergent to the sample. Traces of SDS in the sample cause altered migration profiles during LC separation. Muramidase is inactivated by boiling the sample and precipitation by centrifugation.
44. For sample reduction of Gram-positive bacteria or high reaction volumes long glass tubes are preferred, since bubble formation can lead to loss of sample.
45. Do not use the indicator strip on the sample directly. In order to save sample, take 0.5 μL and drop it on the indicator strip. Then, check the color/pH on the reference table.
46. Do not leave the sample for a long time in an alkaline pH because it leads to β-elimination (the disaccharide loses the peptide).
47. $NaBH_4$ is very reactive and needs to be prepared immediately before use. H_2 is produced and there will be bubble formation. Prevent gas accumulation by leaving the lids opened. Due to the constant bubble formation, it is necessary to ensure

that the right volume of $NaBH_4$ is pipetted. Improper sample reduction will result in peaks and profiles with altered retention times. The reaction takes place during the first minutes, so it is better to add an excess of $NaBH_4$ than to leave the reaction stand for a long time.

48. This step is critical for adequate separation of the muropeptides by UPLC. pH higher than 5 units will result in muropeptides eluting with altered retention times, especially when using the organic LC separation method. Add 4 μL orthophosphoric acid 25 % (v/v) to the sample and check pH using indicator strips as indicated before. Carefully add acid μL by μL to ensure that the sample has the desired pH.
49. Prior to injection, samples need to be filtered to remove impurities (*see* **Note 10**).
50. For long-term storage, samples are preferably stored in glass vials (minimizing solvent evaporation through pre-slit cap mats and avoiding potential leaking of contaminants from well plates). If analysis is undertaken soon after preparation, prepared samples should be kept at 4 °C (on ice or in the refrigerator) until they are transferred to the auto-sampler. If necessary, prepared samples can be stored frozen at −20 °C before analysis. However, this may lead to the formation of insoluble precipitates, which should be removed immediately before injection via centrifugation or heating the sample.
51. For muropeptide profiling, reverse-phase (RP) columns, typically C18-bonded silicas that are able to retain and separate medium-polar and nonpolar metabolites, provide a good separation pattern. For RPLC, maximum retention of analytes is ensured by loading samples onto the column using solvents of low eluotropic strength (i.e., composed mainly or entirely of water). Elution of retained metabolites is accomplished using a gradient of increasing methanol (inorganic method) or acetonitrile (organic) content.
52. In our system, pressure oscillates between ~7300 and ~8600 psi during the inorganic run.
53. To ensure repeatability of the separations, parameters such as flow rate and column temperature need to be carefully controlled, following the manufacturer's indications. With ~12–20-min runs and a 3- to 5-min re-equilibration period, ~50 samples can be run per day. In our experience, we can run ~2000 samples on each chromatographic column before observing pressure issues or substantial degradation in peak quality.
54. The injection volume can be modified depending on the sample concentration. For concentrated samples (e.g., peptidoglycan from Gram-positive bacteria or from large starting cultures), inject less volume. If there are no peaks or absorbance is too low, increase the injection volume.

55. If the chromatograms show poor peak shapes, this is indicative of column degradation or sample overloading. To solve this problem, dilute the sample or improve sample preparation. If the problem persists, consider cleaning or replacing the column.
56. If there are no or few peaks, either the injection failed (try reinjecting the sample) or sample concentration is too low. In this case, inject larger volumes, concentrate the sample using a Speedvac concentrator, or prepare new sample.
57. Changes in the retention time can be due to improper sample pH or presence of detergent (SDS) in the sample. When sample pH > 5, it results in shifted chromatograms in the organic method but not the inorganic one (due to its buffering effect). Adjust pH with orthophosphoric acid 25 % (v/v) and rerun the sample. If there is detergent in the sample, either prepare new sample increasing the amount of washing steps or wash the column after each run to get rid of the retained detergent.
58. Extra peaks appear due to poor sample reduction. Reduce the sample again or prepare new sample. Contamination with other components or short re-equilibration time between runs also contributes to the appearance of ghost peaks.
59. If peak separation is not optimal, optimize the gradients for longer run times. This can be useful for some Gram-positive bacteria peptidoglycan profiles where the amount of peaks overcomes the resolution capability of the detector.
60. In our system, pressure oscillates between ~3800 and ~4100 psi during the organic run.
61. We usually remove the solvent front at the beginning of the run (typically 1–2 min) and the final re-equilibration step (min 22 for the inorganic method and min 12 for the organic method).
62. For new or unknown samples, a reference sample can be run to facilitate peak alignment and muropeptide identification.
63. Divide the area of every peak by the total area of the chromatogram (sum of all individual areas) and multiply by 100 to calculate relative abundance in %.
64. Retention times differ depending on the gradient and UPLC system used. For confirmation of peak identity, individual peaks need to be collected and subjected to MS analysis.

References

1. Vollmer W, Blanot D, de Pedro MA (2008) Peptidoglycan structure and architecture. FEMS Microbiol Rev 32(2):149–167. doi:10.1111/j.1574-6976.2007.00094.x, FMR094 [pii]
2. Alvarez L, Espaillat A, Hermoso JA, de Pedro MA, Cava F (2014) Peptidoglycan remodeling by the coordinated action of multispecific enzymes. Microb Drug Resist 20(3):190–198. doi:10.1089/mdr.2014.0047

3. Cava F, de Pedro MA (2014) Peptidoglycan plasticity in bacteria: emerging variability of the murein sacculus and their associated biological functions. Curr Opin Microbiol 18:46–53. doi:10.1016/j.mib.2014.01.004
4. Glauner B, Holtje JV, Schwarz U (1988) The composition of the murein of Escherichia coli. J Biol Chem 263(21):10088–10095
5. Desmarais SM, De Pedro MA, Cava F, Huang KC (2013) Peptidoglycan at its peaks: how chromatographic analyses can reveal bacterial cell wall structure and assembly. Mol Microbiol 89(1):1–13. doi:10.1111/mmi.12266
6. Ikeda S, Hanaki H, Yanagisawa C, Ikeda-Dantsuji Y, Matsui H, Iwatsuki M, Shiomi K, Nakae T, Sunakawa K, Omura S (2010) Identification of the active component that induces vancomycin resistance in MRSA. J Antibiot 63(9):533–538. doi:10.1038/ja.2010.75
7. Filipe SR, Tomasz A (2000) Inhibition of the expression of penicillin resistance in Streptococcus pneumoniae by inactivation of cell wall muropeptide branching genes. Proc Natl Acad Sci U S A 97(9):4891–4896. doi:10.1073/pnas.080067697
8. Lortal S, Van Heijenoort J, Gruber K, Sleytr UB (1992) S-layer of Lactobacillus helveticus ATCC 12046: isolation, chemical characterization and re-formation after extraction with lithium chloride. Microbiology 138(3):611–618. doi:10.1099/00221287-138-3-611
9. Liang OD, Flock JI, Wadstrom T (1995) Isolation and characterisation of a vitronectin-binding surface protein from Staphylococcus aureus. Biochim Biophys Acta 1250(1): 110–116
10. Regulski K, Courtin P, Meyrand M, Claes IJ, Lebeer S, Vanderleyden J, Hols P, Guillot A, Chapot-Chartier MP (2012) Analysis of the peptidoglycan hydrolase complement of Lactobacillus casei and characterization of the major gamma-D-glutamyl-L-lysyl-endopeptidase. PLoS One 7(2):e32301. doi:10.1371/journal.pone.0032301
11. Hill SA, Judd RC (1989) Identification and characterization of peptidoglycan-associated proteins in Neisseria gonorrhoeae. Infect Immun 57(11):3612–3618

Part II

Genome-Wide Approaches for the Identification of Gene Products with Roles in Cell Wall Homeostasis

Chapter 3

Microarray Analysis to Monitor Bacterial Cell Wall Homeostasis

Hee-Jeon Hong and Andy Hesketh

Abstract

Transcriptomics, the genome-wide analysis of gene transcription, has become an important tool for characterizing and understanding the signal transduction networks operating in bacteria. Here we describe a protocol for quantifying and interpreting changes in the transcriptome of *Streptomyces coelicolor* that take place in response to treatment with three antibiotics active against different stages of peptidoglycan biosynthesis. The results defined the transcriptional responses associated with cell envelope homeostasis including a generalized response to all three antibiotics involving activation of transcription of the cell envelope stress sigma factor σ^E, together with elements of the stringent response, and of the heat, osmotic, and oxidative stress regulons. Many antibiotic-specific transcriptional changes were identified, representing cellular processes potentially important for tolerance to each antibiotic. The principles behind the protocol are transferable to the study of cell envelope homeostatic mechanisms probed using alternative chemical/environmental insults or in other bacterial strains.

Key words Transcriptomics, Streptomyces, Antibiotic, Cell envelope homeostasis

1 Introduction

Microbial cells are particularly vulnerable to their environmental context and have evolved biological systems for sensing and adapting to external changes which are central to their viability. The bacterial cell envelope is at the immediate interface between cell and environment and houses much of the sensory apparatus for detecting environmental insults and initiating the remedial changes in gene expression. Many of these changes in gene expression are involved in cell envelope homeostasis and are directed towards maintaining the integrity of the cell envelope which is essential for bacterial cell survival under all but the most artificial laboratory conditions (Fig. 1). Transcriptomics, the genome-wide analysis of gene transcription, has consequently become a useful approach for identifying and understanding the signal transduction systems important for cell envelope homeostasis in a broad range of

Hee-Jeon Hong (ed.), *Bacterial Cell Wall Homeostasis: Methods and Protocols*, Methods in Molecular Biology, vol. 1440,
DOI 10.1007/978-1-4939-3676-2_3, © Springer Science+Business Media New York 2016

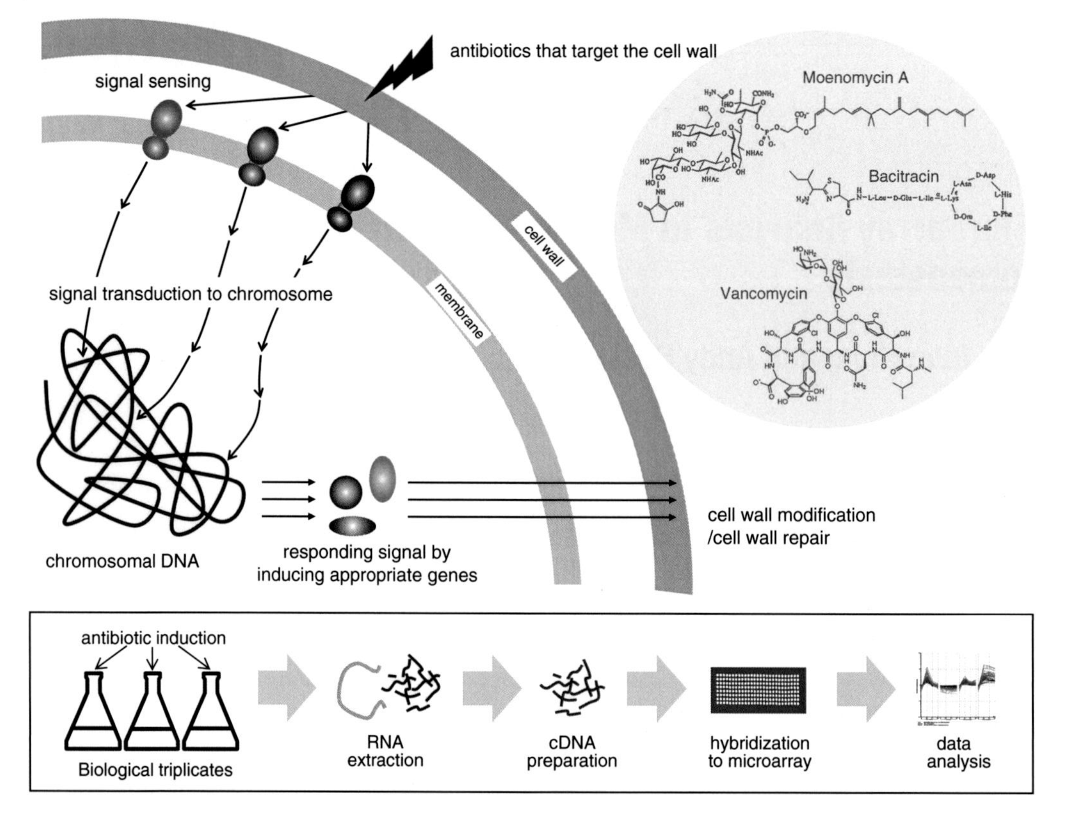

Fig. 1 Sensory systems located in the bacterial cell envelope detect adverse changes in the environment and initiate remedial changes in gene expression, including homeostatic mechanisms for maintaining the functional integrity of the cell envelope. We describe a protocol for analyzing the global transcriptional responses to the antibiotics vancomycin, moenomycin, and bacitracin which inhibit peptidoglycan biosynthesis in the cell wall. Cultures are treated in biological triplicate with each antibiotic and RNA extracted at 30 min intervals after antibiotic addition. Transcripts are quantified by hybridization to DNA microarrays and the major changes taking place revealed by functional analysis of the differentially expressed genes identified

bacterial species. Such studies tend to fall into one of the two broad categories: (1) analysis of the transcriptional response following application of an external stress designed to cause damage to the cell envelope (see for example [1–5]) and (2) characterization of the changes in the transcriptional program associated with strains or mutants which exhibit increased tolerance towards agents causing cell envelope damage (see for example [6–10]).

Here we describe a protocol for collecting and analyzing data on changes induced in the transcriptome of the soil bacterium *Streptomyces coelicolor* following treatment with three antibiotics which target distinct steps in peptidoglycan cell wall biosynthesis. The general scheme—culturing, treatment, extraction of RNA, quantification of transcripts, and data analysis (Fig. 1)—is however transferable to any bacterial species and any type of chemical or stress challenge. The ability to quantitatively detect changes in transcription of all genes in the genome in response to a range of different cell envelope-damaging events provides a global view of the adaptive processes taking place [1]. The collection of time series data allowed the dynamic response to the antibiotics to be observed, and the immediate effects on transcription to be distinguished from the longer term changes. Integration with established knowledge about the control of gene transcription (e.g., transcription factor-binding sites, sigma factor promoter consensus sequences) and about the function of the gene products (e.g., gene ontology annotation, metabolic pathway annotation) further increased the biological understanding obtainable from the transcriptome data collected. The quality and extent of existing genome annotation are crucial for interpreting -omics data sets, and for many bacterial species, and particularly for non-model organisms, this can be a significant limiting factor. Storage of the transcriptome data in public databases provides the potential both for community efforts to improve genome annotation and for the reinterpretation of the datasets at future dates as and when additional information becomes available. Suitable database storage also creates the opportunity for integration with similar data collected from different labs, or with corresponding data on changes in protein and/or metabolite abundance. This can form the basis for beginning to understand the networks of cellular components which interact to determine all cell behavior and physiology.

We have used Affymetrix DNA microarrays to make the transcript abundance measurements in this protocol but these can readily be replaced by a suitable alternative microarray platform, or by RNA sequencing (RNA-seq). The choice of platform for any transcriptome analysis study will be influenced both by the availability of an appropriate microarray for the genome being studied and by the desired level of detail concerning the transcriptome structure. RNA-seq promises to be superior to a standard gene expression microarray in its ability to identify transcription start and stop sites at nucleotide resolution, and to identify and quantify novel regions of transcription including antisense and other noncoding RNA transcripts. Suitable pipelines for the analysis of RNA-seq data are however still evolving, particularly for bacterial genomes which offer some unique challenges (see Creecy and Conway [11] for a review), while mature and reliable methods for analyzing microarray data are already available and in routine use.

2 Materials

2.1 Strain and Culture Media

1. Bacterial strain: *Streptomyces coelicolor* wild-type M600 ($SCP1^-$, $SCP2^-$).
2. 0.05 M Tris[hydroxymethyl]methyl-2-aminoethanesulfonic acid (TES) buffer (pH 8): Dispense 100 mL aliquots and autoclave.
3. Double-strength germination medium (2xGM) (*see* **Note 1**): 1% (w/v) Bacto yeast extract, 1% (w/v) casaminoacids. Add 0.01 M $CaCl_2$ to 2xGM before use.

4. NMMP minimal liquid medium: 0.2 % (w/v) $(NH_4)_2SO_4$, 0.5 % (w/v) casaminoacids (Difco), 0.06 % (w/v) $MgSO_4{\cdot}7H_2O$, 5 % (w/v) PEG6000, 0.1 % (v/v) minor element solution (*see* **Note 2**). Dispense in 80 mL aliquots and autoclave. At the time of use, add 15 mL of NaH_2PO_4/K_2HPO_4 buffer (0.1 M, pH 6.8) (*see* **Note 3**), 2.5 mL of 20 % carbon source, and 2.5 mL of any required growth factors. For this experiment, glucose was used as carbon source and no growth factors were required.
5. Antibiotic stock solutions: Bacitracin (10 mg/mL), moenomycin A (10 mg/mL), vancomycin (10 mg/mL). Filter sterilize all antibiotic solutions and store in aliquots at −20 °C.

2.2 RNA Extraction and Quality Control (*See* Note 4)

1. RNA protect solution.
2. RNaseZAP (Sigma).
3. Egg white lysozyme: Store at −20 °C. Make a fresh solution in TE buffer (15 mg/mL) at the time of use.
4. Phenol/chloroform/isoamyl alcohol: 50 % (v/v) phenol, 50 % (v/v) chloroform, 1 % (v/v) isoamyl alcohol (see **Note 5**). Store at 4 °C.
5. Chloroform.
6. RNase-Free DNaseI Set (Qiagen). Store at −20 °C.
7. 100 % and 70 % ethanol.
8. RNeasy Midi kit.
9. Nanodrop spectrophotometer.
10. Bioanalyzer.
11. Agilent RNA 6000 nano kit.

2.3 cDNA Synthesis and Cleanup

1. Random primers (Invitrogen). Store at −20 °C.
2. 10 mM Deoxynucleoside triphosphate (dNTP) mix (Invitrogen). Store at −20 °C.
3. 100 mM Dithiothreitol (DTT) (Invitrogen).
4. Superscript III reverse transcriptase and accompanying reagents (Invitrogen). Store at −20 °C.
5. Superase In (Ambion): A strong RNase inhibitor. Store at −20 °C.
6. 1 N NaOH.
7. 1 N HCl.
8. QIAquick PCR purification kit (Qiagen).

2.4 cDNA Fragmentation, Terminal Labeling, and Quality Control

1. DNase I (Amersham Biosciences).
2. 10× One Phor-All buffer (Pharmacia Biotech).
3. BioArray Terminal Labeling kit (Enzo Life Sciences).
4. 0.5 M Ethylenediaminetetraacetic Acid (EDTA).

5. NeutrAvidin (Pierce).
6. Phosphate-buffered saline (PBS) solution, pH 7.2 (Sigma-Aldrich).
7. SYBR Gold (Sigma-Aldrich).

2.5 Affymetrix GeneChip Hybridization, Washing, and Staining (See Note 6)

1. Affymetrix *Streptomyces* diS_div712a GeneChip arrays (Affymetrix). Store at 4 °C until use.
2. 12× MES (2-(*N*-Morpholino)EthaneSulfonic Acid) stock solution (*see* **Note 7**): 1.22 M MES, 0.89 M [Na^+].
3. 2× Hybridization buffer (*see* **Note 8**): Final 1× concentration is 100 mM MES, 1 M [Na^+], 20 mM EDTA, 0.01 % Tween-20.
4. 1× Hybridization buffer: 100 mM MES, 1 M [Na^+], 20 mM EDTA, 0.01 % Tween-20. To make 50 mL of 1× hybridization buffer, dilute 25 mL of 2× hybridization buffer with the equal volume of water.
5. Wash A non-stringent wash buffer (*see* **Note 9**): 6× Saline-sodium phosphate-EDTA (SSPE), 0.01 % Tween-20.
6. Wash B stringent wash buffer (*see* **Note 10**): 100 mM MES, 0.1 M [Na^+], 0.01 % Tween-20.
7. 2× Stain buffer (*see* **Note 11**): 100 mM MES, 1 M [Na^+], 0.05 % Tween-20.
8. Hybridization cocktail (*see* **Note 12**).
9. Streptavidin phycoerythrin (SAPE) solution (*see* **Note 13**).
10. Antibody solution mix (*see* **14**).

2.6 Affymetrix GeneChip Data Analysis (See Note 15)

1. R software environment for statistical computing and graphics [12].
2. R software packages from Bioconductor (http://bioconductor.org/). The packages required are named in Subheading 3 and can be downloaded and installed directly into R using the biocLite ("packagename") function.
3. GeneSpring (Agilent Technologies).
4. Ontologizer [13].
5. Gene ontology annotation for *Streptomyces coelicolor* downloadable from ftp://ftp.ebi.ac.uk/pub/databases/GO/goa/proteomes/84.S_coelicolor.goa.

3 Methods

3.1 Culture and Induction with Antibiotics

1. To culture the strains, first germinate aliquots of spores of *S. coelicolor* M600 by heat-shock treatment in 5 mL TES buffer (0.05 M, pH 8) at 50 °C for 10 min, and then incubate (with

shaking) at 37 °C for 2–3 h following dilution with an equal volume of 2× GM. Collect the germlings by centrifugation for 2 min at 4000 × *g* and use to inoculate 50 mL of NMMP broth to produce an initial OD_{450} ≈0.025. Do this for three flasks for each condition being investigated to produce triplicate measurements suitable for the statistical analysis of the transcript abundance data to be collected (*see* **Note 16**). Incubate at 30 °C with 250 rpm shaking until an OD_{450} of 0.4–0.6 is reached (approximately 10–12 h), and then treat by addition of a sub-inhibitory concentration (10 μg/mL) of antibiotic. Also perform an untreated control experiment to characterize the natural changes in expression taking place over the 90 min duration of the studies (*see* **Note 17**).

2. Cultures are sampled at times immediately before (defined as 0 min) and at 30, 60, and 90 min after antibiotic treatment by rapidly harvesting cells from 10 mL aliquots by centrifugation for 1 min at 4000 × *g*. Immediately discard the supernatant and quickly resuspend the pelleted cells in twice the volume of RNA protect solution and then leave for 5 min at room temperature. Spin down for 3 min at 4000 × *g*, decant (making sure that most of the liquid is removed), and freeze at −80 °C. The pelleted cells can be stored like this for up to 1 month until use.

3.2 RNA Isolation (*See* Note 18)

1. Defrost the frozen pellet on ice and resuspend in 1 mL TE buffer containing 15 mg/mL lysozyme. Incubate for 1 h at room temperature. Add 4 mL RLT buffer and then sonicate (full 18 μm amplitude) on ice for three cycles of 20 s allowing the cells to rest on ice for 20 s between cycles. Wash the sonicator probe with 70 % ethanol and RNaseZAP before and after use. The solution may go cloudy while standing on ice but this is a natural property of the RLT buffer.
2. Add 4 mL phenol/chloroform and vortex for 1 min (see **Note 5**). Spin for 1 min at 4 °C with maximum speed to separate the layers. Take the clear supernatant and repeat. Take the clear supernatant and repeat the extraction procedure but this time using just 4 mL chloroform.
3. Take 4 mL of the cleaned supernatant from **step 2** above and mix with 2.24 mL of ethanol. Apply this to a Qiagen RNeasy midi column. Spin to bind. Discard the flow-through.
4. Perform on-column DNaseI digestion. Wash the RNA on-column by applying 2 mL of RW1 buffer to the column, spin down, and discard the flow-through. Prepare a DNaseI stock by adding 550 μL of RNase-free water (included in the kit) and aliquoting to 120 μL portions (enough for 6 digests each). Mix one aliquot with 840 μL RDD buffer and apply 160 μL of this to each column. Allow to digest for 60 min at room

temperature. Wash the column again by applying 2 mL of RW1 buffer, spin down, and discard the flow-through.

5. Wash the column twice by applying 2.5 mL of RPE buffer, spin down, and discard the flow-through. Dry the column of ethanol by spinning dry for one last time (see **Note 19**).
6. To elute the RNA into a fresh RNase-free tube, add 300 μL of RNase-free water, and let stand for 1 min. Spin down for 3 min at maximum speed. Reapply the eluate to the column and repeat the elution step. This gives a final volume of about 250 μL.
7. Quantify the RNA using the NanoDrop spectrophotomer, checking the purity by ensuring that the A260/280 and A260/230 values for each sample are between 1.8 and 2.2.
8. Check the integrity of the RNA with the Bioanalyzer (Agilent Technologies) using the RNA 6000 nano kit according to the manufacturer's instructions (*see* **Note 20**) (Fig. 2a, b).

3.3 cDNA Synthesis

1. RNA primer hybridization: For each sample prepare the following reaction mixture in a 0.2 mL PCR tube: 10 μg of RNA (prepared as detailed in Subheading 3.2), 10 μL of 75 ng/μL random primers, and RNase-free water in a final total volume of 30 μL. Incubate the reaction mixture for 10 min at 70 °C and 10 min at 25 °C and then cool at 4 °C for at least 2 min.
2. First-strand cDNA synthesis: Prepare enough master mix for all samples (*see* **Note 21**). The reaction mixture for a single reaction contains 12 μL of 5× first-strand reaction mix (provided from Superscript III reverse transcriptase kit; Invitrogen), 6 μL of 0.1 M DTT, 3 μL of 10 mM dNTP, 1.5 μL of Superase In, and 7.5 μL of Superscript III in a final volume of 30 μL. Incubate for 2 h at 42 °C, 30 min at 50 °C, 30 min at 55 °C, and 10 min at 70 °C. Add 20 μL of 1 N NaOH and incubate at 65 °C for 30 min to degrade the template RNA. Cool the samples for at least 2 min at 4 °C and neutralize with 20 μL of 1 N HCl.
3. cDNA cleanup: Use a QIAquick PCR purification kit (Qiagen) to clean up the cDNA prepared as in **step 2** above. Add 500 μL of buffer PB to the cDNA reaction mix. Place a QIAquick spin column in a provided 2 mL collection tube. Apply the sample to the column, centrifuge for 1 min at 13,000 × *g*, and discard flow-through. Place the column back in the same tube, then add 0.75 mL of Buffer PE to the column, and spin down for 1 min. Discard the flow-through, place the column back in the tube, and then spin down once again for 1 min to get rid of ethanol completely from the column (*see* **Note 19**). Place the column in a clean 1.5 mL tube and apply 30 μL of Buffer EB to the center of the membrane in column. Let the column stand for 1 min and then spin down for 1 min. Add another 30 μL of Buffer EB to the center of the membrane in column.

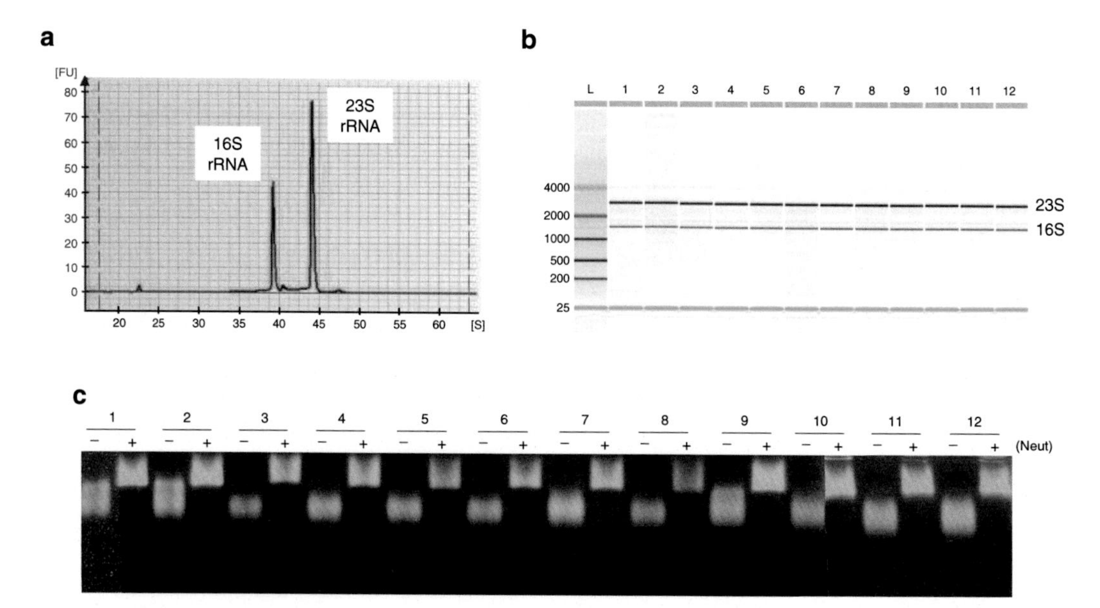

Fig. 2 Checking the quality of the RNA and labeled cDNA samples prior to hybridization to the DNA microarrays. (**a**) A Bioanalyzer electropherogram showing the separation and relative quantitation of the 23S and 16S ribosomal RNAs in a sample of total RNA extracted from *Streptomyces coelicolor.* RNA samples exhibiting a ratio of 23S to 16S peak areas of >1.8 are accepted for further analysis. (**b**) A pseudogel generated from the Bioanalyzer electropherograms produced in the analysis of the 12 RNA samples (*lanes 1–12*) from the moenomycin induction experiment. *Lane L* corresponds to separation of a nucleotide size marker ladder (sizes indicated to the *left*). If a Bioanalyzer is not available for checking the samples, RNA quality can be similarly visually assessed by separation using RNA gel electrophoresis. (**c**) Agarose gel electrophoresis to assess the fragment size and biotin labeling of the cDNA generated from the RNA samples. Lanes marked + or – correspond to biotinylated cDNA samples which have and have not been incubated with neutravidin (Neut), respectively, prior to loading onto the gel. The complete shift to a higher apparent molecular weight observed in the presence of neutravidin indicates effective labeling of the cDNA with biotin

Let the column stand for 1 min and then spin down for 1 min with maximum speed. Quantify the purified cDNA using the NanoDrop spectrophotometer.

3.4 cDNA Fragmentation and Terminal Labeling

1. Fragmentation: Prepare enough master mix for all samples. The reaction mixture for a single reaction contains 6 μL of 10× One Phor-All buffer, 3–7 μg of cDNA, 3–7 U of 10 U/μL DNaseI, and water in a final volume of 60 μL. Incubate for 10 min at 37 °C and then inactivate the enzyme by incubating at 98 °C for 10 min. Keep the fragmented cDNA in the same tube for the terminal labeling reaction. The reaction solution can be stored at −20 °C. Check the fragmentation by 2 % agarose gel electrophoresis (in 1× TBE). Good-quality fragmented cDNA should appear as a broad band centered between 50 and 100 bp in size (e.g., Fig. 2c).
2. Terminal labeling: Prepare a master mix using reagents provided from the labeling kit. The reaction mixture for a single reaction contains 18 μL of 5× reaction buffer, 9 μL of $CoCl_2$, 1.5 μL of

Biotin-ddUTP, 3 μL of TDT, and 59 μL of fragmented cDNA (1.5–6 μg) in a final volume of 90 μL. Incubate the reaction for 60 min at 37 °C and then stop the reaction by adding 2 μL of 0.5 M EDTA. The sample is now ready to be hybridized to the GeneChip microarrays or it may be stored at −20 °C.

3.5 Quality Control of Labeled and Fragmented cDNA Using a Gel Shift Assay (*See* Note 22)

1. Prepare 2 mg/mL NeutrAvidin solution in PBS.
2. For each sample to be tested, remove two 150–200 ng aliquots of fragmented and biotinylated sample to fresh tubes.
3. Add 5 μL of 2 mg/mL NeutrAvidin to one of the two tubes (Neut+) for each sample tested.
4. Mix and incubate at room temperature for 30 min.
5. Add loading dye to all samples to a final concentration of 1× loading dye.
6. Prepare 100 bp DNA ladders (1 μL of ladder + 7 μL of water + 2 μL of loading dye for each lane).
7. Carefully load the samples, Neut+ and Neut−, side by side onto a 2 % agarose gel. Include the DNA size ladder.
8. Run the gel. Do not run for too long or the fragmented DNA bands will be difficult to see.
9. While the gel is running, prepare at least 10 mL of a 1× staining solution of SYBR Gold. Stain gel for 10 min. Alternately SYBR Gold can be added to the gel loading dye.
10. Visualize and photograph the gel on a UV light box (Fig. 2c).

3.6 Hybridization, Washing, Staining, and Scanning of Arrays

1. Equilibrate the GeneChip microarrays to room temperature immediately before use.
2. Make a master mix of hybridization cocktail reagents and add to each individual fragmented labeled cDNA sample in separate 1.5 mL tubes.
3. Heat the hybridization cocktail to 90 °C for 5 min in a thermomixer.
4. Wet the array by filling it through one of the septa with 1× hybridization buffer using a pipette and appropriate tips. It is necessary to use two pipette tips when filling the microarray cartridge: one for filling and the second to allow venting of air from the hybridization chamber.
5. Incubate the microarray filled with 1× hybridization buffer at 50 °C for 10 min with rotation at 60 rpm.
6. Transfer the hybridization cocktail that has been heated at 99 °C to a 50 °C thermomixer for 5 min.
7. Spin the hybridization cocktails at maximum speed in a microcentrifuge for 5 min to remove any insoluble material from the hybridization mixture.

8. Remove the buffer solution from the probe array cartridge.
9. Fill each microarray GeneChip with the clarified hybridization cocktail, avoiding any insoluble matter at the bottom of the tube. Label the array (see **Note 23**).
10. For hybridization place all loaded GeneChips into the hybridization oven set to 50 °C. Load the arrays in a balanced configuration to minimize stress to the motor. Hybridize for 16 h with rotation at 60 rpm.
11. Immediately after hybridization, proceed to the washing, staining, and scanning steps using the Affymetrix Fluidics Station and Scanner and the protocol recommended by Affymetrix for the GeneChip microarray being used (*see* **Note 24**).
12. Using the Affymetrix workstation software, process the scanned .dat image files recorded for each sample to corresponding . CEL files for export and data analysis.

3.7 Data Analysis (*See* Note 25)

1. To assess the quality of the microarray data produced, import the .CEL files from each scanned microarray into R for analysis using the "affyPLM" and "affyQCReport" packages (*see* **Note 26**). If necessary omit any .CEL files with poor-quality control metrics from the subsequent statistical analysis. Data from four microarrays corresponding to replicate three of the bacitracin antibiotic treatments were not considered further in this analysis.
2. Process the probe-level measurements from the Affymetrix microarrays into normalized measurements at the gene level using the robust multi-array average (rma) algorithm (see **Note 27**). This can be achieved either in GeneSpring or in R using the "affy" package. Perform principal component analysis (PCA) to view the major trends in the data (Fig. 3a).
3. Filter out data for any uninformative genes (those that are not expressed, or unchangingly expressed, across all samples in the experiment) using the filtering tools incorporated into GeneSpring, or the varFilter function in the R package "genefilter."
4. Using the filtered data, test for transcripts whose abundance is significantly altered by the experimental conditions, contrasting the expression values in each antibiotic-treated sample with the corresponding time point from the untreated control. This can be achieved using two-way ANOVA in GeneSpring or using linear models in the R package "limma." Include a suitable multiple testing correction (e.g., Benjamini and Hochberg) in the statistical analysis to control the false discovery rate in the large number of statistical tests being performed.
5. Analyze the sets of significantly differently expressed genes identified in the experiment using a Venn diagram (intrinsic to GeneSpring, or achievable using the R package "VennDiagram")

to visualize the results (Fig. 3b) and to produce lists of genes for functional enrichment testing (*see* **step** 7 below).

6. Quality threshold (qt) clusters the expression profiles of the significantly differently expressed genes to identify those which may be co-regulated (see **Note 28**). The promoter regions of co-clustered genes can be further analyzed to search for over-represented DNA-binding motif sequences that may be associated with a known regulator or RNA polymerase sigma factor. Lists of genes from the same expression clusters can also be tested for functional enrichment (*see* **step** 7 below). Algorithms for qt clustering are included in GeneSpring and in the "flexclust" R package.
7. Lists of significantly differently expressed sets of genes identified in **steps 5** or **6** above are analyzed to determine whether they are significantly associated with any particular cellular function. This is accomplished using gene ontology (GO) analysis with the open-source Ontologizer program (*see* **Note 29**) and the *Streptomyces coelicolor* GO annotation, testing for functions which are significantly over-represented in the gene lists under test. The majority of bacterial genomes have only computationally created GO annotation where functions are inferred from protein sequence homology to gene products from better characterized organisms, and it is often also desirable to perform additional statistical comparisons of the lists

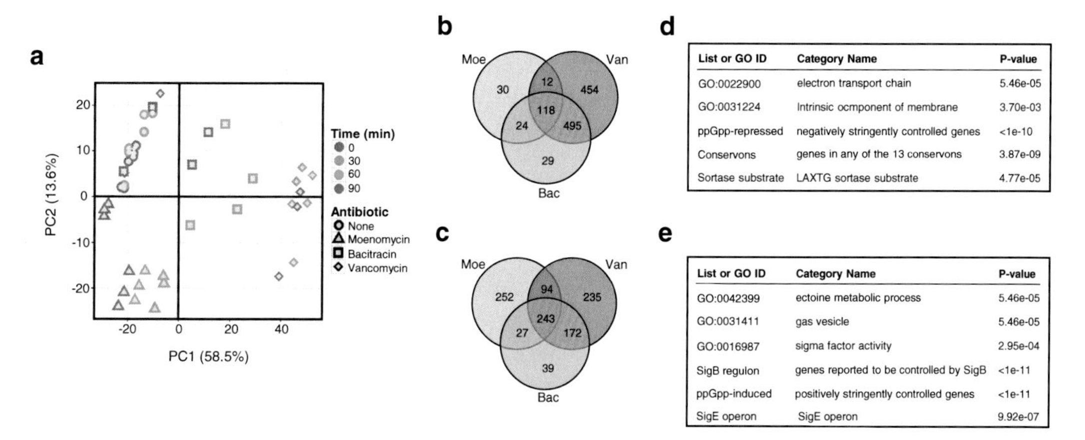

List or GO ID	Category Name	P-value
GO:0022900	electron transport chain	5.46e-05
GO:0031224	Intrinsic ocmponent of membrane	3.70e-03
ppGpp-repressed	negatively stringently controlled genes	<1e-10
Conservons	genes in any of the 13 conservons	3.87e-09
Sortase substrate	LAXTG sortase substrate	4.77e-05

List or GO ID	Category Name	P-value
GO:0042399	ectoine metabolic process	5.46e-05
GO:0031411	gas vesicle	5.46e-05
GO:0016987	sigma factor activity	2.95e-04
SigB regulon	genes reported to be controlled by SigB	<1e-11
ppGpp-induced	positively stringently controlled genes	<1e-11
SigE operon	SigE operon	9.92e-07

Fig. 3 Analysis of the DNA microarray data. (**a**) Principal component analysis (PCA) showing the general trends in the data. The samples for all time points in the control cultures cluster together with each other and with the 0 min samples from the antibiotic-treated cultures. The 30, 60, and 90 min samples from the antibiotic-treated cultures tend to cluster in groups according to the antibiotic used. (**b**) Venn diagram classifying the genes identified as being significantly downregulated following antibiotic treatment into sets according to their response to moenomycin (Moe), vancomycin (Van), and bacitracin (Bac). (**c**) Similar Venn classification of the significantly up-regulated genes. (**d**) Functional analysis of the 118 genes identified as being significantly downregulated in response to all three of the antibiotics used. (**e**) Functional analysis of the 243 genes identified as being significantly up-regulated in response to all three of the antibiotics used

with in-house manually curated lists of functionally related genes gleaned from the literature. This is achieved using the "find similar entity lists" tool in GeneSpring or by running Fisher's exact tests for over-representation with the "stats" package in R. For illustration, Fig. 3 contains a summary of the functional enrichment analysis results obtained using the genes found to be significantly downregulated (Fig. 3d) and up-regulated (Fig. 3e) in response to all of the different antibiotics tested. The method effectively reduces the complexity of the results, highlighting a small number of biological processes that are represented by the hundreds of genes identified as being significantly differently expressed in the experiments.

3.8 Database Submission

Transcriptome data is both expensive to produce and information rich. To maximize the usefulness of the data and to create a long-term record, submit the data to a suitable open-access database repository, e.g., ArrayExpress at EMBL-EBI (https://www.ebi.ac.uk/arrayexpress/) or NCBI's Gene Expression Omnibus (http://www.ncbi.nlm.nih.gov/geo/). The microarray data from this study is available from ArrayExpress under accession number E-MEXP-3032.

4 Notes

1. Prepare $CaCl_2$ as 1 M stock solution and autoclave. To make the final concentration of 0.01 M $CaCl_2$ in 2xGM, add 100 μL of 1 M $CaCl_2$ solution in total 10 mL of 2xGM.
2. Minor element solution: 0.1% (w/v) $ZnSO_4 \cdot 7H_2O$, 0.1% (w/v) $FeSO_4 \cdot 7H_2O$, 0.1% (w/v) $MnCl_2 \cdot 4H_2O$, 0.1% (w/v) $CaCl_2$ anhydrous. Make a fresh solution every 2–4 weeks and store at 4 °C.
3. Prepare 0.1 M solutions of NaH_2PO_4 and K_2HPO_4 separately, and then mix together in equal volumes. Adjust pH to 6.8.
4. RNase enzymes are a significant problem when handling RNA under laboratory conditions. All reagents, tubes, and pipette tips used in RNA work should be RNase free, and suitable powder-free gloves should be worn at all times. Care should be taken not to contaminate the gloves by touching skin or hair. RNase-free aqueous solutions and 1.5 mL tubes are most conveniently prepared by double-autoclaving at 115 °C. Clean any surfaces/instruments which will come directly into contact with the RNA samples using RNaseZAP.
5. Phenol is toxic and hazardous. Handle carefully according to local safety rules.
6. All buffers and solutions for hybridization should be prepared either in Milli-Q water or molecular biology-grade

diethylpyrocarbonate (DEPC)-treated water. For this study, DEPC-treated water was used.

7. For 250 mL of stock solution, dissolve 16.15 g of MES hydrate and 48.25 g of MES sodium salt in 200 mL of molecular biology-grade water. The pH should be adjusted to between 6.5 and 6.7 and then sterilize the solution by filtering through a 0.2 μm filter. Store at 4 °C and shield from light. Discard the solution if it turns yellow on storage.
8. For 50 mL, add 8.3 mL of 12× MES stock solution, 17.7 mL of 5 M NaCl, 4.0 mL of 0.5 M EDTA, and 0.1 mL of 10% Tween-20 into 19.9 mL water. Do not autoclave. Store at 4 °C and shield from light. Discard the solution if it turns yellow on storage.
9. For 1 L, add 300 mL of 20× SSPE and 1 mL of 10% Tween-20 into 699 mL of water. Sterilize the solution by filtering through a 0.2 μm filter.
10. For 250 mL, add 20.83 mL of 12× MES stock solution, 1.3 mL of 5 M NaCl, and 0.25 mL of Tween-20 into 227.5 mL of water. Sterilize the solution by filtering through a 0.2 μm filter. Store at 4 °C and shield from light. Discard the solution if it turns yellow on storage.
11. For 200 mL, add 41.7 mL of 12× MES stock solution, 92.5 mL of 5 M NaCl, and 2.5 mL of Tween-20 into 113.3 mL of water. Sterilize the solution by filtering through a 0.2 μm filter. Store at 4 °C and shield from light. Discard the solution if it turns yellow on storage.
12. Calculate the total volume required based on the number of microarrays being used. To make 220 μL (the volume required per chip in this study), mix 85 μL of biotin-labeled cDNA, 2.2 μL of oligo B2, 2.2 μL of Herring sperm DNA, 2.2 μL of 50 mg/mL bovine serum albumin (BSA) solution, 15.4 μL of DMSO, 110 μL of 2× hybridization buffer, and 3 μL of water. Only prepare this master mix immediately prior to use.
13. For 2× 700 μL volume (the given volumes are sufficient for one probe array), add 700 μL of 2× stain buffer, 56 μL of 50 mg/mL BSA, 14 μL of 1 mg/mL SAPE, and 630 μL of water. SAPE is light sensitive and should be stored in the dark at 4 °C, foil-wrapped. Do not freeze SAPE. Mix SAPE well before preparing the stain solution. Always prepare the stain solutions fresh, on the day of use.
14. For 700 μL, add 350 μL of 2× stain buffer, 56 μL 28 μL of 50 mg/mL BSA, 7 μL of 10 mg/mL goat IgG (ImmunoGlobulin G) stock, 4.2 μL of 0.5 mg/mL anti-streptavidin goat antibody, and 310.8 μL of water.
15. There are a variety of software tools available which will ultimately accomplish the same or similar statistical analysis. The

publication first reporting this data used a combination of the commercial tool GeneSpring (Agilent Technologies) and freely available packages from Bioconductor (http://www.bioconductor.org/) for use in the R statistical environment.

16. The statistical advice for the optimum number of biological replicates to perform is often "the more the better." Triplicates are a common practical compromise between the statistical power required to identify the majority of changes in gene expression taking place and the affordability of the experiment.
17. For this study, we treated cultures with three different cell wall-specific antibiotics, i.e., bacitracin, moenomycin A, and vancomycin, and ran an untreated control set of cultures. Prepare four times more germinated *S. coelicolor* M600 spores (20 mL in total) and inoculate equally into all flasks of NMMP medium used.
18. The protocol for RNA isolation in this study is essentially according to the manufacturer's instruction for use of the RNeasy Midi kit but modified to include an additional phenol/chloroform cleanup. Once started, proceed to the end of the protocol avoiding unnecessary delays.
19. Ethanol must be completely removed for the efficient elution of bound nucleotides from the column.
20. If a Bioanalyzer is not available for checking the integrity of the samples, RNA quality can be visually assessed following separation using RNA gel electrophoresis.
21. Prepare sufficient master mix solution to perform all reactions. When there are more than two samples, it is prudent to prepare ca. 10% more than is needed to compensate for potential pipetting inaccuracies or losses.
22. The efficiency of the labeling procedure can be assessed using the procedure described in Subheading 3.5. This quality control protocol prevents hybridizing poorly labeled target onto the probe array. The addition of biotin residues is monitored in a gel shift assay, where the fragments are incubated with avidin prior to electrophoresis. The nucleic acids are then detected by staining. The procedure takes approximately 90 min to complete. The absence of a shift pattern indicates poor biotin labeling. The problem should be addressed before proceeding to the hybridization step.
23. Choose a concise, unique name as once assigned it should be used to track the microarray all the way through to data analysis and storage.
24. The Fluidics Station 450 and GeneChip Scanner 3000 (Affymetrix) were used to automate the washing, staining, and scanning of the GeneChip expression probe arrays. After completing the procedures described in Subheading 3.6, the scanned probe array image

(.dat file) is converted to a .CEL file ready for downstream quality control and analysis.

25. The publication first reporting this data used a combination of the commercial tool GeneSpring (Agilent) and freely available packages from Bioconductor (http://www.bioconductor.org/) for use in the R statistical environment. The description here focuses on the original analysis but also provides open-source alternatives to the commercial tools. Detailed manuals are available for all the software packages used, so only an outline of the data processing is provided here.
26. Importing the .CEL file data into R requires an R package made from the Affymetrix chip description file (CDF) for the microarray used. Packages for commonly used microarrays are called automatically from within R when using the "affy" package. Packages for custom microarrays such as diSdiv712a can be made using the R package "makecdfenv."
27. Alternative methods are available for the normalization and summarization of microarray data but rma is one of the top performers [14].
28. Gene expression data can also be clustered using many other different algorithms and approaches [15].
29. Other popular tools for GO analysis include the BINGO app for cytoscape [16] and the "GOSTATS" package in R.

Acknowledgments

This work has been supported by funding from the Medical Research council, UK (G0700141) and the Royal Society, UK (516002.K5877/ROG).

References

1. Hesketh A, Hill C, Mokhtar J, Novotna G, Tran N, Bibb M et al (2011) Genome-wide dynamics of a bacterial response to antibiotics that target the cell envelope. BMC Genomics 12:226
2. Muthaiyan A, Silverman JA, Jayaswal RK, Wilkinson BJ (2008) Transcriptional profiling reveals that daptomycin induces the *Staphylococcus aureus* cell wall stress stimulon and genes responsive to membrane depolarization. Antimicrob Agents Chemother 52:980–990
3. Song Y, Lunde CS, Benton BM, Wilkinson BJ (2012) Further insights into the mode of action of the lipoglycopeptide telavancin through global gene expression studies. Antimicrob Agents Chemother 56:3157–3164
4. Liu X, Luo Y, Mohamed OA, Liu D, Wei G (2014) Global transcriptome analysis of *Mesorhizobium alhagi* CCNWXJ12-2 under salt stress. BMC Microbiol 14:1
5. Lechner S, Prax M, Lange B, Huber C, Eisenreich W, Herbig A et al (2014) Metabolic and transcriptional activities of *Staphylococcus aureus* challenged with high-doses of daptomycin. Int J Med Microbiol 304:931–940
6. Scherl A, François P, Charbonnier Y, Deshusses JM, Koessler T, Huyghe A et al (2006) Exploring glycopeptide-resistance in *Staphylococcus aureus*: a combined proteomics and transcriptomics approach for the identification of resistance-related markers. BMC Genomics 7:296

7. Delauné A, Dubrac S, Blanchet C, Poupel O, Mäder U, Hiron A et al (2012) The WalKR system controls major staphylococcal virulence genes and is involved in triggering the host inflammatory response. Infect Immun 80: 3438–3453
8. Falord M, Mäder U, Hiron A, Débarbouillé M, Msadek T (2011) Investigation of the *Staphylococcus aureus* GraSR regulon reveals novel links to virulence, stress response and cell wall signal transduction pathways. PLoS One 6:e21323
9. Shikuma NJ, Davis KR, Fong JN, Yildiz FH (2013) The transcriptional regulator, CosR, controls compatible solute biosynthesis and transport, motility and biofilm formation in *Vibrio cholerae*. Environ Microbiol 15:1387–1399
10. Reyes LH, Abdelaal AS, Kao KC (2013) Genetic determinants for n-butanol tolerance in evolved *Escherichia coli* mutants: cross adaptation and antagonistic pleiotropy between n-butanol and other stressors. Appl Environ Microbiol 79:5313–5320
11. Creecy JP, Conway T (2015) Quantitative bacterial transcriptomics with RNA-seq. Curr Opin Microbiol 23:133–140
12. R Core Team (2014) R: a language and environment for statistical computing. R Foundation for Statistical Computing, Vienna, Austria, http://www.R-project.org/
13. Bauer S, Grossman S, Vingron M, Robinson PN (2008) Ontologizer 2.0—a multifunctional tool for GO term enrichment analysis and data exploration. Bioinformatics 24:1650–1651
14. McCall MN, Almudevar A (2012) Affymetrix GeneChip microarray preprocessing for multivariate analyses. Brief Bioinform 13:536–546
15. Do JH, Choi DK (2008) Clustering approaches to identifying gene expression patterns from DNA microarray data. Mol Cells 25:279–288
16. Maere S, Heymans K, Kuiper M (2005) BiNGO: a Cytoscape plugin to assess overrepresentation of gene ontology categories in biological networks. Bioinformatics 21:3448–3449

Chapter 4

Cell Shaving and False-Positive Control Strategies Coupled to Novel Statistical Tools to Profile Gram-Positive Bacterial Surface Proteomes

Nestor Solis and Stuart J. Cordwell

Abstract

A powerful start to the discovery and design of novel vaccines, and for better understanding of host-pathogen interactions, is to profile bacterial surfaces using the proteolytic digestion of surface-exposed proteins under mild conditions. This "cell shaving" approach has the benefit of both identifying surface proteins and their surface-exposed epitopes, which are those most likely to interact with host cells and/or the immune system, providing a comprehensive overview of bacterial cell topography. An essential requirement for successful cell shaving is to account for (or minimize) cellular lysis that can occur during the shaving procedure and thus generate data that is biased towards non-surface (e.g., cytoplasmic) proteins. This is further complicated by the presence of "moonlighting" proteins, which are proteins predicted to be intracellular but with validated surface or extracellular functions. Here, we describe an optimized cell shaving protocol for Gram-positive bacteria that uses proteolytic digestion and a "false-positive" control to reduce the number of intracellular contaminants in these datasets. Released surface-exposed peptides are analyzed by liquid chromatography (LC) coupled to high-resolution tandem mass spectrometry (MS/MS). Additionally, the probabilities of proteins being surface exposed can be further calculated by applying novel statistical tools.

Key words Cell shaving, Gram-positive bacteria, Mass spectrometry, Surface proteomics, Surfaceome

Abbreviations

CID	Collision-induced dissociation
DTT	Dithiothreitol
ESI	Electrospray ionization
IAA	Iodoacetamide
LC-MS/MS	Liquid chromatography-tandem mass spectrometry
MeCN	Acetonitrile

Hee-Jeon Hong (ed.), *Bacterial Cell Wall Homeostasis: Methods and Protocols*, Methods in Molecular Biology, vol. 1440, DOI 10.1007/978-1-4939-3676-2_4,

1 Introduction

1.1 Bacterial Surface Topography

Interactions between an organism and its environment are initially mediated by the interplay of surface-exposed proteins and other macromolecular structures on the exterior face of the cell. As a major example, it is the ability of a pathogen to recognize its environment and respond to it that enables colonization and infection in the host. Interactions between bacterial and host surfaces and extracellular matrix further facilitate pathogenesis. As such, surface-exposed proteins are critical for understanding adherence, colonization, and disease progression caused by bacterial pathogens. Furthermore, surface structures including lipopolysaccharide, capsule, and proteins are the first markers recognized by the human immune response and thus knowledge of the topography of bacterial cells is crucial for vaccine design.

The cell surface in Gram-positive bacteria consists of a thick peptidoglycan wall and an inner cytoplasmic membrane. There are four major groups of cell wall/envelope proteins, including those anchored to the cytoplasmic membrane by hydrophobic domains, lipoproteins, cell wall proteins anchored by sortase via an LPXTG signal, and non-covalently cell wall-associated proteins [1–4]. Some of these proteins may remain buried within the envelope and are thus not truly "surface exposed." Membrane-embedded proteins are characterized by the Ala-X-Ala *N*-terminal signal peptidase I recognition sequence, while lipoproteins are covalently anchored to phospholipid and contain the signal peptidase II recognition sequence, Leu-Ala-Ala-Cys. Cell wall-anchored proteins contain an *N*-terminal *Sec* signal and a *C*-terminal LPXTG motif. Such proteins are retained in the membrane by a hydrophobic *C*-terminal domain and cleaved by sortase.

1.2 Cell Shaving Proteomics

Identification of those proteins representing the true "surfaceome" of an organism represents a rich reservoir of information that can be utilized in the production of novel therapeutics and vaccines, based on either individual proteins or multiple combined peptide epitopes. Substantial technical progress has been made in subcellular proteomics of bacterial pathogens, with several studies describing methods for enriching outer membrane, periplasmic, and secreted proteins. These studies however do not provide specific assignment of surface-exposed proteins, nor those peptide epitopes located outside the cell. Due to their low abundance and hydrophobic transmembrane regions, surface proteins are generally considered very difficult to enrich from among complex protein mixtures. Several methods for analysis of membrane-associated proteins have been proposed, including (1) surface labeling by biotinylation and capture through streptavidin affinity; (2) precipitation, density gradient ultracentrifugation, and detergent extraction; and (3) detergent-phase partitioning. Such methods are useful for enrichment of membrane-associated proteins prior to separation and analysis using gel-based and gel-free approaches; however false positives may occur in these protocols due to cell lysis or residual biotin resulting in labeling of cytoplasmic contaminants. Additionally, specific analysis of surface proteins and their surface-exposed epitopes has remained challenging.

A novel method for better understanding bacterial surface protein topology involves cell "shaving," where a proteolytic enzyme is incubated with whole cells to release their exposed peptide epitopes while maintaining cell integrity (Fig. 1) [5–12]. This approach, combined with the resolution of LC-MS/MS, provides a peptide repertoire of surface-exposed epitopes belonging to surface and membrane-associated proteins. The method provides a simple and fast route for the gentle digestion of entire cells, purification of released peptides, and proteomic identification. A false-positive control strategy [9] can also be employed to better control for cell lysis and the release of intracellular proteins. Furthermore, using novel statistical tools and bioinformatic predictions, a probability can be calculated for the likelihood of any identified protein being surface exposed [11]. A final high-confidence list of proteins can then be functionally validated.

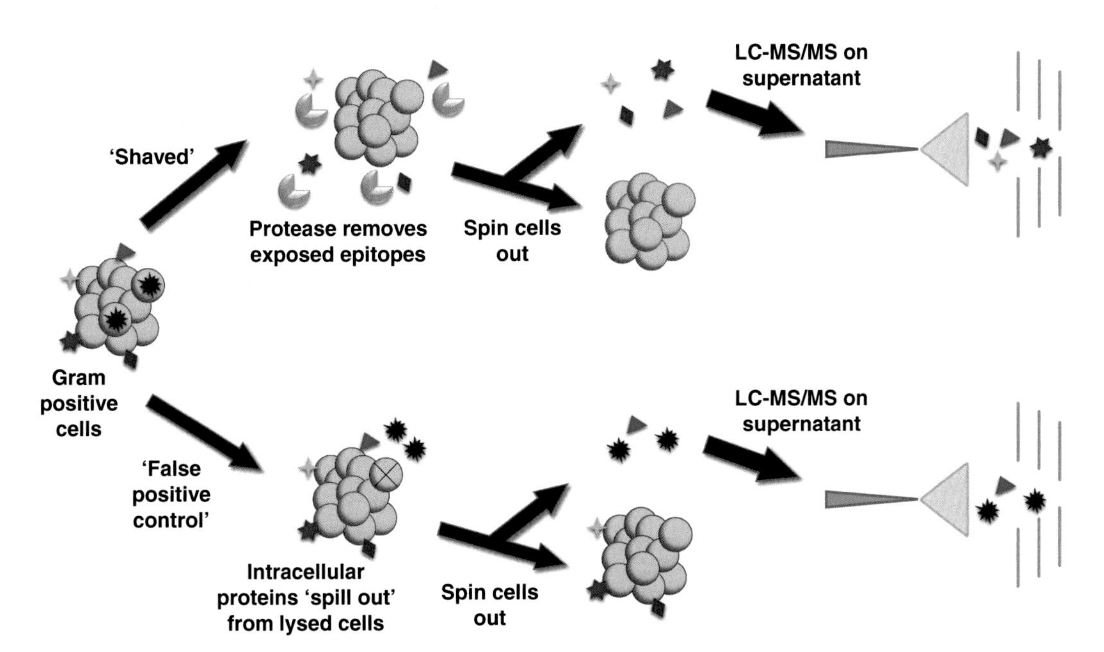

Fig. 1 Cell shaving and false-positive control strategy for Gram-positive cell surfaceomics. (*Upper*) Cell shaving; whole cell-shaved fractions are generated by high concentration, short-duration incubation with a protease (generally trypsin) in isotonic buffer. Released peptides are collected and analyzed by LC-MS/MS. (*Lower*) False-positive control; false-positive control is used to identify proteins released by cell lysis. Cells are incubated as for the shaving protocol but no protease is included. Whole cells are removed by centrifugation and the supernatants then digested with a protease. Any identified peptides are present as a result of lysis and are analyzed by LC-MS/MS

2 Materials

2.1 Growth of Microorganisms

1. Frozen stock of pure culture of Gram-positive organism to be analyzed.
2. Agar plates (12%) with media of choice (e.g., Luria-Bertani Broth, Tryptic Soya Broth).
3. Liquid media of choice (e.g., Luria-Bertani Broth, Tryptic Soya Broth) with supplements as required.

2.2 Sample Preparation for Cell Shaving

1. Conical bottom sterile 50 mL volume Falcon tubes.
2. Wash buffer (150 mM NaCl, 20 mM Tris–HCl, pH 7.5).
3. Digestion buffer (150 mM NaCl, 20 mM Tris–HCl, 1 M D-arabinose, 10 mM $CaCl_2$, pH 7.5) (*see* **Note 1**).
4. Sequencing-grade trypsin (vial of 20 μg).
5. Formic acid (HPLC grade).
6. 2 mL Tubes with 1 kDa dialysis membrane cutoffs (e.g., Mini-Dialysis Kit with 1 kDa cutoff, GE Life Sciences, Uppsala, Sweden).
7. 0.22 μm Filters (suitable for use with a handheld 2 mL syringe).
8. 4 L Buckets of cold distilled water for dialysis.
9. Vacuum centrifuge.
10. Dithiothreitol (DTT; 1 M stock).
11. Iodoacetamide (IAA; 0.5 M stock).
12. Acetonitrile (MeCN; 100% stock).
13. C_{18} material for peptide purification (e.g., POROS R2 resin) using home-packed columns.

2.3 Mass Spectrometry and Data Analysis

1. LC-MS/MS system capable of high-speed, high-sensitivity data-dependent acquisition.
2. Nanoflow HPLC system (e.g., Agilent 1100/1200 series or Thermo Scientific EasyLC system).
3. Full genome sequence of organism converted to translated proteome (FASTA format).
4. Database search engine (e.g., MASCOT).

3 Methods

3.1 Cell Shaving and False-Positive Control

1. Inoculate from a stock onto an agar plate to generate a pure culture of the Gram-positive organism to be examined. Grow at the desired temperature until colonies are visible.
2. Inoculate a single colony into the desired broth and grow until $OD_{600} > 1$.

3. For each experiment (shaved and false-positive control), aliquot 200 μL of turbid culture into 19.8 mL of fresh media (1:100 dilution) in a conical shaped sterile tube (50 mL Falcon tube) and grow under desired conditions until mid-log phase. Precool a swing-bucket rotor to 4 °C (*see* **Note 2**).
4. Following growth to mid-log phase, place each tube on ice for 5 min and then centrifuge in a precooled swing-bucket rotor at 1000 × *g* for 15 min at 4 °C.
5. Carefully decant supernatant into waste.
6. Resuspend the cell pellet with ice-cold wash buffer (*see* **Note 3**).
7. Centrifuge in the precooled swing-bucket rotor centrifuge at 1000 × *g* for 15 min at 4 °C and then carefully decant the supernatant into waste.
8. Repeat **steps 6** and 7 another two times, for a total of three washes (*see* **Note 4**). During spin steps prepare fresh 4 mL of digestion buffer. Resuspend 10 μg of sequencing-grade trypsin with 10 μL digestion buffer immediately before the next step and keep on ice.
9. Carefully resuspend the cell pellets for the control and shaved experiments in a total of 4 mL digestion buffer. Slowly invert and keep a homogenous mixture (*see* **Note 3**).
10. Split 2 mL each into two separate large low-protein-binding microfuge (2 mL sized) tubes.
11. To one of the tubes add the ice-cold trypsin—this will be the cell-shaved fraction.
12. Place both tubes on a rotator and spin slowly in a 37 °C controlled room or incubator for 15 min (*see* **Notes 5** and **6**).
13. Immediately after digestion, place the tubes on ice and centrifuge at 1000 × *g* for 10 min at 4 °C. During this centrifugation, prepare new 2 mL low-protein-binding tubes labeled "shaved" and "control," as well as two dialysis filter membranes (1 kDa cutoff) by gently rinsing in water.
14. Remove supernatants carefully by pipetting into fresh microfuge tubes (*see* **Note** 7).
15. Pipette each fraction into separate 2 mL syringes each with a 0.22 μm filter at the end. Pass the solution through the filter directly into separate pre-washed dialysis tubes.
16. Screw the dialysis membranes onto the tubes and dialyze in 4 L of water at 4 °C for 3 h and then replace the water with 4 L fresh cold water and dialyze overnight. Next morning, replace the water one more time for a further 3-h dialysis.
17. Recover samples from dialysis tubes into 2 mL microfuge tubes and concentrate by vacuum centrifugation to 100 μL.

18. Reduce samples with DTT to a final concentration of 10 mM for 1 h at 37 °C and then alkylate with IAA to a final 15 mM concentration at room temperature in the dark. Quench with additional DTT to a final 20 mM.
19. To the false-positive "control" fraction add 1 μg trypsin in digestion buffer and digest overnight at 37 °C.
20. Acidify both samples to a final 0.1% formic acid.
21. Activate a C_{18} micro-column in 70% MeCN and 0.1% formic acid. Equilibrate the column twice with 50 μL 0.1% formic acid and load with 50 μL sample. Wash twice with 50 μL 0.1% formic acid and elute peptides with 50 μL 70% MeCN and 0.1% formic acid.
22. Lyophilize purified peptides to complete dryness and store at −20 °C until required for mass spectrometric analysis.

3.2 LC-MS/MS (See Note 8)

1. Purified peptide supernatants are separated by reversed-phase nanoflow LC (e.g., using an EASY-nLC [Thermo Scientific, San Jose CA]).
2. Peptides are resolved using a one column reversed-phase (3 μm particle size, 50 cm × 50 μm inner diameter [I.D.], C_{18}) setup over a linear gradient of 0–40% buffer B (80% MeCN, 0.1% formic acid) at 250 nL/min over 103 min (*see* **Note 9**).
3. Peptides are eluted into the mass spectrometer via electrospray ionization (ESI).
4. Operate the mass spectrometer in data-dependent acquisition mode, which automatically switches between MS and MS/MS. Depending on the mass spectrometer, for each MS scan, the 3–30 most intense peptide ions are automatically selected for fragmentation by collision-induced dissociation (CID) (*see* **Note 10**).

3.3 Data Analysis (See Notes 11 and 12)

1. Search raw MS files in a database search engine of choice. RAW files generated by an LTQ Orbitrap XL are searched in the Proteome Discoverer environment using SEQUEST with an MS1 tolerance of 10 ppm and an MS2 tolerance of 0.8 Da. Allow for four missed cleavages as discussed in [7]. Variable modifications should include oxidation of methionine and carbamidomethylation of cysteines. Semi-tryptic protease specificity can also be employed to maximize coverage of surface-exposed peptides (*see* **Note 13**).
2. Determine the predicted localization of all protein hits identified by database searching. This can be done with a variety of tools such as PSORTb [13], SurfG+ [14], or LocateP [15] (Fig. 2) (*see* **Note 14**).

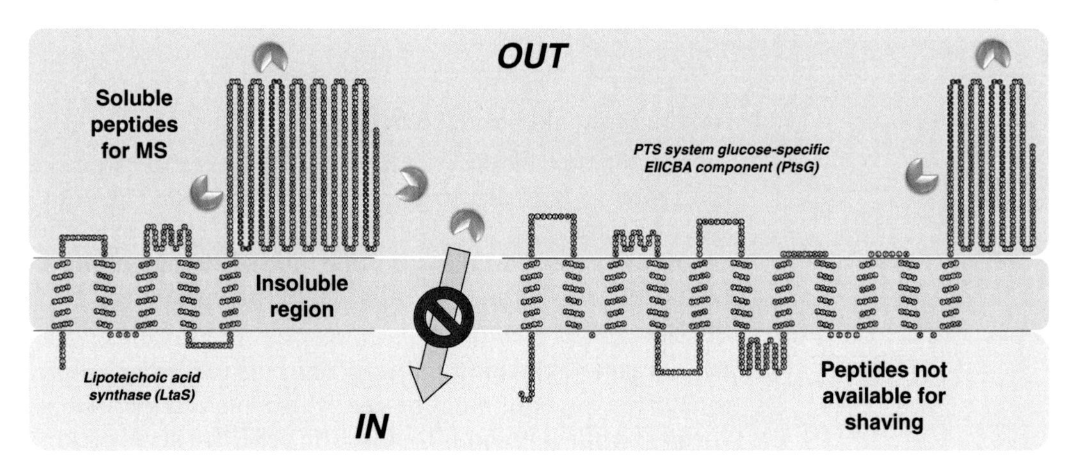

Fig. 2 Cell shaving data and surface-exposed peptide topology prediction (SurfG+) from 2 *Staphylococcus aureus* proteins. (*Left*) Lipoteichoic acid synthase (LtaS). (*Right*) PTS system glucose-specific EIICBA component (PtsG). Amino acid sequences shown in red were identified by cell shaving of *S. aureus* cells. Transmembrane and internal amino acid sequences are intractable to shaving, as they are not exposed

3. Once protein and peptide lists have been compiled for shaved and false-positive control fractions, a statistical methodology can be employed to determine the likelihood of a protein being surface exposed based on the number of peptides identified in the shaved and false-positive control fraction [11] using this equation:

$$p_{\text{experimental}} = 1 - \sum_{k=\frac{m}{2}}^{m} \frac{\binom{n_c}{k}\binom{n_s}{m-k}}{\binom{n_T}{m}}$$

where n_c = number of control peptides, n_s = number of shaved peptides, $n_T = n_c + n_s$, and $m = 0.4 \times n_T$ (to closest higher integer):

$$p_{\text{adjusted}} = \frac{(p_{\text{experimental}})(p_{\text{predicted}})}{(p_{\text{experimental}})(p_{\text{predicted}}) + (1 - p_{\text{experimental}})(1 - p_{\text{predicted}})}$$

4. Protein lists with number of peptides per identified protein should be compiled and the probabilities of each protein being surface exposed calculated from **step 2**. These can be used as an input to calculate (a) experimental probability of the protein being surface exposed and (b) the adjusted probability (accounting for the predictions made in **step 2** and based on sequence predictions) of the protein being surface exposed. These can be run directly on https://github.com/mehwoot/cellshaving to provide a final score for each protein.

4 Notes

1. Arabinose should be added fresh before use.
2. Mid-log phase typically contains cells that are most robust and less prone to lysis due to active cellular division. The conical shape of the Falcon tube is important for pelleting cells at low centrifugal speeds. At this step, one tube of 20 mL culture will be for cell shaving and one tube of 20 mL culture will be for the false-positive control.
3. Do not pipette the buffer directly onto the cell pellet—pipette against the inside of the tube gently to avoid cell lysis. Do not vortex. Gentle inversions to wash the pellet are appropriate.
4. Additional washes ensure removal of media components and any loosely bound, non-surface-specific proteins on the exterior of the cell surface.
5. For optimizing incubation periods, obtain 10 μL aliquots every 5 min for each treatment and perform cell counts using a hemocytometer under a phase-contrast microscope. This will give an estimate of the number of intact cells.
6. Digestion times are preferably short to minimize cell lysis and as such require higher amounts of trypsin to achieve proteolysis. However, lower amounts of trypsin (2–5 μg) may be required for cells with less rigid cell walls.
7. It is best to leave a small volume close to the cell pellet for higher purity. At this stage, it is also optional to acidify with formic acid to a final 0.1 % to stop any proteolytic digestion.
8. Cell shaving protocols are reliant on comprehensive peptide coverage in the relatively simple fractions generated by tryptic digestion of cell surfaces. Therefore, access to a high-speed, high-sensitivity mass spectrometer is essential for proper determination of peptide sequences representing surface-exposed proteins.
9. Any typical nanoflow reversed-phase LC setup will be compatible with these peptide analyses.
10. Settings on the mass spectrometer should be optimized and will be instrument dependent.
11. Raw MS data being analyzed for cell shaving experiments are best searched against the strain of the organism in question to overcome issues associated with point mutations or posttranslational modifications.
12. Data analysis in many cases depends on the acquisition mass spectrometer and proprietary software. Here, we describe an analysis workflow used for cell shaving data acquired on an LTQ Orbitrap XL and the use of Proteome Discoverer (Thermo Scientific) software.

13. It is important to consider that the identified proteins may not necessarily generate a large sequence coverage (as typically found in many proteins during shotgun/bottom-up proteomics experiments), since only their surface-exposed regions should be cleaved and these may not have amino acid sequences flanked by lysine and arginine residues suitable for tryptic cleavage and MS analysis. Hence, one peptide per protein is the minimum requirement for identification and we suggest manual verification of these "one-hit wonders" to ensure correct sequence assignation.
14. It is important to either keep these localizations as a reference for later validation or preferably have a probability score associated with their surface localization that can be used in the next step.

References

1. Schneewind O, Missiakas D (2014) Sec-secretion and sortase-mediated anchoring of proteins in Gram-positive bacteria. Biochim Biophys Acta 1843:1687–1697
2. Buist G, Steen A, Kok J, Kuipers OP (2008) LysM, a widely distributed protein motif for binding to (peptido)glycans. Mol Microbiol 68:838–847
3. Scott JR, Barnett TC (2006) Surface proteins of gram-positive bacteria and how they get there. Annu Rev Microbiol 60:397–423
4. Navarre W, Schneewind O (1999) Surface proteins of gram positive bacteria and mechanisms of their targeting to the cell wall envelope. Microbiol Mol Biol Rev 63: 174–229
5. Dreisbach A, van der Kooi-Pol MM, Otto A, Gronau K, Bonarius HP, Westra H, Groen H, Becher D, Hecker M, van Dijl JM (2011) Surface shaving as a versatile tool to profile global interactions between human serum proteins and the *Staphylococcus aureus* cell surface. Proteomics 11:2921–2930
6. Olaya-Abril A, Jimenez-Munguia I, Gomez-Gascon L, Rodriguez-Ortega MJ (2013) Surfomics: shaving live organisms for a fast proteomic identification of surface proteins. J Proteomics 97:164–176
7. Rodriguez-Ortega MJ, Norais N, Bensi G, Liberatori S, Capo S, Mora M, Scarselli M, Doro F, Ferrari G, Garaguso I, Maggi T, Neumann A, Covre A, Telford JL, Grandi G (2006) Characterization and identification of vaccine candidate proteins through analysis of the group A *Streptococcus* surface proteome. Nat Biotechnol 24:191–197
8. Solis N, Cordwell SJ (2011) Current methodologies for proteomics of bacterial surface-exposed and cell envelope proteins. Proteomics 11:3169–3189
9. Solis N, Larsen MR, Cordwell SJ (2010) Improved accuracy of cell surface shaving proteomics in *Staphylococcus aureus* using a false-positive control. Proteomics 10:2037–2049
10. Tjalsma H, Lambooy L, Hermans PW, Swinkels DW (2008) Shedding & shaving: disclosure of proteomic expressions on a bacterial face. Proteomics 8:1415–1428
11. Solis N, Parker BL, Kwong SM, Robinson G, Firth N, Cordwell SJ (2014) *Staphylococcus aureus* surface proteins involved in adaptation to oxacillin identified using a novel cell shaving approach. J Proteome Res 13:2954–2972
12. Severin A, Nickbarg E, Wooters J, Quazi SA, Matsuka YV, Murphy E, Moutsatsos IK, Zagursky RJ, Olmsted SB (2007) Proteomic analysis and identification of *Streptococcus pyogenes* surface-associated proteins. J Bacteriol 189:1514–1522
13. Yu NY, Wagner JR, Laird MR, Melli G, Rey S, Lo R, Dao P, Sahinalp SC, Ester M, Foster LJ, Brinkman FS (2010) PSORTb 3.0: improved protein subcellular localization prediction with refined localization subcategories and predictive capabilities for all prokaryotes. Bioinformatics 26:1608–1615
14. Barinov A, Loux V, Hammani A, Nicolas P, Langella P, Ehrlich D, Maguin E, van de Guchte M (2009) Prediction of surface exposed proteins in *Streptococcus pyogenes*, with a potential application to other Gram-positive bacteria. Proteomics 9:61–73
15. Zhou M, Boekhorst J, Francke C, Siezen RJ (2008) LocateP: genome-scale subcellular-location predictor for bacterial proteins. BMC Bioinformatics 9:173

Chapter 5

Differential Proteomics Based on Multidimensional Protein Identification Technology to Understand the Biogenesis of Outer Membrane of *Escherichia coli*

Alessandra M. Martorana, Sara Motta, Paola Sperandeo, Pierluigi Mauri, and Alessandra Polissi

Abstract

Cell envelope proteins in bacteria are typically difficult to characterize due to their low abundance, poor solubility, and the problematic isolation of pure surface fraction with only minimal contamination. Here we describe a method for cell membrane fractionation followed by mass spectrometry-based proteomics to analyze and determine protein abundance in bacterial membranes.

Key words MudPIT, Membrane proteins, Gel-free, Gram-negative cell envelope, Outer membrane, Cell fractionation

1 Introduction

The traditional proteomics approach is based on two-dimensional gel electrophoresis (2DG), where the protein spots of interest are isolated and identified by mass spectrometry (MS) via in-gel digestion. The 2DG approach has a relatively high resolution, which is limited however by the difficulty in detecting certain classes of proteins. These include membrane proteins due to their low solubility in gel electrophoresis buffer, proteins with either low (<10 kDa) or high (>200 kDa) molecular weight (MW), as well as those with an extreme isoelectric point (pI<4 or >9). Moreover, this approach has a limited dynamic range, which impairs the analysis of low-abundance proteins and it is tedious and time consuming.

These problems are overcome by "mass spectrometry-based proteomics," using approaches such as Multidimensional Protein Identification Technology (MudPIT), which is a shotgun proteomics methodology [1]. MudPIT represents a fully automated technology based on two-dimensional capillary chromatography coupled to tandem mass spectrometry (2DC-MS/MS).

Hee-Jeon Hong (ed.), *Bacterial Cell Wall Homeostasis: Methods and Protocols*, Methods in Molecular Biology, vol. 1440, DOI 10.1007/978-1-4939-3676-2_5, © Springer Science+Business Media New York 2016

2DC-MS/MS combines ion exchange with reversed-phase separation of peptide mixtures obtained from direct digestion of total (or pre-fractioned) proteins by means of two micro- or nano-HPLC columns and direct analysis of eluted peaks by data-dependent fragmentation (MS/MS) [2]. This technology provides a significant improvement over gel-based proteomic analysis; in fact, MudPIT approach enables the quantitative determination of proteins [3] and the identification of the protein mixtures in wide pI and MW ranges. In addition, mass spectrometry-based proteomics approach allows the characterization of hydrophobic proteins making possible to analyze the proteome of membranes, from both eukaryotic [4] and prokaryotic cells [5].

The outer membrane (OM) is the hallmark of gram-negative bacteria. The cell envelope of these diderm organisms consists two membranes that delimit an aqueous space, the periplasm, containing a thin layer of peptidoglycan. The inner (cytoplasmic) membrane (IM) is composed by phospholipids whereas the OM is an atypical asymmetric membrane composed of phospholipids in the inner leaflet and a unique glycolipid, lipopolysaccharide (LPS), in the outer leaflet [6]. IM and OM contain many integral and peripheral proteins as well as lipoproteins to fulfil the numerous functions played by the bacterial envelope including metabolic activities, signal transduction, interaction with the environment, adhesion, and immune evasion [7–9]. Based on their different structure and composition, IM and OM possess different densities and these properties allow their separation onto a discontinuous sucrose gradient [10, 11].

Here, we combine cell fractionation procedures to separate IM and OM membranes and therefore their respective proteins, with the MudPIT technology to analyze and characterize the membranome of *Escherichia coli*.

2 Materials

Prepare all solutions using ultrapure water and analytical grade reagents.

Diligently follow all waste disposal regulations when disposing waste materials.

Prepare and store all reagents at room temperature (unless indicated otherwise).

2.1 Membrane Purification and Fractionation

1. LD medium: Dissolve 10 g bacto-tryptone, 5 g yeast extract, and 5 g NaCl in 900 mL deionized water. Adjust pH to 7.5 with NaOH. Make up to 1 L with deionized water. Sterilize by autoclaving.
2. Potassium phosphate buffer, pH 8: Prepare solution A dissolving 27.8 g $KH_2PO_4{\cdot}H_2O$ in 1 L deionized water, and solution

B dissolving 53.65 g $K_2HPO_4{\cdot}7H_2O$ in 1 L deionized water. To obtain potassium phosphate buffer 0.1 M mix 5.3 mL of solution A and 94.7 mL of solution B, and make up to 400 mL with deionized water.

3. Tris–HCl pH 8.0, 1 M: Dissolve 121.1 g of Tris base in 800 mL of water. Adjust pH 8 by adding concentrated HCl. Make up to 1 L with water and sterilize by autoclaving.
4. EDTA 0.5 M: Weigh 186.12 g EDTA, transfer it to 2 L beaker, and add 800 mL deionized water. While stirring vigorously on a magnetic stirrer, add NaOH pellet or 10 N NaOH to adjust the solution to pH 8.0 (*see* **Note 1**). Make up to 1 L with water. Filtration can be used to remove any undissolved material.
5. Phenylmethylsulfonyl fluoride (PMSF): 0.1 M solution. Store at 4 °C.
6. Lysozyme: Dissolve solid lysozyme at a concentration of 10 mg/mL in 10 mM Tris–HCl pH 8, immediately before use.
7. DNase: Dissolve solid DNase at 50 mg/mL in MilliQ water. Store at −20 °C.
8. Sucrose solution, 0.25 M: Weigh 85.57 g sucrose, transfer it to 2 L beaker, and add 800 mL of 10 mM Tris–HCl pH 8.0. Mix vigorously on a magnetic stirrer until the sucrose is dissolved. Make up to 1 L with 10 mM Tris–HCl pH 8.0.
9. Solutions of increasing sucrose concentration: Weigh 855.6 g sucrose, transfer it to 2 L beaker, add 460 mL of 10 mM Tris–HCl pH 8 and 5 mM EDTA, and mix vigorously on a magnetic stirrer until the sucrose is dissolved to obtain a 65 % (w/w) sucrose solution. 55, 50, 45, 40, 35, 30, and 25 % sucrose solutions are obtained by dilution in 10 mM Tris–HCl pH 8 and 5 mM EDTA. Store at 4 °C.

2.2 Resolution of Inner and Outer Membranes

1. β-Nicotinamide adenine dinucleotide (NADH), 50 mM solution: Weigh 35 mg of solid NADH and dissolve it in 1 mL of Tris–HCl 50 mM Tris–HCl pH 7.5. Make it fresh each time and keep on ice and away from light.
2. Pierce™ Coomassie Protein Assay: Store at 4 °C.
3. Albumin standard ampules (BSA), 2 mg/mL: Prepare by serial dilutions (2×) of BSA standards: 2, 1, 0.5, 0.25, and 0.125 mg/mL.
4. Loading buffer (5×): 0.12 M Tris–HCl (pH 6.8), 5.5 % SDS, 9 % β-mercaptoethanol, 0.01 % bromophenol blue (BPB), 15 % glycerol.
5. Acrylamide/Bis-acrylamide 37.5/1, 30 % solution.
6. SDS-PAGE running buffer (10×): Weigh 30 g Trizma base, 144 g glycine, 10 g SDS. Transfer Trizma base and SDS to 2 L

beaker and add 800 mL deionized water. While stirring vigorously on a magnetic stirrer, add glycine gradually. Make up to 1 L with water.

7. Resolving gel buffer: 1.37 M Tris–HCl, pH 8.8, 0.4% SDS. Dissolve 187 g Trizma base in 800 mL water. Add 40 mL of 10% SDS solution and adjust solution pH 8.8 with HCl. Make up to 1 L with water.
8. Stacking gel buffer: 0.45 M Tris–HCl, pH 8.8, 0.4% SDS. Dissolve 60.6 g Trizma base in 800 mL water. Add 40 mL of 10% SDS solution and adjust solution pH 6.8 with HCl. Make up to 1 L with water.
9. Ammonium persulfate: 10% solution in water. Store at 4 °C.
10. *N,N,N′,N′*-Tetramethylethylenediamine (TEMED): Store at 4 °C.
11. Prestained protein MW marker: Broad range 10–180 kDa. Coomassie blue staining solution: 0.15% w/v Coomassie Brilliant Blue R-250, 50% ethanol, 10% acetic acid solution in water.
12. Destaining solution: 30% ethanol, 10% acetic acid solution in water

2.3 MudPIT Analysis

1. Ultrapure water.
2. Ammonium bicarbonate (NH_4HCO_3) 0.1 M pH 8.0: Add about 50 mL water to a 250 mL graduated cylinder. Weigh 1.97 g ammonium bicarbonate and transfer to the 250 mL graduated cylinder. Add water to a volume of about 200 mL. Mix and adjust pH if necessary. Make up to 250 mL with water. Store at room temperature.
3. RapiGest SF (Waters Corporation, Milford, MA, USA): Store at 4 °C.
4. SPN™—Protein Assay (G-Biosciences, St Louis, MO, USA).
5. Trypsin: Split into aliquots of 1 μg and store at −20 °C.
6. Acetonitrile (CH_3CN), HPLC gradient grade.
7. Trifluoroacetic acid (TFA) eluent additive for LC-MS (CF_3COOH).
8. PepClean™ C-18 Spin Columns.
9. Vacuum system.
10. Formic acid (HCOOH), eluent additive for LC-MS, ~98%.
11. Evolution 60S, UV–visible spectrophotometer.

2.4 MS Analysis

1. 0.1% Formic acid in water (eluent A).
2. 0.1% Formic acid in acetonitrile (eluent B).
3. Ammonium chloride (NH_4Cl).

4. Solutions of increasing ammonium chloride concentration: 20, 40, 80, 120, 200, 400, and 700 mM. Prepare 1 mL for each concentration. In each 1 mL glass vial put 20, 40, 80, 120, 200, 400, and 700 μL of ammonium chloride and add, respectively, 980, 960, 920, 880, 800, 600, and 300 μL of 0.1% formic acid in water.
5. Surveyor AS (Thermo Finnigan Corp., San Jose, CA, USA).
6. Biobasic SCX column, 0.32 i.d. × 100 mm, 5 μm (Thermo Electron Corporation, Bellefonte, PA, USA).
7. Peptide trap (Zorbax 300 SB-C18, 0.3 mm × 5 mm, 5 μm) (Agilent Technologies, Santa Clara, CA, USA).
8. Biobasic C18 column (0.180 i.d. × 100 mm, 5 μm) (Thermo Electron Corporation, Bellefonte, PA, USA).
9. Nano-LC electrospray ionization source (nano-ESI) (Thermo Finnigan Corp., San Jose, CA, USA).
10. LCQ Deca XP plus or LTQ (Thermo Finnigan Corp., San Jose, CA, USA).

2.5 Computational Analysis

1. Cluster PC.
2. Non-redundant *Escherichia coli* protein sequence database downloaded from the NCBI website (http://www.ncbi.nlm.nih.gov/).
3. Bioworks version based on SEQUEST algorithm (University of Washington licensed to Thermo Fisher Scientific Inc., USA).
4. Proteome Discoverer version (Thermo Fisher Scientific Inc., USA).
5. MAProMa software (Multidimensional Algorithm Protein Map) [12].
6. EPPI (Experimental Proteotypic Peptide Investigator) [13].

3 Methods

Perform all procedures at room temperature unless otherwise specified.

The use of low-retention pipette tips and the use of glass vials or inserts are suggested in order to minimize sample loss, in particular for peptide mixtures.

The whole procedure is depicted in Fig. 1.

3.1 Bacterial Growth

Grow the *E. coli* bacterial strain static at 37 °C in 5 mL LD for 16–18 h. Set up the control and the treated cultures diluting the static pre-culture to an $OD_{600} = 0.05$ in 500 mL of suitable LD medium. Grow at 37 °C with aeration (160–180 rpm) for 300–330 min and harvest 125 OD_{600} of each culture by centrifugation

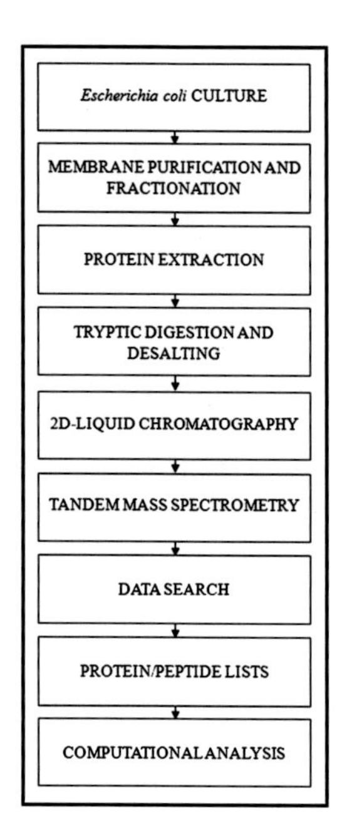

Fig. 1 Experimental procedure flow chart. General steps for the differential analysis of membrane proteome

at 3000 × *g* at 4 °C. Wash cells with 10 mL of 0.1 M potassium phosphate buffer pH 8.0 and collect cells by centrifugation at 3000 × *g* at 4 °C. Cell pellets can be stored at −20 °C before further processing.

3.2 Whole Membrane Purification

1. Resuspend cell pellets (125 OD_{600}) in 3 mL of 10 mM Tris (pH 8.0), 1 mM EDTA, 1 mM PMSF, and 0.2 mg/mL lysozyme and incubate on ice for 30 min. Add 0.2 mg/mL DNase and disrupt cells by a single passage through a cell disruptor at 25,000 psi (*see* **Note 2**).
2. Unbroken cells are removed by centrifugation at 3000 × *g* for 20 min at 4 °C. The cleared lysates were then subjected to ultracentrifugation at 100,000 × *g* for 60 min at 4 °C. Discard the supernatant, resuspend pellets that contain whole-cell membranes, in 0.4 mL of MilliQ water, and lyophilize for 16–18 h.
3. Lyophilized samples are then ready to be processed according to the method in Subheading 3.5 below.

3.3 Membrane Fractionation

1. Resuspend cell pellets (125 OD_{600}) in 12 mL of 10 mM Tris (pH 8.0), 1 mM EDTA, 1 mM PMSF, and 0.2 mg/mL lysozyme and incubate on ice for 30 min. Add 0.2 mg/mL DNase and disrupt cells by a single passage through a cell disruptor at 25,000 psi (*see* **Note 3**).
2. Unbroken cells are removed by centrifugation at 3000 × *g* for 20 min at 4 °C. The cleared lysates are then subjected to ultracentrifugation at 100,000 × *g* for 90 min in an ultracentrifuge.

Discard the supernatant and resuspend pellets that contain whole-cell membranes, in 9 mL of 0.25 M sucrose and 3.3 mM Tris pH 8.0.

3. Whole-cell membranes are ultracentrifuged at 100,000 ×*g* for 90 min in an ultracentrifuge. Discard supernatant and resuspend membranes in 0.8 mL 25 % sucrose. Save 50 μL for total protein quantification (see below).
4. Prepare a discontinuous sucrose gradient in a 12 mL polyallomer tube by layering the following sucrose solutions: 0.5 mL 55 %, 2 mL 50 %, 2 mL 45 %, 2 mL 40 %, 2 mL 35 %, and 2 mL 30 % (*see* **Note 4**). Load 750 μL of membranes in 25 % sucrose on the top of the sucrose gradient. Membranes are ultracentrifuged for 13 h at 240,000 ×*g* at 4 °C in a TH-641 swinging bucket rotor in an ultracentrifuge (*see* **Note 5**).
5. 0.3 mL Fractions (usually 36 fractions) are collected manually from the top of each tube and stored at −20 °C. Usually odd fractions are analyzed.

3.4 Resolution of Inner and Outer Membranes from Sucrose Gradient

The distribution of inner and outer membranes along the sucrose gradient is analyzed as follows: total protein profile along the gradient is assessed by the Bradford assay, the NADH activity is assayed as a marker of the inner membrane, and the profile of OmpC/F and OmpA porins, assessed by SDS-polyacrylamide gel electrophoresis, is used as a marker of outer membrane.

3.4.1 Total Protein Profile: Bradford Assay

Dilute the Bradford reagent twofold in deionized water (1 part Bradford:1 part water). Add 5–10 μL of each fraction to 1 mL of the diluted reagent and mix. Incubate at room temperature for 5 min and measure the blue color formed at 595 nm. Prepare a standard curve with samples of known protein concentration using a serial dilution series (0.125–2.0 mg/mL) of BSA. Using the standard curve extrapolate the amount of BSA in a given fraction; determine protein concentration in the fraction by dividing the amount of protein by the sample volume.

3.4.2 Inner Membrane Profile: NADH Activity Assay

Incubate mixtures containing 50 mM Tris–HCl pH 7.5, 0.3 mM NADH, and 50 μL of each fraction in a volume of 1 mL. The decrease in absorbance at 340 nm in 1 min is measured at room temperature in a JASCO V-550 spectrophotometer. The specific activity was calculated using an extinction coefficient of 6.22 mM/cm (*see* **Note 6**).

3.4.3 Outer Membrane Profile: SDS-Polyacrylamide Gel Electrophoresis and Coomassie Blue Staining

1. Prepare 12.5 % running gel by mixing 2.5 mL of resolving buffer, 4.16 mL of acrylamide solution, and 3.34 mL of water in a 50 mL conical flask. Add 50 μL of 10 % ammonium persulfate and 10 μL of TEMED, mix gently, and cast gels within a 1.0 mm spaced gel cassette (4.5 mL for each gel). Gently overlay with ethanol.

2. Prepare 4.5 stacking gel by mixing 2.5 mL of stacking buffer, 1.5 mL of acrylamide solution, and 6 mL water. Remove ethanol from the top of stacking gel with filter paper. Add 50 μL of 10% ammonium persulfate and 15 μL of TEMED to the mix and layer it on stacking gel. Insert a 10-well gel comb avoiding air bubble formation.
3. Add 5 μL of loading buffer 5× to 20 μL aliquot from odd fractions and heat in boiling water for 5 min. Centrifuge the heated samples for 0.5 min at max speed to remove insoluble material and load 20 μL of each fraction on two gels. Prestained protein ladder is used as marker. Run the gels at 100 V until the BFB has reached the bottom.
4. At the end of the run, separate the gel plates and remove the stacking gel. Put the gel in a glass bowl.
5. Rinse the gel with water to remove traces of running buffer.
6. Stain the gel with Coomassie blue staining solution for 1 h.
7. Destain the gel by soaking for at least 2 h in destaining solution with at least two changes of this solvent. If the gel still has a Coomassie blue background then continue destaining until the background is nearly clear. Rinse the gel with water and store destained gel in MilliQ water. OmpC/F and OmpA are visible on SDS-PAGE gel as polypeptides migrating approximately at 40 and 37 kDa.
8. Fractions containing the inner membrane and those containing the outer membrane are separately pooled. Inner membrane and outer membrane pools are ultracentrifuged at 100,000 × *g* for 2 h in an ultracentrifuge at 4 °C, to remove the sucrose. The pellets are resuspendend in 2 mL of water and lyophilized for 16–18 h.
9. Lyophilized samples are then ready to be processed according to the method in Subheading 3.5 below.

3.5 Sample Preparation for MudPIT Analysis

1. Resuspend the lyophilized sample in 100 μL of 0.1 M ammonium bicarbonate, pH 8.0. Control pH and make sure that pH > 7 (*see* **Note 7**).
2. Reconstitute the 1 mg lyophilized RapiGest™ SF (Waters Corporation) in 100 μL of ammonium bicarbonate 0.1 M pH 8.0 to reach the concentration of 1% w/v (*see* **Note 8**).
3. Add RapiGest™ SF to sample to give 0.2% (w/v) (*see* **Note 9**).
4. Boil the protein/RapiGest™ SF mixture at 100 °C for 15 min.
5. Cool the sample to room temperature for 5 min and then quantify protein content using SPN™-Protein Assay (G-Biosciences) (*see* **Note 10**).
6. Put 50 μg of sample in a glass insert and add acetonitrile (ACN) at the final concentration of 10% (w/v).

7. Digest protein sample adding sequencing-grade trypsin in a ratio of 1:50 (1 μg enzyme:50 μg substrate) and incubate at 37 °C overnight, at 300 rpm (*see* **Note 11**).
8. To improve the digestion efficiency, add a second aliquot of trypsin in a ratio of 1:100 (0.5 μg enzyme:50 μg substrate) and then incubate the sample at 37 °C for 4 h, at 300 rpm.
9. Stop the tryptic digestion adding trifluoroacetic acid (TFA); the final TFA concentration should be 0.5 % (w/v). Make sure that pH is 2 using a litmus paper.
10. Evaporate ACN in a vacuum system (60 °C for 5 min). Prevent loss of peptides, trying not to dry the sample.
11. Incubate peptide mixture at 37 °C for 45 min (300 rpm) and then centrifuge acid-treated sample at 13,000 × *g* for 10 min; a pellet may be observed.
12. Collect the supernatant and discard the pellet.
13. Desalt and concentrate the sample using PepClean™ C18 Spin Column (*see* **Note 12**).
14. Gently dry the sample in a vacuum system at 60 °C and wash it three times with 20 μL of 0.1 % formic acid in water.
15. Reconstitute sample in 50 μL of 0.1 % formic acid to obtain a final protein concentration of 1 μg/μL.

3.6 MS Analysis

Trypisn-digested peptides are analyzed by two-dimensional micro-liquid chromatography coupled to ion trap mass spectrometry (MudPIT).

1. By means of an autosampler, load 10 μL of the digested peptide mixtures onto a strong cation-exchange column and then elute using eight steps of increasing ammonium chloride concentration (0, 20, 40, 80, 120, 200, 400, and 700 mM). Eluted peptides, obtained by each salt step, are at first captured in turn onto two peptide traps mounted on a 10-port valve, for concentration and desalting, and subsequently loaded on a reversed-phase C-18 column for separation with an acetonitrile gradient. Set the gradient profile as shown: 5–10 % eluent B in 5 min, 10–40 % eluent B in 40 min, 40–80 % eluent B in 8 min, 80–95 % eluent B in 3 min, 95 % eluent B for 10 min, 95–5 % eluent B in 4 min, and 5 % eluent B for 15 min (eluent A, 0.1 % formic acid in water; eluent B, 0.1 % formic acid in acetonitrile). The flow rate is 100 μL/min split in order to achieve a final flux of 1.5 μL^{-1} min^{-1} (*see* **Note 13**).
2. The peptides eluted from the C-18 column are directly analyzed with an ion trap mass spectrometer equipped with a nano electrospray ionization source (nano-ESI). Nanospray is achieved with an un-coated fused silica emitter held to 1.5 kV and the heated capillary is held at 185 °C (*see* **Note 14**).

3. Acquire full mass spectra in positive mode and over a 400–2000 *m/z* range, followed by three MS/MS events sequentially generated in a data-dependent manner on the first, second, and third most intense ions selected from the full MS spectrum, using dynamic exclusion for MS/MS analysis (collision energy 35%) (*see* **Note 15**). MudPIT analysis generates files in .raw format, containing the experimental mass spectra (full Ms and MS/MS); the number of files created is equivalent to the salt step number set in the analysis.

3.7 Computational Analysis

1. From National Central for Biotechnology Information (NCBI) website (http://www.ncbi.nlm.nih.gov/) download the non-redundant (NR) protein sequence database in FASTA format.
2. From "NR" retrieve the *Escherichia coli* protein sequence database.
3. Save .raw files and protein sequence database into computer equipped with SEQUEST algorithm (*see* **Note 16**).
4. Using software based on SEQUEST algorithm (University of Washington), such as Bioworks (Thermo Fisher Scientific Inc.) or Proteome Discoverer (Thermo Fisher Scientific Inc.), correlate the experimental mass spectra to tryptic peptide sequences by comparing with theoretical mass spectra, obtained by in silico digestion of *Escherichia coli* protein database. Set the following parameters: no enzyme mode (*see* **Note 17**), tolerance on the mass measurements of 2.00 amu for precursor peptide, and 1.00 amu for fragment ions.
5. Combine the files generated by applying stringent filters: Xcorr scores greater than 1.5 for singly charged peptide ions and 2.0 and 2.5 for doubly and triply charged ions, respectively, peptide probability $\leq$0.001, and protein consensus score value $\geq$10.
6. Export and save in Excel format the protein list and the peptide list (*see* **Note 18**).
7. From the protein list generate the 2D map of the sample using software such as an in-house algorithm called MAProMa (Multidimensional Algorithm Protein Map) [12]. Proteins are plotted according to their theoretical pI and MW.
8. Compare the protein lists obtained from the analysis of all samples with MAProMa or similar software; the final file is a list of all the proteins identified in the samples and for each protein is reported the values of SEQUEST Spectral Count (SpC) and of SEQUEST score.
9. Verify the repeatability of the analysis by plotting SEQUEST scores of each identified protein in the first technical replicate versus the second technical replicate. Evaluate the linear correlation (R_2) and compare the slope value obtained with the theoretical value (1.00).

10. Identify proteins with significant differences in level by other two tools of MAProMA: DAve (Differential Average) and DCI (Differential Coefficient Index) (*see* **Note 19**).
11. Characterize proteotypic peptides by means of another in-house software called EPPI (Experimental Proteotypic Peptide Investigator) [13].

4 Notes

1. EDTA powder dissolves completely when solution reaches pH 8.0. About 20 g NaOH pellet is required to adjust to pH 8.0.
2. Alternatively, cell lysis could be performed by six cycles of sonication at 20% amplitude for 10 s.
3. Alternatively, membranes can be prepared by conversion of the cells to spheroplasts by the lysozyme–EDTA treatment and disruption of spheroplast by sonication. Add 100 μg/mL lysozyme to the cell suspension and leave on ice for 2 min. Keep the sample in ice (put a small beaker in a bigger one filled with ice) and add 12 mL of 0.3 mM EDTA over a period of 10 min with a peristaltic pump by mixing gently with a magnetic stirrer. EDTA solution must be added dropwise by the side of the beaker. Add 10 μM PMSF and 0.2 mM DTT. Take an aliquot and read absorbance at 450 nM (A_{450}). Disrupt spheroplasts by sonication at 10% amplitude for 15 s. Repeat this step four times leaving the samples on ice for 1 min after every cycle of sonication. Read A_{450} of the spheroplast suspension and check whether it has decreased to approximately 5% of its original value; if not sonication should be continued for 1–2 cycles. Remove the unbroken cells by centrifugation at 3000 × *g* for 20 min at 4 °C and continue with the protocol.
4. Carefully layer the sucrose solutions 1 mL at the time. Once the sucrose gradient is poured discrete layers of sucrose should be visible in the tube.
5. Immediately after the run the tube should be removed from the rotor, taking great care not to disturb the layers of sucrose. Two discrete membrane layers should be visible.
6. The following formula can be used to calculate the enzyme activity: $(|OD_{430}f - OD_{430}i| / \text{time} \times \varepsilon) \times$ cuvette vol.; $OD_{430}f$, $OD_{430}i$: final and initial absorbance values at 430 nm; time: expressed in minutes; ε: 6.22 mM^{-1} cm^{-1}; cuvette vol 1000 μL. We find that storing fractions at −20 °C reduces NADH oxidase activity.
7. The lyophilized sample needs to be completely covered by ammonium bicarbonate. If 100 μL is not enough, keep on adding 10 μL until the sample is not completely resuspended. If the final volume exceeds 200 μL, concentrate the sample in a vacuum system to a final volume of 200 μL.

8. Rapigest is an enzyme-friendly and a mass spectrometry-compatible reagent. It is able to help the solubilization of proteins and thus the enzymatic digestion. The lyophilized powder is stable at room temperature until the expiration date written on package, but once reconstituted the solution is stable for 1 week at 2–8 °C.
9. The recommended concentration is 0.1 % (w/v), but hydrophobic proteins, as membrane proteins, may require higher concentrations.
10. SPN™-Protein Assay is a protein estimation method which is fast, efficient, and compatible with laboratory agents and detergents, such as sodium dodecyl sulfate (up to 2 %). The kit is stored at ambient temperature and requires only 0.5–10 g of proteins, polystyrene cuvettes, and deionized water as blank. Protein concentration is determined by comparing the optical density (OD_{595}) obtained from sample treated with the reference OD_{595}. The absorbance can be read either using spectrophotometer or using microplate reader.
11. Trypsin is the most common protease used for digestion due to its well-defined specificity. Many factors and parameters could affect the effectiveness of protein digestion, in particular pH of the reaction and temperature. A slightly alkaline environment (pH 8) is optimal and a temperature of 37 °C is recommended. The use of modified trypsin is also necessary to avoid autolysis.
12. PepClean™ C-18 Spin Column is useful for removing contaminants and for concentrating peptides, realizing them in MS-compatible solutions. Each column can bind up to 30 μg of total peptide from 10 to 150 μL of sample volume. Mix 3 parts sample to 1 part sample buffer (2 % TFA in 20 % ACN); prepare column adding 200 μL of activation solution (50 % ACN), centrifuge at 1500 × *g* for 1 min, discard flow-through, and repeat. Repeat this procedure using equilibration solution (0.5 % TFA in 5 % ACN) instead of activation solution. Load sample on column, and centrifuge at 1500 × *g* for 1 min. To ensure complete binding, recover flow-through and reload sample on column. After centrifugation, add 200 μL wash solution (0.5 % TFA in 5 % ACN) and centrifuge (1.5 g for 1 min). Repeat this step three times. Elute sample by adding 20 μL of elution buffer (70 % ACN) twice.
13. The flow is maintained constant and equal to that set by a splitter to pulse dampener. If the pressure and the flow increase, part of the flow is conveyed toward the waste in order to maintain the outlet flow predetermined.
14. Eluted peptides can be analyzed also using an LTQ mass spectrometer (Thermo Finnigan Corp.) equipped with a nano-LC electrospray ionization source. The values of the parameters set are the same between LCQ Deca XP plus and LTQ.

15. With LTQ, acquire full mass spectra in positive mode and over a 400–2000 *m/z* range, followed by five MS/MS events sequentially generated in a data-dependent manner on the five most intense ions selected from the full MS spectrum, using dynamic exclusion for MS/MS analysis (collision energy 35 %).
16. Cluster PC or multi-processor is useful for reducing processing time.
17. "No enzyme mode" guarantees the identification of peptides generated by nonspecific cuts. This parameter ensures the identification of a greater/larger number of peptides.
18. Using the Autoformat tool of an in-house algorithm called MAProMa (Multidimensional Algorithm Protein Map) [12], arrange the raw protein list to have for each protein identified: reference, accession, spectral count (SpC), SEQUEST-based SCORE, isoelectric point (pI), and molecular weight (MW).
19. These two algorithms are based on score values assigned by SEQUEST software to each identified protein in two samples compared. Specifically, DAve is an index of the relative ratio between the two samples and DCI evaluates the confidence of DAve. Briefly, using MAProMA each identified protein in the two samples is aligned and then DAve and DCI indexes are calculated for all proteins. The threshold values imposed could be very stringent (DAve > 0.4 and DAve < –0.4, DCI > 400 and DCI < –400) or less stringent (DAve > 0.2 and DAve < –0.2, DCI > 200 and DCI < –200). It is necessary that both indexes, DAve and DCI, satisfy these thresholds [14].

Acknowledgements

We thank CNR Project "FaReBio di Qualità," MIUR PRIN 2012WJSX8K, and Regione Lombardia Progetti di Ricerca Industriale e Sviluppo Sperimentale ID n. 30190679 for the financial support provided for this study. We thank Marta G. Bitonti for providing the MAProMA software.

References

1. Washburn MP, Wolters D, Yates JR 3rd (2001) Large-scale analysis of the yeast proteome by multidimensional protein identification technology. Nat Biotechnol 19:242–247
2. Park SK, Venable JD, Xu T, Yates JR 3rd (2008) A quantitative analysis software tool for mass spectrometry-based proteomics. Nat Methods 5:319–322
3. Mauri P, Scigelova M (2009) Multidimensional protein identification technology for clinical proteomic analysis. Clin Chem Lab Med 47: 636–646
4. De Palma A, Roveri A, Zaccarin M, Benazzi L, Daminelli S, Pantano G, Buttarello M, Ursini F, Gion M, Mauri PL (2010) Extraction methods of red blood cell membrane proteins for

multidimensional protein identification technology (MudPIT) analysis. J Chromatogr A 1217:5328–5336

5. Martorana AM, Motta S, Di Silvestre D, Falchi F, Dehò G, Mauri P, Sperandeo P, Polissi A (2014) Dissecting *Escherichia coli* outer membrane biogenesis using differential proteomics. PLoS One 9:e100941
6. Polissi A, Sperandeo P (2014) The lipopolysaccharide export pathway in *Escherichia coli*: structure, organization and regulated assembly of the Lpt machinery. Mar Drugs 12:1023–1042
7. Silhavy TJ, Kahne D, Walker S (2010) The bacterial cell envelope. Cold Spring Harb Perspect Biol 2:a000414
8. Kinnebrew MA, Pamer EG (2011) Innate immune signaling in defense against intestinal microbes. Immunol Rev 245:113–131
9. Thanassi DG, Bliska JB, Christie PJ (2012) Surface organelles assembled by secretion systems of Gram-negative bacteria: diversity in structure and function. FEMS Microbiol Rev 36:1046–1082
10. Osborn MJ, Gander JE, Parisi E (1972) Mechanism of assembly of the outer membrane of *Salmonella typhimurium*. Site of synthesis of lipopolysaccharide. J Biol Chem 247:3973–3986
11. Sperandeo P, Cescutti R, Villa R, Di Benedetto C, Candia D, Dehò G, Polissi A (2007) Characterization of *lptA* and *lptB*, two essential genes implicated in lipopolysaccharide transport to the outer membrane of *Escherichia coli*. J Bacteriol 189:244–253
12. Mauri P, Dehò G (2008) A proteomic approach to the analysis of RNA degradosome composition in *Escherichia coli*. Methods Enzymol 447:99–117
13. Di Silvestre D, Brunetti P, Vella D, Brambilla F, De Palma A, Mauri P (2015) Automated extraction of proteotypic peptides by shotgun proteomic experiments: a new computational tool and two actual cases. Curr Biotechnol 4:39–45
14. Mauri P, Scarpa A, Nascimbeni AC, Benazzi L, Parmagnani E, Mafficini A, Della Peruta M, Bassi C, Miyazaki K, Sorio C (2005) Identification of proteins released by pancreatic cancer cells by multidimensional protein identification technology: a strategy for identification of novel cancer markers. FASEB J 19:1125–1127

Chapter 6

Random Transposon Mutagenesis for Cell-Envelope Resistant to Phage Infection

Ruth Reyes-Cortés, Emma S. Arguijo-Hernández, Marco A. Carballo-Ontiveros, Eva Martínez-Peñafiel, and Luis Kameyama

Abstract

In order to identify host components involved in the infective process of bacteriophages, we developed a wide-range strategy to obtain cell envelope mutants, using *Escherichia coli* W3110 and its specific phage mEp213. The strategy consisted in four steps: (1) random mutagenesis using transposon miniTn10Km^r; (2) selection of phage-resistant mutants by replica-plating; (3) electroporation of the phage-resistant mutants with mEp213 genome, followed by selection of those allowing phage development; and (4) sequencing of the transposon-disrupted genes. This strategy allowed us to distinguish the host factors related to phage development or multiplication within the cell, from those involved in phage infection at the level of the cell envelope.

Key words Cell-envelope, Infective process, Random mutagenesis, Bacterial receptor, Bacteriophages

1 Introduction

We are proposing a genome-wide screening strategy to identify host genes whose products are involved in the phage infection process at the cell-envelope level. This strategy combines four methods that were performed as follows: (1) random transposon mutagenesis using the phage λNK1316, which contains the miniTn10Km^r transposon [1]; (2) screening and selection of phage-resistant mutants by replica-plating method; (3) electroporation of phage mEp213 entire genome into the phage-resistant mutants, and selection of mutants that produced mature phage particles. This was the key step for selecting mutants resistant to phage infection at the cell-envelope level, different from those inhibiting phage development within the cell. And (4) sequencing of the disrupted genes using a Y-linker method based on PCR, where the gene regions flanking the transposon insertion were amplified using a Y-linker primer and a miniTn10Km^r transposon primer [2] (Fig. 1).

Hee-Jeon Hong (ed.), *Bacterial Cell Wall Homeostasis: Methods and Protocols*, Methods in Molecular Biology, vol. 1440, DOI 10.1007/978-1-4939-3676-2_6,

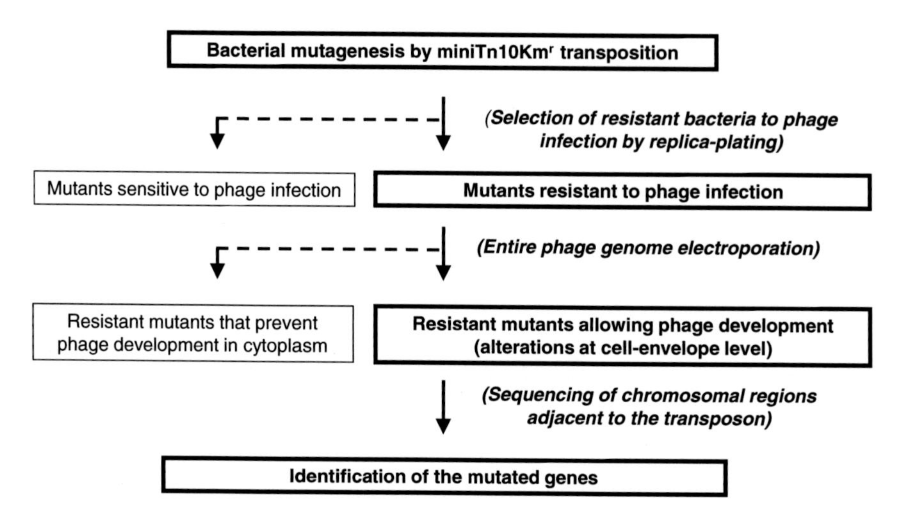

Fig. 1 Experimental strategy for the selection of bacterial mutants resistant to phage infection at the cell-envelope level. For details see the text

To better illustrate the advantages provided by this strategy, we briefly describe the results obtained with *E. coli* W3110 and its phage mEp213 [3]. Firstly, we generated kanamycin-resistant mutants with miniTn10Km^r transposon, and then we selected 12 phage-resistant mutants by replica-plating. Electroporation of the viral genome into these yielded nine mutants that were able to produce viral progeny detected in the supernatant, indicating that such mutants prevented the phage infection at the cell-envelope level. Six of these mutants allowed the phage infection when they were complemented with the plasmid pUCJA (containing the gene *fhu*A, the usual receptor for mEp213 phage). The other three mutants did not show FhuA complementation, and harbored transposon mutations at two different positions of *waa*C and another one at the *gmh*D gene, respectively. The products of both genes are involved in lipopolysaccharide (LPS) biosynthesis.

All of the obtained mutants were related to cell envelope. The positive results observed in the complementation test using pUCJA (*fhu*A$^+$) indicated that the strategy was in the right way and that the LPS (together with the FhuA receptor) increased the efficiency of phage infection.

We propose that the use of this strategy may facilitate the elucidation of novel cell-envelope factors, and will enrich the current knowledge of the phage infection process.

2 Materials

2.1 Strains, Plasmids, and Phages

1. Strains: *Escherichia coli* W3110 [4, 5], *E. coli* C600 [6], *E. coli* LE392 [7], and *E. coli* DH5α [8] (*see* **Note 1**).
2. Plasmids: pUCJA [9] and pPROEX-d [9] (*see* **Note 2**).

3. Bacteriophages: λ wild type, λNK1316 [1], and mEp213 [10] (*see* **Note 3**).

2.2 Bacterial Media

1. Luria-Bertani broth (LB): 10 g/L Tryptone, 5 g/L yeast extract, and 5 g/L NaCl. For semisolid LB, in addition add 15 g/L of agar (*see* **Note 4**).
2. Tryptone broth (Tϕ): 10 g/L Tryptone and 5 g/L NaCl. For Tϕ soft-agar media, in addition add 7.5 g/L of agar.
3. GYT broth: 25 g/L Tryptone, 12.5 g/L yeast extract, 10% glycerol at final concentration [11] (*see* **Note 5**).
4. SOB broth: 20 g/L Tryptone, 5 g/L yeast extract, 2 mL of NaCl 5 M (0.58 g/L), 2.5 mL of KCl 1 M (0.186 g/L), 10 mL of $MgCl_2$ 1 M (2.03 g/L), 10 mL of $MgSO_4$ 1 M (1.2 g/L) [12] (*see* **Note 6**).

2.3 Buffers

1. SM buffer: 5.8 g/L NaCl, 2 g/L $MgSO_4 \cdot 7H_2O$, 50 mL Tris–HCl 1 M (7.88 g/L), pH 7.5, and 5 mL of 2% gelatin solution (0.1 g/L) [11].
2. TE buffer: 10 mM Tris–HCl pH 7.4 and 1 mM EDTA [7] (*see* **Note 7**).

2.4 Antibiotics and Inductor

1. 100 μg/mL Ampicillin (Amp), 50 μg/mL kanamycin (Km), 50 μg/mL chloramphenicol (Cm), 1 mM isopropyl β-D-1-thiogalactopyranoside (IPTG).

2.5 Oligonucleotides

Oligonucleotide Y-linker 1: 5′-TTTCTGCTCGAATTCAAGCTT CTAACGATGTACGGGGACACATG-3′, Y-linker 2: 5′-TGTCC CCGTACATCGTTAGAACTACTCGTACCATCCA CAT-3′, Y-linker primer: 5′-CTGCTCGAATTCAAGCTTCT-3′, Tn10 primer: 5′-TGACAAGATGTGTATCCACCTTAAC-3′, and Tn10Km^r primer: 5′-TTCATTTGATGCTCGATGAG-3′. For *waa*C gene, forward primer was designed as follows: waaC-EcoFwd 5′-GGAATTCAAGAGGAAGCCTGACGGATG-3′, and reverse was waaC-HindRev 5′-CCCAAGCTTTAAAGGATGTTAGCAT GTTTTACC-3′. For *gmh*D gene, forward primer was designed as gmhD-EcoFwd 5′-GGAATTCGAAGGTTACAGTTATGATC-3′, and reverse as gmhD-HindRev 5′-CCCAAGCTTCATGCAGA GCTCTTATGC-3′.

2.6 DNA-Related Kits

1. GenElute Bacterial Genomic DNA Kit, High Pure Plasmid Isolation Kit.
2. MinElute PCR Purification Kit.
3. QIAquick Gel Extraction Kit.
4. ABI P_{RISM}® Big Dye® Terminator v3.1 Cycle Sequencing Ready Reaction Kit.

2.7 Enzymes

1. DNA cloning: Restriction enzymes *Eco*RI, *Hin*dIII, and *Nla*III with their respective 10× buffers *Eco*RI, number 2 and 4, T4 DNA ligase and its 10× buffer, T4 polynucleotide kinase, 10× kinase buffer, and 10× of BSA solution (1 mg/mL).
2. DNA amplification: AmpliTaq DNA Polymerase, Platinum Taq DNA Polymerase High Fidelity.

2.8 Other Reagents

1. Low-melting-point agarose and agarose.
2. Polyethylene glycol (PEG 8000-ES).
3. Chloroform.
4. Phenol.
5. Isoamyl alcohol.
6. Ethanol.

3 Methods

3.1 Random Transposon Mutagenesis

1. Mutagenesis was carried out infecting *E. coli* W3110 with λNK1316 phage (*see* **Note 8**). For adsorption, 10 μL of λNK1316 phage at 10^8 PFU/mL was mixed with 100 μL of W3110 bacteria at 10^7 CFU/mL, and incubated at room temperature (~26 °C) for 10 min (*see* **Note 9**).
2. After adsorption, 1 mL of LB broth was added to the mixture, and then it was centrifuged at 5500 × *g* for 5 min to remove non-adsorbed phages. This step was repeated two times, in order to avoid reinfection.
3. The infected bacterial pellet was resuspended in 1 mL of LB broth and incubated at 37 °C for 1 h with uniform shaking (~150 rpm).
4. 100 μL of culture was spread on LB plates supplemented with kanamycin (LB-Km), and incubated overnight at 37 °C.
5. After ~18 h of incubation, kanamycin-resistant mutants were observed and counted.

   Approximately, ~10^2 CFU/plate were obtained (*see* **Note 10**).

3.2 Selection of Host Mutants Resistant to mEp213

1. Km-resistant mutants were replicated on two LB plates: one supplemented only with kanamycin (LB-Km) and the other LB-Km plate in which virulent phage mEp213 (mEp213_V) was spread homogeneously at 10^9 PFU/plate (*see* **Note 11**). Both plates were incubated overnight at 37 °C (*see* **Note 12**).
2. Bacteria colonies with uniform rounded shape that developed in the LB-Km plate spread with virulent phage were selected as candidates for phage-resistant mutants. The candidate colonies were taken from the plates having only the antibiotic as selecting agent.

3. For colony purification, the phage-resistant mutants were streaked at least three times on LB-Km plates. All of the selected host mutants should accomplish the following conditions: (a) they should be able to grow on LB-Km plates, indicating the presence of miniTn10Km^r; (b) they should be sensitive to λ phage infection, indicating that no integration of λNK1316 occurred; and c) they should be non-lysogenic for mEp213, which indicates that the observed resistance to phage infection is not due to the mEp213 repressor, but due to the transposon insertion.
4. Sensitivity test was performed using an overlay assay. A colony of each mutant was grown overnight in LB-Km broth at 37 °C, then 0.5 mL of the overnight culture was added to 3 mL of Tϕ soft agar (~42–45 °C), and the mixture was poured onto LB-Km agar plates and left at room temperature until it solidified. Then, 10 μL of serial dilutions (10^{-2} to 10^{-8}) of mEp213 phage were spotted onto the overlay plates and were incubated overnight at 37 °C.
5. If the candidates were resistant to mEp213, the formation of plaques would not be expected. Lysogenic test: A colony of W3110 was grown in LB broth overnight at 37 °C, and then 0.5 mL of the overnight culture was mixed with 3 mL of Tϕ soft-agar (~40 °C) and poured onto an LB plate. When the mixture solidified, 10 μL of serial dilutions of the supernatant of each mutant candidate were spotted onto the overlay plates, and then were incubated overnight at 37 °C. If the candidates are indeed not lysogenic for mEp213, the absence of phage in the supernatant is expected and therefore no formation of plaques should be observed. We obtained 12 host mutants that fulfilled the above conditions (*see* **Note 13**).

3.3 Electroporation of Phage-Resistant Host Mutants with Viral DNA

Electroporation was the key step to select host mutants related to the initial step of phage infection. All mutants were electroporated with the entire mEp213 genome, and screening for those allowing phage development was then performed. Bacterial burst indicated that the transposon mutations inhibited phage infection at the cell-surface level, during the previous experiments.

3.3.1 Phage DNA Extraction

This step was performed according to Sambrook et al. [11] with minor modifications.

1. NaCl and polyethylene glycol (PEG 8000-ES) were added to the phage lysate to a final concentration of 1 M and 10%, respectively, and were then mixed thoroughly. The mixture was incubated overnight at room temperature without agitation.
2. The next day, the mix was centrifuged at 10,000 ×*g* for 20 min, the supernatant was discarded, and the pellet was resuspended in 1 mL of SM buffer.

3. Polyethylene glycol was removed by adding an equal volume of chloroform and shaking vigorously.
4. The aqueous phase was recovered by centrifugation at 10,000 × *g* for 5 min, and then an equal volume of phenol:chloroform:isoamyl alcohol solution (25:24:1) was added and mixed with vortex.
5. The aqueous phase was recovered and mixed with an equal volume of chloroform:isoamyl alcohol (ratio 24:1).
6. The mix was centrifuged at 10,000 × *g* for 5 min and then the aqueous phase was gently mixed with 100 μL of 3 M sodium acetate and 2 volumes of absolute ethanol. The precipitation reaction was kept on ice for 30 min.
7. DNA was pelleted by centrifugation at 12,000 × *g* for 10 min and washed with 1 mL of 70 % ethanol. The ethanol was then removed by centrifugation and the pellet was allowed to dry.
8. The DNA was resuspended in 100 μL of TE buffer. The integrity of phage DNA was evaluated by electrophoresis in 1 % agarose gel stained with 0.5 μg/mL of ethidium bromide.

3.3.2 Preparation of Electro-Competent Cells

1. A fresh single colony was inoculated into 50 mL of LB broth. It was incubated overnight at 37 °C with agitation (~150 rpm).
2. The overnight culture was diluted 1:200 in fresh LB broth (2.5–500 mL, respectively) and incubated at 37 °C with vigorous shaking until it reached an OD_{600} of 0.4–0.6 (approximately 2–4 h).
3. The culture was transferred to centrifuge tubes (from this point, the culture temperature should never exceed 4 °C), and it was centrifuged at 3500 × *g* for 15 min at 4 °C. The supernatant was discarded and the pellet was resuspended in 1 volume of cold pure MilliQ water.
4. Washing steps were repeated; initially the pellet was resuspended in 0.5 and later in 0.1 volumes of cold 10 % glycerol.
5. Finally, the pellet was resuspended in 1 mL of GYT medium, and the OD_{600} was measured and adjusted (1:100) to a concentration approximately of 2×10^7 cells/mL. Aliquots of 100 μL were taken and stored at −70 °C, until use.

3.3.3 Electroporation

1. The electroporation assay was performed by mixing 1 μg of phage mEp213 DNA with ~4×10^5 competent bacterial cells [phage-resistant mutants (PRM mutants)], in a total volume of 20 μL. The Cell Porator® Electroporation System (Gibco BRL) was used with the following settings: capacitance at 330 μF, voltage at 200 V, and resistance at low 4 kΩ.
2. The transformed cells were cultured in 3 mL of LB broth for 4 h at 37 °C and centrifuged at 13,000 × *g* for 5 min at room temperature.

3. Serial dilutions of the supernatant were spotted onto overlay plates to test for plaque production (*see* **Note 14**).
4. For this assay, nine mutants that produced mature virions were detected from the supernatant assay, strongly suggesting that their resistance to phage infection was at the cell-envelope level. In contrast, the other three host mutants that did not produce virions indicate that the mutation affected the phage development within the bacterial cytoplasm (*see* **Note 15**).

3.3.4 Mutants Related to FhuA Receptor

We previously reported that mEp213 requires the FhuA receptor for infection [9]. Therefore, once we identified bacterial mutants at the cell-envelope level, the next step was to differentiate between mutants in the *fhu*A gene and those affected in genes different to *fhu*A.

For this purpose, the mutants were transformed with the plasmid pUCJA (*fhu*A$^+$) and their susceptibility to mEp213 phage infection was analyzed by double-layer assay (*see* **Note 16**). Six out of nine mutants, as well as the control *E. coli* strain C600 (*fhu*A$^-$) with the plasmid pUCJA (*fhu*A$^+$), showed complementation for the phage mEp213 infection process. Complementation failed for the other three mutants, suggesting that the transposon insertion possibly affected genes other than *fhu*A.

3.4 Sequencing the Host Chromosome Regions Adjacent to the Transposon

Mutants not complemented in trans by *fhu*A$^+$ were selected for sequencing. Chromosomal regions adjacent to the miniTn10Km^r insertion site were amplified by the Y-linker method described by Kwon and Ricke [2], which we describe briefly:

1. DNA extraction was performed using the GenElute Bacterial Genomic DNA Kit. Two micrograms (2 μL) of bacterial genomic DNA were digested with 10 U (1 μL) of restriction enzyme *Nla*III, 1 μL of 10× restriction buffer No. 4 (NEB), and nuclease-free water to a final volume of 10 μL, at 37 °C for 3 h.
2. On the other hand, 9 μL of linker 2 (350 ng/μL) was phosphorylated with 10 U (1 μL) of T4 polynucleotide kinase (PNK), 2 μL of 10× PNK buffer, 2 μL of ATP 10 mM, and nuclease-free water to a final volume of 20 μL. The mixture was incubated at 37 °C during 1 h. The PNK was denatured at 65 °C for 20 min and then it was mixed with 9 μL of linker 1 (350 ng/μL). Both linkers were heated at 95 °C for 2 min and then allowed to cool gradually at room temperature to make the "Y-linker."
3. The digested chromosomal DNA was cleaned by phenol-chloroform extraction. Approximately, 10 μL of DNA (equivalent to 50 ng) was ligated with 1 μg of Y-linker using 20 U of T4 DNA ligase (1 μL), 2 μL of 10× T4 ligase buffer, and

nuclease-free water up to 20 μL, at 22 °C for 24 h. The mixture was diluted with distilled water to a final volume of 200 μL and then heated to 65 °C for 10 min to denature the ligase. Two microliters of this DNA solution was used as template for PCR amplification.

4. AmpliTaq DNA Polymerase, the adaptor primer (5′-CTGCTCGAATTCAAGCTTCT-3′), and the specific primer designed for the Km resistance cassette of the transposon (5′-TTCATTTGATGCTCGATGAG-3′) were used to amplify the transposon-flanking sequence. The reaction contained 5 μL of 10× PCR buffer, 3.5 μL of each primer (~350 ng), 3 μL of dNTP mix (25 mM of each dNTP), 3 μL DMSO, and 2 μL template DNA in a 49 μL reaction volume. It was incubated at 95 °C for 2 min. As an additional step, 1 μL of Taq DNA Polymerase (5 U) was added during hot-start incubation at 80 °C to prevent nonspecific priming.
5. Reactions were performed using a Flexigen PCR System under the following conditions: a denaturing step at 94 °C/2 min, followed by 30 amplification cycles of 94 °C/30 s, 56 °C/60 s, and 72 °C/60 s for each cycle, and a final extension step of 72 °C/2 min.
6. The amplified fragments were separated by electrophoresis in a 2% low-melting-point agarose gel, and then purified using the MinElute PCR Purification Kit, according to the manufacturer's specifications.
7. The fragments were amplified and sequenced with the ABI P_{RISM}® Big Dye® Terminator v3.1 Cycle Sequencing Ready Reaction Kit, using the adaptor primer and the specific primer (5′-TGACAAGATGTGTATCCACCTTAAC-3′), designed for the insertion sequence (IS) region of miniTn10Km^r. The reaction conditions were as follows: a denaturing step of 94 °C/2 min, followed by 30 cycles of amplification of 96 °C/30 s, 53 °C/20 s, and 60 °C/2 min for each cycle, and a final extension step of 60 °C/2 min. Sequencing reactions were resolved in a Perkin Elmer ABI P_{RISM} Genetic Analyzer 310.
8. The gene sequences were aligned against the *E. coli* W3110 genome reference sequence (GenBank accession No. AP009048.1, GI: 85674274), using the BLAST program and the GenBank database at the NCBI website (http://blast.ncbi.nlm.nih.gov/Blast.cgi).

Sequence mapping indicated that the transposon insertion in each of the three mutants occurred at two different sites of *waa*C and the other at the *gmh*D gene, respectively. The products of both genes are involved in the biosynthesis of LPS.

3.4.1 Plasmid Constructions

*waa*C and *gmh*D genes were amplified from the *E. coli* W3110 genome and were directly cloned into the expression vector pPROEX-d in *Eco*RI and *Hin*dIII restriction sites [9, 13]. The *Eco*RI and *Hin*dIII recognition sequences were included in the forward (Fwd) and reverse (Rev) primers, respectively (*see* Subheading 2.5). The PCR conditions for both genes were the same: a denaturing step at 94 °C/2 min, followed by 30 amplification cycles of 94 °C/30 s, 55 °C/60 s, and 72 °C/60 s, followed by a final extension step of 72 °C/2 min. The amplified fragments of the *waa*C and *gmh*D genes were separately restricted in 20 μL reactions with 1 μL (20 U) of *Eco*RI and 1 μL (20 U) of *Hin*dIII enzymes, 2 μL *Eco*RI buffer 10×, and 14 μL of total DNA (1 μg) and filled up with nuclease-free water. The reactions were incubated at 37 °C for 3 h, resolved by electrophoresis in agarose gels, and then purified using QIAquick Gel Extraction Kit. Restriction conditions for plasmids were as follows: 2 μL of DNA (1 μg), 1 μL (20 U) of *Eco*RI, 1 μL (20 U) of *Hin*dIII, 2 μL of *Eco*RI buffer 10×, and 14 μL of nuclease-free water. Restriction efficiency was evaluated by 1 % agarose gel electrophoresis. The PCR products and digested plasmids were purified from gel using the QIAquick Gel Extraction Kit and observed on a 1 % agarose gel for quantification. For ligation, the plasmid:insert ratio was 1:3 (~50 ng/vector and ~30 ng/insert) in 20 μL reactions containing 2 μL of ligase buffer 10×, 1 μL of T4 ligase (20 U), and distilled water. The reaction was gently mixed and incubated at 16 °C for 24 h.

The new plasmids pWaaC and pGmhD were used for complementation assay (*see* **Note 17**).

4 Notes

1. *E. coli* strain W3110 (F^- λ^- *rph*$^-$) was used to multiply the lambdoid phage mEp213 [4, 5]. The C600 strain (*leu*B6 *thi*-1 *lac*Y1 *sup*E44 *thr*-1 *rfb*D1 *fhu*A21) was used to expand the λ phage [6]. The *E. coli* LE392 strain (F^- *e*14$^-$ ($MrcA^-$) *hsd*R514(r^-m^+) *sup*E44 *sup*F58 *lac*Y1 *gal*K2 *gal*T22 *met*B1 *trp*R55) was used to grow the phage λNK1316, which requires suppressor mutations [7]. The *E. coli* DH5α (*end*A1 *hsd*R17(r^-m^+) *sup*E44 *thi*-1 *rec*A1 *gyr*A96 *rel*A1 Δ*lac*U169 [ϕ80d*lac*ZΔM15]) [8] was used for transformation with the plasmid derived from pPROEX-d.
2. The plasmid pUCJA presents the following relevant characteristics: *cat fhu*A^+. This construction derived from pPROEX-d [(*bla lac*I^Q pTrc Δ(80 bp)] and lacks the nucleotide region associated to 6xHis [9].
3. The λ phage corresponds to the wild type. The genotype of phage λNK1316 is λ(miniTn10Km^r *c*I857 P*am*80 *nin*5 b522 att2).

Transposon miniTn10Km^r was used for mutagenesis and it cannot replicate in the strain W3110. The phage mEp213 is a lambdoid phage classified in the immunity group IX [9] by Kameyama et al. [10].

4. For Luria-Bertani broth (LB): Weigh 10 g of Bacto™ Tryptone, 5 g of NaCl, and 5 g of Bacto™ Yeast Extract. Mix with ~300 mL of distiller water until complete dissolution. Adjust pH to 7.0 with 5 N NaOH, and fill up to a final volume of 1 L with distiller water. Sterilize by autoclaving for 15 min at 15 psi. For semisolid LB, add the above components and 15 g of Bacto™ Agar.
5. For GYT medium weigh 25 g Bacto™ Tryptone and 12.5 g Bacto™ Yeast Extract, and mix with ~300 mL of distiller water until complete dissolution. Add 100 mL glycerol, and adjust to a final volume of 1 L with distilled water [11].
6. For SOB medium weigh 20 g Bacto™ Tryptone and 5 g Bacto™ Yeast Extract, and mix with ~300 mL of distilled water until complete dissolution. Add 2 mL of 5 M NaCl, 2.5 mL of 1 M KCl, 10 mL of 1 M $MgCl_2$, and 10 mL of 1 M $MgSO_4$. Adjust to a final volume of 1 L with distilled water [12].
7. The TE buffer contains 10 mM Tris–HCl pH 7.4 and 1 mM EDTA. Preparation of TE buffer: First, it is required to prepare a stock solution of 1 M Tris–HCl; for this, weigh 121 g of Tris base and add ~300 mL of distiller water, adjust pH to 8.0, and fill up to a final volume of 1 L with distiller water. Second, prepare a stock solution of 250 mM EDTA; for this, weigh 93 g $Na_2 \cdot EDTA \cdot 2H_2O$, adjust pH to 8.0, and fill up to a final volume of 1 L with distiller water. Sterilize by autoclaving.
8. Phage λNK1316 cannot replicate nor integrate in *E. coli* W3110, because it harbors a mutation in gene *P* (the P protein is required for λ replication) and also a deletion of the site-specific recombination sequence (att2) [1].
9. The bacteria-to-phage ratio used for mutagenesis was 1:1. However, a 5:1 ratio is desirable, in order to avoid multiple phage infection in one host.
10. Mutagenesis efficiency was around 0.12 %, which resulted sufficient to obtain mutants. Thus, we considered that there was no need to modify the conditions of the assay to improve the mutagenesis efficiency. We conducted a large number of independent mutagenesis assays.
11. The classical replica-plating technique is efficient; however, for our application it was not as precise as required. A modified replica-plating method can be used with liquid cultures on 96-well plates. Each mutant could be cultured by duplicate in 300 μL of LB-Km or LB-Km plus 10 μL of phage suspension

(10^8 PFU/mL). The cultures are incubated for 4–8 h at 37 °C with constant shaking. Finally, the OD_{600} of both cultures is measured, expecting higher OD for candidates than for their respective lysate controls.

12. Virulent mEp213 phage displays only the lytic replication cycle. Its use as a selecting agent consisted in eliminating all bacteria with mutations that do not affect the development of phages.

    Phage mEp213_V construction: The temperate phage mEp213 was mutagenized with *N*-methyl-*N*-nitro-*N*--nitrosoguanidine (MNNG) [14]. Mutagenesis was performed by mixing 0.1 mL of an overnight *E. coli* W3110 culture (10^9 CFU/mL), 0.1 mL of phage lysate (2000–4000 PFU), and 40 μL of MNNG fresh solution (1 mg/mL). Then, 2.5 mL of soft agar was added and the mixture was poured onto an LB agar plate and incubated overnight at 37 °C. The next day, a clear plaque was selected and this phage was denominated mEp213_C [14]. The mEp213_C phage was subjected to a second round of chemical mutagenesis under the conditions described before, but now we used the lysogenic *E. coli* W3110 (mEp213) to select the virulent mEp213 phage.

13. Lysogenic bacteria are those in which the genome of a temperate phage is inserted in the bacterial genome. When these bacteria multiply, mature virions are eventually released and can be detected in the overnight cultures. To quantify the phages in the supernatant, 1 mL of overnight culture is centrifuged for 5 min at 6000 × *g*, then the bacterial pellet is discarded, and serial dilutions of the supernatant are tested for phage presence.

14. For a good electroporation, quality and integrity of viral DNA are essential. DNA phage extraction by CsCl discontinuous gradient can be used [11].

15. The 3:1 mutation ratio observed between cell surface and cytoplasmic components could be biased, since any mutations related to cytoplasmic factors affecting the survival of bacteria would be underestimated.

16. Identification of host mutants altered in FhuA receptor: Plasmid DNA was isolated from the C600/pUCJA strain using the High Pure Plasmid Isolation Kit (following the supplier's instructions). Chemical competent cells were prepared according to Hanahan [8]. For the transformation assay: (1) 100 ng of DNA plasmid was mixed with ~10^6 competent cells. The mixture was incubated on ice for 30 min, heat-shocked at 42 °C for 1 min, and left on ice for two additional minutes. (2) After this, 1 mL of LB broth was added and incubated at 37 °C for 1 h under constant shaking. (3) 100 μL of the culture was

spread on LB chloramphenicol (LB-Cm) plate, and then incubated at 37 °C for 16–18 h. The next day, colonies from each plate were isolated.

For the complementation assay, one colony of each plate was cultured overnight in LB-Cm broth at 37 °C. Next day, 3 mL of Tϕ soft agar (~40 °C) was mixed with 0.5 mL of the overnight culture and 3.5 μL of 1 M IPTG was added to a final concentration of 1 mM. The mixture was poured onto LB-Cm agar plates. When the mixture solidified, serial dilutions of mEp213 phage were spotted onto it.

17. Complementation assay: To carry out the complementation assay, plasmids pWaaC and pGmhD were transformed into their respective mutants: PRM4, PRM11, and PRM12 [3]. Then, their susceptibility to phage mEp213 was tested by overlay assay. The phage mEp213 efficiency of plating (e.o.p.) values in PRM12, PRM4, and PRM11 were 0.045, 0.068, and 0.004, respectively, contrasting to that of W3110 (which had a value of 1). The negative control strain C600 (*fhuA*[+]) displayed an e.o.p. value of 0. Susceptibility to mEp213 infection was completely recovered when these mutants were complemented with their respective plasmids: pWaaC for mutants PRM4 and PRM12, and pGmhD for PRM11, showing e.o.p. values of 1. These results suggest that the gene products involved in biosynthesis of the LPS of the outer membrane, together with the FhuA receptor, are necessary for the efficient development of mEp213 phage.

    The e.o.p. value was calculated by determining the ratio of phage titer on the hosting strain (e.g., PRMs) versus the phage titer in the phage-sensitive strain W3110.

Acknowledgements

We specially thank to Dr. F. Fernández-Ramírez (Hospital General de México, SSA) for proofreading of the manuscript and Dr. M. de la Garza (CINVESTAV-IPN) for providing access to her laboratory facilities. We acknowledge M.Sc., M.A., Guadalupe Aguilar-González (CINVESTAV-IPN) for technical assistance in DNA sequencing. This work was supported by Secretaría de Ciencia, Tecnología e Innovación de la Ciudad de México (SeCyTI), grant No. PICSA11-107.

References

1. Kleckner N, Bender J, Gottesman S (1991) Uses of transposons with emphasis on Tn10. Methods Enzymol 204:139–180
2. Kwon YM, Ricke SC (2000) Efficient amplification of multiple transposon-flanking sequences. J Microbiol Methods 41:195–199

3. Reyes-Cortes R, Martínez-Peñafiel E, Martínez-Pérez F, De la Garza M, Kameyama L (2012) A novel strategy to isolate cell-envelope mutants resistant to phage infection: bacteriophage mEp213 requires lipopolysaccharides in addition to FhuA to enter *Escherichia coli* K-12. Microbiology 158:3063–3071
4. Bachmann BJ (1972) Pedigrees of some mutant strains of *Escherichia coli* K-12. Bacteriol Rev 36:525–557
5. Jensen KF (1993) The *Escherichia coli* K-12 "wild types" W3110 and MG1655 have an *rph* frameshift mutation that leads to pyrimidine starvation due to low *pyrE* expression levels. J Bacteriol 175:3401–3407
6. Appleyard RK (1954) Segregation of new lysogenic types during growth of a doubly lysogenic strain derived from *Escherichia coli* K-12. Genetics 39:440–452
7. Silhavy TJ, Berman ML, Enquist LW (1984) Experiments with gene fusions. Cold Spring Harbor Laboratory, Cold Spring Harbor, NY
8. Hanahan D (1983) Studies on transformation of *Escherichia coli* with plasmids. J Mol Biol 166:557–580
9. Uc-Mass A, Loeza EJ, De la Garza M, Guarneros G, Hernández-Sánchez J, Kameyama L (2004) An orthologue of the *cor* gene is involved in the exclusion of temperate lambdoid phages. Evidence that Cor inactivates FhuA receptor functions. Virology 329: 425–433
10. Kameyama L, Fernández L, Calderón J, Ortiz-Rojas A, Patterson TA (1999) Characterization of wild lambdoid bacteriophages: detection of the wide distribution of phage immunity groups and identification of the Nus-dependent, nonlambdoid phage group. Virology 263:100–111
11. Sambrook J, Russell DW (2001) Molecular cloning: a laboratory manual, 3rd edn. Cold Spring Harbor Laboratory, Cold Spring Harbor, NY
12. Hanahan D, Jessee J, Bloom FR (1991) Plasmid transformation of *Escherichia coli* and other bacteria. Methods Enzymol 204:63–113
13. Polayes D, Huges AJ Jr (1974) Efficient protein expression and simple purification using the pPROEX-1 super(TM) system. Focus 16:81–84
14. Dhillon EK, Dhillon TS (1974) N-methyl-N′-nitro-N-nitrosoguanidine and hydroxylamine induced mutants of the rII region of phage T4. Mutat Res 22:223–233

Part III

Functional Analysis of Cell-Wall Associated Proteins

Chapter 7

Zymographic Techniques for the Analysis of Bacterial Cell Wall in *Bacillus*

Tatsuya Fukushima and Junichi Sekiguchi

Abstract

Zymography of cell wall hydrolases is a simple technique to specifically detect cell wall or peptidoglycan hydrolytic activity. The zymographic method can be used for assessing the hydrolytic activities of purified target proteins, cell surface proteins, and proteins secreted to culture. Here, methods of cell wall and peptidoglycan purification, extraction of cell surface proteins containing cell wall hydrolases, and zymographic analysis are described. The purified or extracted proteins are separated by electrophoresis using an SDS gel containing cell wall or peptidoglycan material and then the proteins are renatured in the gel. The renatured cell wall hydrolases in the gel hydrolyze the material around the proteins. The cell wall or peptidoglycan in the gel is stained by methylene blue and the hydrolyzed material cannot be stained, resulting in the detection of cell wall hydrolytic activities of the enzymes on the gel.

Key words Endopeptidase, Carboxypeptidase, Muramidase, Glucosaminidase, Lytic transglycosylase, Wall teichoic acid, Lipoteichoic acid, Gram-positive bacteria

1 Introduction

Cell wall metabolism in Gram-positive bacteria is an essential process to adjust to various environmental conditions. Bacterial cell wall consists of peptidoglycan and the other polymers such as wall/lipoteichoic acid and teichuronic acid [1]. Recently many scientists have discovered that modifications of peptidoglycan by cell wall hydrolases are essential for cell elongation, sporulation, and germination [2–4].

Zymography using cell wall or peptidoglycan is a straightforward technique and several cell wall hydrolases: *N*-acetylmuramidase/lytic transglycosylase, *N*-acetylglucosaminidase, L-alanine amidase, L,D-endopeptidase, D,L-endopeptidase, and D,D-endopeptidase (Fig. 1) can be detected by zymographic analyses [5–11]. The procedure of zymographic analysis consists of two parts: separation of proteins by electrophoresis using an SDS-gel containing cell wall or peptidoglycan material and hydrolysis of the material by

Hee-Jeon Hong (ed.), *Bacterial Cell Wall Homeostasis: Methods and Protocols*, Methods in Molecular Biology, vol. 1440,
DOI 10.1007/978-1-4939-3676-2_7,

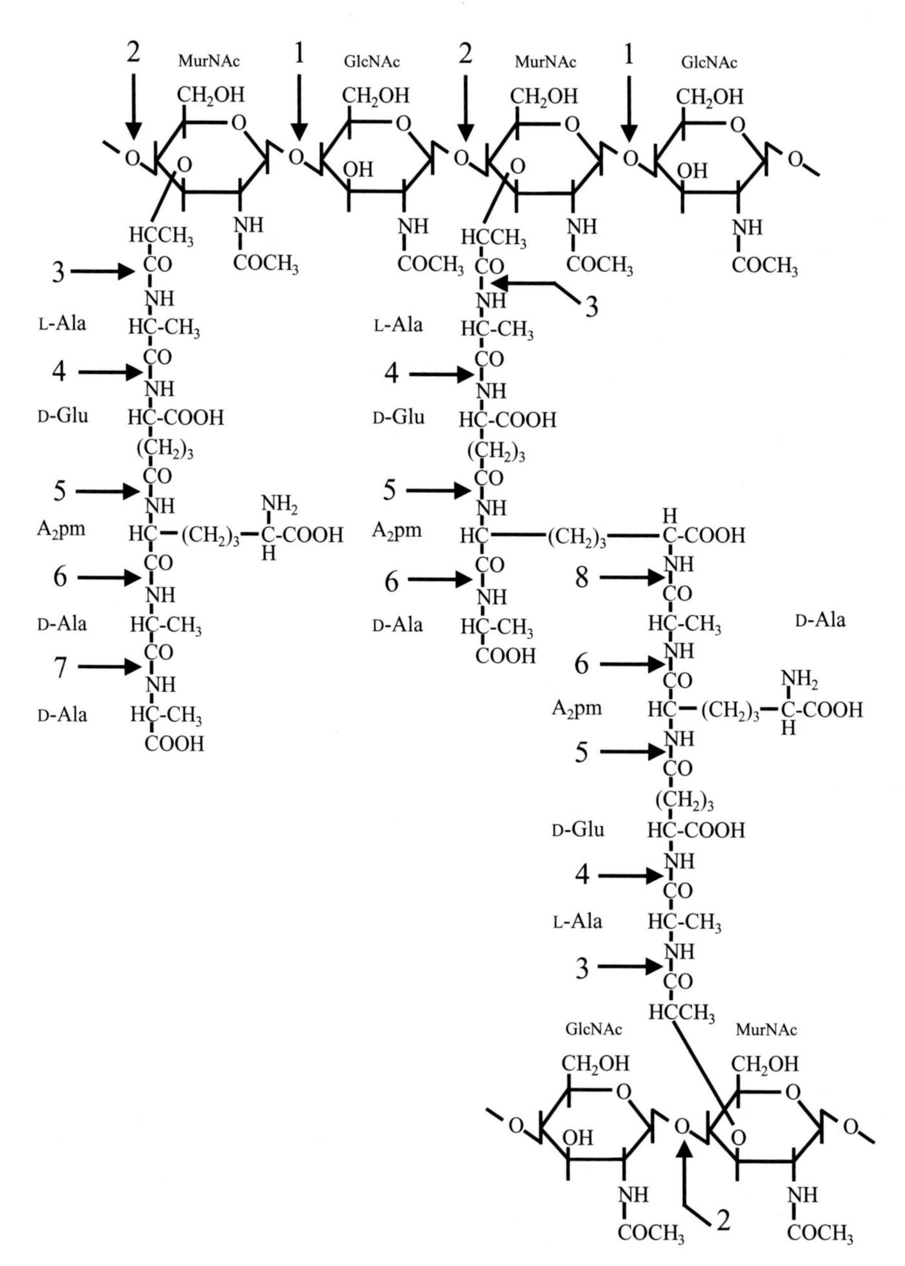

Fig. 1 Structure of *B. subtilis* peptidoglycan. *Arrow 1*, cleavage sites by *N*-acetylmuramidases (e.g., lysozyme, *B. subtilis* CwlT [N-terminal domain] [5]) or lytic transglycosylases (e.g., *B. subtilis* CwlQ [6]); *arrow 2*, cleavage sites by *N*-acetylglucosaminidases (e.g., *B. subtilis* LytD [CwlG] [7]); *arrow 3*, cleavage sites by L-alanine amidases (e.g., *B. subtilis* CwlB [LytC] [8], CwlC [12]); *arrow 4*, cleavage sites by L,D-endopeptidases (e.g., *B. subtilis* CwlK [9]); *arrow 5*, cleavage sites by D,L-endopeptidases (e.g., *B. subtilis* LytF [CwlE] [10], CwlO [13], CwlS [14], and CwlT [C-terminal domain] [5]); *arrow 6*, cleavage sites by L,D-carboxypeptidases (e.g., *B. subtilis* LdcB [YodJ] [15]); *arrow 7*, cleavage sites by D,D-carboxypeptidases (e.g., *B. subtilis* DacA [16]); *arrow 8*, cleavage site by D,D-endopeptidases (e.g., *B. subtilis* CwlP [C-terminal domain] [11]). The figure is modified from Fig. 1 in ref. [9]

renatured cell wall hydrolases in the gel. One of the benefits of zymography is to detect weak activity of cell wall hydrolysis. The other benefit of the analysis is that hydrolytic activities of the wild-type cell wall hydrolase and its point-mutated hydrolases can be simply compared like SDS-PAGE [5, 12]. Zymography can also be applied for analyzing cell wall hydrolytic activities of cell surface proteins and proteins secreted to culture [13]. The analysis informs the expression periods and locations of the cell wall hydrolases. We have described the zymographic technique using *Bacillus subtilis* cell wall and peptidoglycan. The method can also be applied to the cell wall hydrolase studies in the other Gram-positive bacteria such as *Staphylococcus* and *Streptococcus* [11].

2 Materials

Prepare all solutions using ultrapure water (Milli-Q water) and perform the following procedures at room temperature unless otherwise noted.

2.1 SDS-Polyacrylamide Gel Electrophoresis

1. Components of separation gel solution containing cell wall or peptidoglycan: 3–4 mL of 40% (w/v) acrylamide (acrylamide:bis-acrylamide = 29:1), 2.5 mL of separating gel buffer (1.5 M Tris–HCl, 0.4% [w/v] SDS [pH 8.8]), 5–10 mg purified cell wall or peptidoglycan, 10 mg fresh ammonium persulfate (APS), up to 10 mL of Milli-Q water (the final concentrations of the components in the gel solution are 12–16% acrylamide, 375 mM Tris–HCl, 0.1% SDS, 0.5–1 mg/mL cell wall or peptidoglycan [pH 8.8]).
2. Components of stacking gel solution: 1 mL of 40% (w/v) acrylamide (acrylamide:bis-acrylamide = 29:1), 2.5 mL of stacking gel buffer (0.5 M Tris–HCl, 0.4% [w/v] SDS [pH6.8]), 10 mg fresh APS, 6.5 mL of Milli-Q water (the final concentrations of the components in the gel solution are 4% acrylamide, 125 mM Tris–HCl, 0.1% SDS [pH 6.8]).
3. Components of 2× loading dye: 20% (w/v) glycerol, 150 mM Tris–HCl, 4% (w/v) SDS, 100 μL/mL β-mercaptoethanol, 0.1 mg/mL bromophenol blue (pH 6.8).
4. Components of SDS running buffer: 14.4 g/L of glycine, 3 g/L of Tris, and 1 g/L of SDS in Milli-Q water.

2.2 Zymography

1. Components of renature buffer: 25 mM Tris–HCl, 1% (v/v) Triton X-100 (pH 7.2) (the buffer component and the buffer pH can be changed to optimize the target cell wall hydrolytic activity).

2. Components of methylene blue staining solution: 0.01% (w/v) methylene blue and 0.01% (w/v) KOH in Milli-Q water.

2.3 Protein Staining

1. Components of Coomassie Brilliant Blue staining solution: 0.02% (w/v) Coomassie Brilliant Blue R-250, 40% (v/v) methanol, and 10% (v/v) acetic acid in Milli-Q water.
2. Components of destaining solution: 10% (v/v) methanol and 10% (v/v) acetic acid in Milli-Q water.

2.4 Extraction of Cell Surface Proteins

1. Component of cold cell washing buffer: 25 mM Tris–HCl (pH 7.2).
2. Components of cold extraction solution: 3 M LiCl, 25 mM Tris–HCl (pH 7.2).

3 Methods

3.1 Purification of B. subtilis Cell Wall (Peptidoglycan Containing [Lipo/Wall] Teichoic Acid)

1. Incubate *B. subtilis* 168 in 1 L of LB medium at 37 °C until the cells enter the stationary phase (*see* **Note 1**). Collect and centrifuge the culture (10,000 *g* 10 min, room temperature). Remove the supernatant, resuspend the pellet in 20–30 mL of 4 M LiCl, and then boil the suspension for 15 min in order to remove cell wall-binding proteins from cell wall.
2. Cool down the suspension at room temperature and replace it to a homogenizer (e.g., ACE Homogenizer, Nissei). Add glass beads (ϕ 0.1 mm) into the suspension until it becomes highly condensed and then perform homogenization for 15 min four times (total for 60 min) (since the homogenizer becomes warm during the process, the homogenizer is cooled down after each 15-min homogenization). Confirm that the cells are broken by microscopic observation.
3. Completely transfer the homogenized sample into a flask (500 mL or 1 L flask) (the sample is named "the glass beads sample"). Keep the sample for 15–20 min in order to precipitate glass beads (the supernatant should have some turbidity).
4. Collect the supernatant (containing cell wall, solubilized materials such as proteins and DNA, and small amount of beads) of the sample and then centrifuge it (1500 *g*, 10 min) to separate the small amount of beads, followed by collection of the supernatant.
5. Centrifuge the supernatant (11,000 *g*, 10 min) to precipitate crude cell wall.
6. Add 100 mL of Milli-Q water in "the glass beads sample" in the procedure in **step 3** and mix them well. Keep the sample for 15–20 min.

7. Repeat the procedure in **steps 4–6** until the supernatant of "the glass beads sample" of the procedure in **step 6** is clear.
8. (Optional) Resuspend the crude cell wall precipitation (procedure in **step 5**) in 20 mL of Milli-Q water and sonicate it for 15 min. Centrifuge the sample and discard the supernatant (this procedure helps obtaining better quality of cell wall).
9. Add 40 mL of 4% (w/v) SDS in the crude cell wall precipitation (procedure in **steps 5** or **8**) and boil the sample for 15 min.
10. Centrifuge the sample (11,000 *g*, 10 min, room temperature [no 4 °C due to avoiding SDS precipitation]) and then discard the supernatant.
11. Resuspend the pellet with 40 mL of 1 M NaCl. Centrifuge the suspension (11,000 *g*, 10 min, room temperature) and then discard the supernatant. Repeat the washing procedure (resuspension → centrifugation → discarding the supernatant) until the cell wall pellet becomes white.
12. Wash the cell wall pellet by centrifugation with 40 mL of Milli-Q water at least three times and then resuspend the cell wall with 5–10 mL of Milli-Q water (the cell wall resuspension should be white) (Fig. 2a).
13. Measure absorbance of the cell wall suspension at 540 nm ($A_{540} = 1$ of the suspension indicates approximately 1 mg cell wall/mL). Pick up some amount of the suspension (e.g., pick up predicted 10 mg of cell wall based on the absorbance measured at 540 nm) and dry up the cell wall, followed by measurement

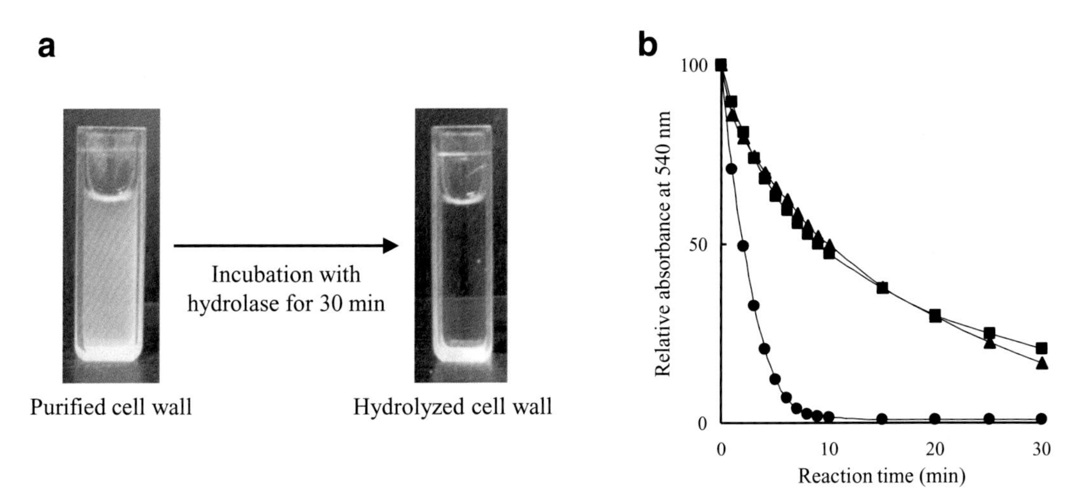

Fig. 2 Cell wall/peptidoglycan hydrolysis assay. (**a**) Pictures of purified cell wall material (*left panel*) and hydrolyzed material using cell wall hydrolase, CwlP (*right panel*). (**b**) Time course of hydrolysis assay of the N-terminal domain of CwlT (*N*-acetylmuramidase) (*squares*), the C-terminal domain of CwlT (D,L-endopeptidase) (*triangles*), or the full length (both the domains) of CwlT (*circles*). Fig. (**b**) is modified from Fig. (**b**) in ref. [5]

of weight of the dried cell wall. Calculate the final concentration of the cell wall suspension.

14. Store the cell wall suspension at 4 °C (do not freeze). If the suspension is needed to keep for a long time (over 1 month), add 0.02–0.05 % (w/v) sodium azide.

3.2 Purification of *B. subtilis* Peptidoglycan Without Lipo/Wall Teichoic Acids

1. Add 1 mL of 10 % (w/v) TCA in 10 mg of purified cell wall and then incubate the mixture at 37 °C for 12 h to remove lipo/wall teichoic acid. Centrifuge the sample (11,000 *g*, 10 min) and discard the supernatant.
2. Repeat the procedure in **step 1** and wash the pellet with 1 mL of Milli-Q water five times to eliminate TCA. Resuspend the pellet (this pellet should be peptidoglycan) in 1 mL of Milli-Q water and measure the pH using a pH indicator to confirm that TCA is completely removed (*see* **Note 2**).
3. Measure absorbance of the peptidoglycan suspension at 540 nm (A_{540} = 1 of the suspension indicates approximately 3 mg cell wall/mL). Pick up some amount of the suspension (e.g., pick up predicted 10 mg of peptidoglycan based on the absorbance measured at 540 nm) and dry up the peptidoglycan, followed by measurement of weight of the dried peptidoglycan. Calculate the final peptidoglycan concentration of the suspension.
4. Store the peptidoglycan suspension at 4 °C (do not freeze). If the suspension is needed to keep for a long time (over 1 month), add 0.02–0.05 % (w/v) sodium azide.

3.3 Confirmation of Quality of *B. subtilis* Cell Wall and Peptidoglycan

Since quality of the purified cell wall and peptidoglycan is very important for zymographic analysis, cell wall/peptidoglycan hydrolysis assay using lysozyme (muramidase) and/or known hydrolases such as LytF (D,L-endopeptidase [10]) and CwlT (muramidase and D,L-endopeptidase [5]) is performed.

1. Resuspend 0.99 of OD_{540} of the purified cell wall or peptidoglycan in 3 mL of buffer such as 50 mM sodium phosphate buffer (pH 6.0) or 50 mM HEPES-NaOH buffer (pH 7.0). Final absorbance of the suspension at 540 nm is 0.33 (Fig. 2).
2. Add 1–10 nmol of lysozyme or known hydrolase in 3 mL of the suspension and incubate the sample at 37 °C.
3. Measure absorbance at 540 nm during the incubation (e.g., 0-, 5-, 10-, 20-, 30-min incubation) to confirm that the purified material is actually hydrolyzed (Fig. 2).

3.4 Zymographic Analysis

1. Add 0.5–1 mg/mL of purified cell wall or peptidoglycan to 10 mL of separating gel solution and mix them (*see* Subheading 2.1).

2. Add 10 μL of *N,N,N′,N′*-tetramethylethane-1,2-diamine (TEMED) to the separation gel mixture, pour the mixture into a cast gel, and gently overlay water or isopropanol. Remove the water or isopropanol from the cast gel after the separating gel becomes solid.
3. Make 10 mL of stacking gel solution and add 10 μL of TEMED to the solution. Pour the solution in the cast gel and insert a well comb.
4. Make 10 μL of target protein, positive control protein (cell wall hydrolase such as lysozyme or CwlT), and negative control protein (e.g., BSA) dissolved in SDS-loading dye (approximately 1 μg of protein is dissolved) (*see* **Note 3**) and keep the samples for 10 min at 96 °C.
5. Perform SDS-PAGE using the cast gel.
6. Open glass plates of the cast gel and remove the stacking gel from the separating gel. Cut the separating gel to split the gel for protein staining by Coomassie Brilliant Blue and for zymographic assay (Fig. 3a) (following procedure in **steps 7–11** is for zymographic analysis and procedure in **steps 12** and **13** is for protein staining).
7. (Optional) Put the separating gel for zymographic analysis in a tray containing 20–50 mL of Milli-Q water and gently shake it for 5 min to remove SDS.
8. Transfer the separating gel into a tray containing 20–50 mL of renature buffer and incubate the gel at 37 °C for 20–60 min (Fig. 3b).
9. Check the gel whether a white band appears or not (the white band indicates the position where cell wall/peptidoglycan is hydrolyzed by cell wall hydrolases). Go to the procedure in **step 10** if the white band appears or if the renature process is performed for a long time (over 3 h); otherwise go to the procedure in **step 8** to repeat renaturation.
10. Stain the gel with 20–50 mL of the methylene blue staining solution until the entire gel becomes blue (Fig. 3c).
11. Put the gel to 20–50 mL of Milli-Q water and gently shake it until a destained band appears in the blue background gel (destaining treatment, Fig. 3d) (if the observed band is weak or nothing, perform renaturation procedure as in **step 8** again).
12. Put the separating gel for Coomassie Brilliant Blue staining in the procedure in **step 6** in 20–40 mL of Coomassie Brilliant Blue staining solution and gently shake the gel in the solution for 30–60 min (Fig. 3e).
13. Destain the gel using 20–40 mL of the destaining solution (Fig. 3e).

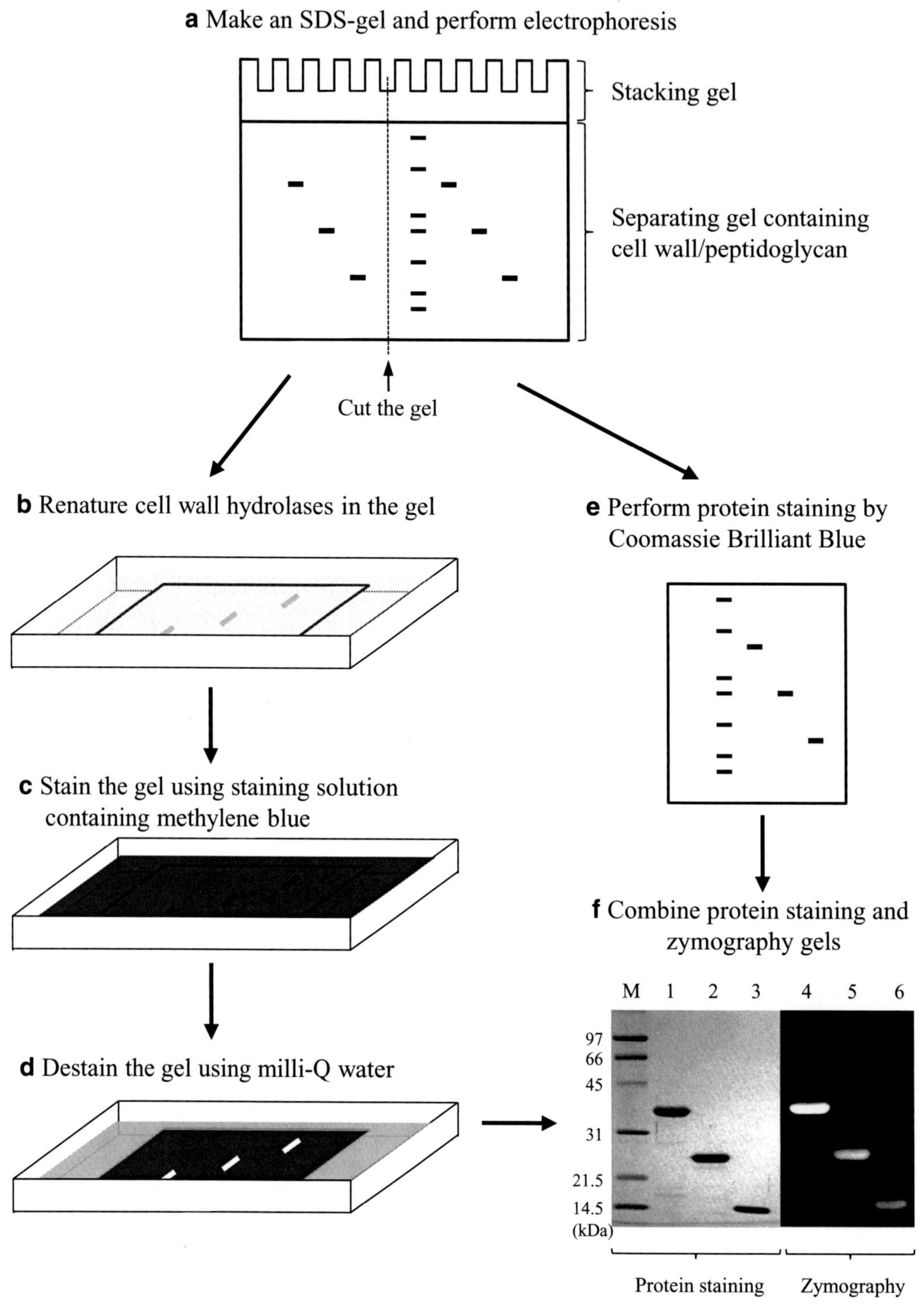

Fig. 3 Zymographic procedure. (**a**) SDS-PAGE (*see* Subheading 3.4, **steps 1–6**). After an SDS-gel containing cell wall or peptidoglycan is made, target protein(s), cell wall hydrolase (the positive control), and a protein for the negative control such as BSA are applied onto the gel followed by SDS-PAGE. The gel is cut for protein staining and zymography. (**b**) Renaturation on zymography (*see* Subheading 3.4, **steps 8** and **9**). The gel is put in renature buffer to renature cell wall hydrolases. (**c**) Gel staining on zymography (*see* Subheading 3.4, **step 10**). (**d**) Gel destaining on zymography (*see* Subheading 3.4, **step 11**). (**e**) Protein staining using Coomassie Brilliant Blue (*see* Subheading 3.4, **step 12**). (**f**) Final result of zymography and protein staining. Fig. (**f**) is modified from Fig. 2a in ref. [5]

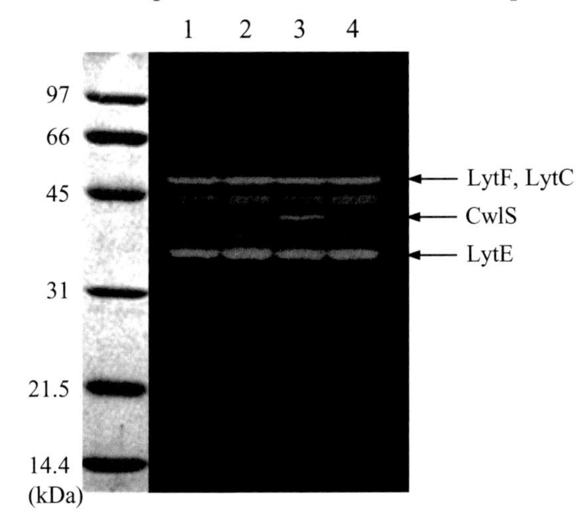

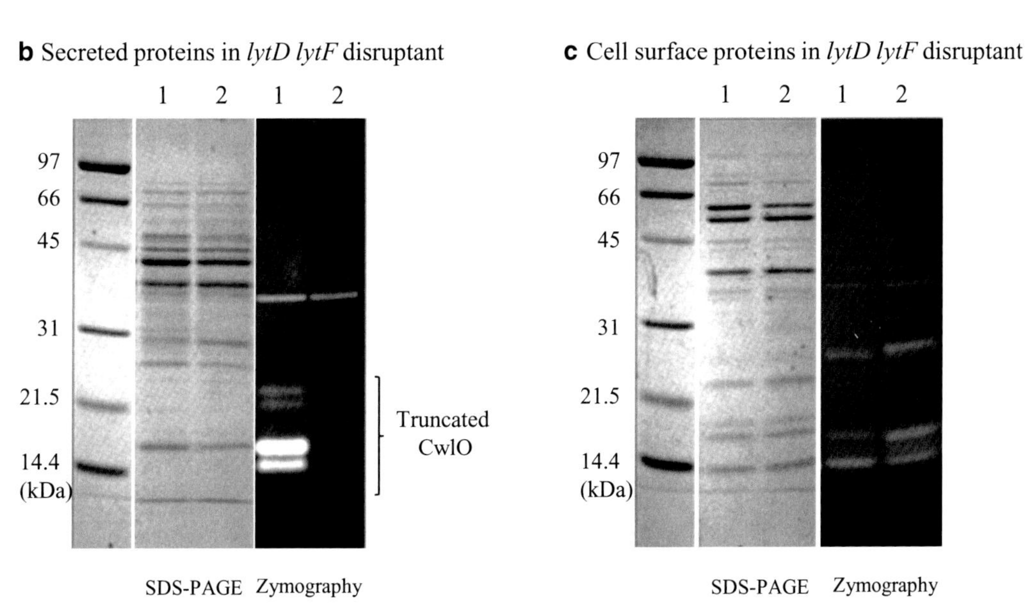

Fig. 4 (**a**) Zymographic analysis of cell surface proteins extracted from *B. subtilis* strains with or without CwlS expression. *Lane 1*, cell surface protein extracted from *B. subtilis* 168 (wild type); *lane 2*, cell surface proteins extracted from the *cwlS* disruptant strain; *lanes 3* and *4*, cell surface proteins extracted from the *cwlS* conditional depletion strain with and without the *cwlS* expression (*lane 3* and *4*, respectively). The result clearly shows that D,L-endopeptidase, CwlS, is a cell wall hydrolase located on the cell surface. (**b**, **c**) Zymographic analysis of proteins secreted to culture (**b**) and cell surface proteins (**c**) in a *lytC lytF*-double disruptant and a *lytC lytF cwlO*-triple disruptant. *Lane 1*, the double disruptant; *lane 2*, the triple disruptant. The result revealed that CwlO is mainly secreted to culture. Fig. (**a**) is modified from Fig. 5 in ref. [14] and Figs. (**b**, **c**) are modified from Fig. 6 in ref. [13]

3.5 Preparation of Cell Surface Proteins for Zymography

Zymography can be used not only for purified cell wall hydrolases but also cell surface proteins extracted from bacteria. Major cell wall hydrolases on cell surface can be detected by the zymographic analysis (Fig. 4).

1. Incubate *B. subtilis* 168 (e.g., *B. subtilis* 168 and a *B. subtilis* 168 strain with no expression or overexpression of a target gene) in 50–200 mL of LB medium. Collect 15 mL of the culture during the vegetative phase and/or stationary phase (*see* **Note 4**), centrifuge the culture (7000 *g*, 5 min, 4 °C), and then separate the supernatant and pellet (the supernatant can be used for preparation of proteins secreted to culture, *see* Subheading 3.6, **step 2**, for the preparation of secreted proteins).
2. Resuspend the cell pellet with 1 mL of cold cell washing buffer, centrifuge the suspension (7000 *g*, 5 min, 4 °C), and then remove the supernatant. Repeat these processes (washing the cell) one more time to remove unbound proteins on cell wall.
3. Resuspend the pellet in 1 mL of cold extraction solution and keep the sample on ice for 20 min to extract cell surface proteins.
4. Centrifuge the sample (7000 *g*, 5 min, 4 °C) and collect 900 μL of the supernatant containing cell surface proteins (to collect the supernatant *without any cells*, only 900 μL of the supernatant is taken for the next step procedure).
5. Add 37.5 μL of 50% (w/v) TCA solution in the supernatant (the final concentration of TCA in the sample should be 2%) and keep the sample on ice for 30 min.
6. Centrifuge the sample 11,000 *g*, 5 min, 4 °C) and discard the supernatant. Wash the sample with 1 mL of 70% (v/v) ethanol five times (add ethanol → centrifugation → discarding the supernatant). Dry up the sample, add SDS-loading dye, and then keep the samples for 10 min at 96 °C (the sample is cell surface proteins).
7. Apply the extracted cell surface proteins on an SDS-gel containing cell wall/peptidoglycan for zymographic analysis (the proteins are extracted from 8 to 10 of OD_{600} of the cells). The following procedures for zymographic analysis are same as the procedures in Subheading 3.4, **step 5**. Examples of zymographic analysis using cell surface proteins are shown in Fig. 4.

3.6 Preparation of Proteins Secreted to Culture on Zymography

1. Incubate *B. subtilis* strain in 50–200 mL of LB medium. Collect 10–30 mL of the culture during the vegetative phase and/or stationary phase (*see* **Note 4**), centrifuge the culture (7000 *g*, 5 min, 4 °C), and then collect the supernatant.
2. Add 50% (w/v) TCA to the supernatant (to make 2% [w/v] TCA solution as a final concentration) and keep the solution for 30 min on ice.

3. Centrifuge the sample (11,000 *g*, 5 min, 4 °C) and discard the supernatant. Wash the sample with 1 mL of 70 % (v/v) ethanol five times (add ethanol → centrifugation → discarding the supernatant). Dry up the sample, add SDS-loading dye, and then keep the samples for 10 min at 96 °C (the sample is proteins secreted to culture).
4. Apply the secreted proteins on an SDS-gel containing cell wall/peptidoglycan for zymographic analysis (the proteins secreted to culture are extracted from 8 to 12 of OD_{600} of the cell culture). The following procedures for zymographic analysis are same as the procedures in Subheading 3.4, **step 5**. An example of zymographic analysis using proteins secreted to culture is shown in Fig. 4b.

4 Notes

1. The culture volume and incubation period can be changed. For example, if the culture volume is 5 L, 100–150 mL of 4 M LiCl is added to the cell pellet.
2. When peptidoglycan derived from the other bacteria is purified, attached materials to the peptidoglycan such as polysaccharides, cell wall-anchoring proteins, and/or poly-γ-glutamic acids are removed using digestion enzymes (e.g., α-amylase, trypsin, and/or poly-γ-glutamic acid hydrolase).
3. Positive control and negative control are important for zymographic analysis. When the gel is incubated in renature buffer for a long time (over 12 h), sometimes the band of negative control protein appears. The negative control protein should not be seen for zymographic analysis.
4. Amount of cell culture harvested is dependent on the experiment. On zymography, normally ten of OD_{600} of the cells or its cell culture is necessary to analyze one sample of cell surface proteins or proteins secreted to culture, respectively.

References

1. Foster SJ, Popham DL (2002) Structure and synthesis of cell wall, spore cortex, teichoic acids, S-layers, and capsules. In: Sonenshein AL, Hoch JA, Losick R (eds) *Bacillus subtilis* and its closest relatives: from genes to cells. American Society for Microbiology, Washington, DC, p 21
2. Bisicchia P, Noone D, Lioliou E, Howell A, Quigley S, Jensen T, Jarmer H, Devine KM (2007) The essential YycFG two-component system controls cell wall metabolism in *Bacillus subtilis.* Mol Microbiol 65:180–200
3. Morlot C, Uehara T, Marquis KA, Bernhardt TG, Rudner DZ (2010) A highly coordinated cell wall degradation machine governs spore morphogenesis in *Bacillus subtilis.* Genes Dev 24:411–422
4. Ishikawa S, Yamane K, Sekiguchi J (1998) Regulation and characterization of a newly deduced cell wall hydrolase gene (*cwlJ*) which affects germination of *Bacillus subtilis* spores. J Bacteriol 180:1375–1380
5. Fukushima T, Kitajima T, Yamaguchi H, Ouyang Q, Furuhata K, Yamamoto H, Shida

T, Sekiguchi J (2008) Identification and characterization of novel cell wall hydrolase CwlT: a two-domain autolysin exhibiting *N*-acetylmuramidase and DL-endopeptidase activities. J Biol Chem 283:11117–11125

6. Sudiarta IP, Fukushima T, Sekiguchi J (2010) *Bacillus subtilis* CwlQ (previous YjbJ) is a bifunctional enzyme exhibiting muramidase and soluble-lytic transglycosylase activities. Biochem Biophys Res Commun 398:606–612
7. Rashid MH, Mori M, Sekiguchi J (1995) Glucosaminidase of *Bacillus subtilis*: cloning, regulation, primary structure and biochemical characterization. Microbiology 141:2391–2404
8. Kuroda A, Sekiguchi J (1991) Molecular cloning and sequencing of a major *Bacillus subtilis* autolysin gene. J Bacteriol 173:7304–7312
9. Fukushima T, Yao Y, Kitajima T, Yamamoto H, Sekiguchi J (2007) Characterization of new L, D-endopeptidase gene product CwlK (previous YcdD) that hydrolyzes peptidoglycan in Bacillus subtilis. Mol Genet Genomics 278: 371–383
10. Ohnishi R, Ishikawa S, Sekiguchi J (1999) Peptidoglycan hydrolase LytF plays a role in cell separation with CwlF during vegetative growth of *Bacillus subtilis*. J Bacteriol 181: 3178–3184
11. Sudiarta IP, Fukushima T, Sekiguchi J (2010) *Bacillus subtilis* CwlP of the SP-β prophage has two novel peptidoglycan hydrolase domains, muramidase and cross-linkage digesting DD-endopeptidase. J Biol Chem 285:41232–41243
12. Shida T, Hattori H, Ise F, Sekiguchi J (2001) Mutational analysis of catalytic sites of the cell wall lytic *N*-acetylmuramoyl-L-alanine amidases CwlC and CwlV. J Biol Chem 276: 28140–28146
13. Yamaguchi H, Furuhata K, Fukushima T, Yamamoto H, Sekiguchi J (2004) Characterization of a new *Bacillus subtilis* peptidoglycan hydrolase gene, *yvcE* (named *cwlO*), and the enzymatic properties of its encoded protein. J Biosci Bioeng 98:174–181
14. Fukushima T, Afkham A, Kurosawa S, Tanabe T, Yamamoto H, Sekiguchi J (2006) A new D, L-endopeptidase gene product, YojL (renamed CwlS), plays a role in cell separation with LytE and LytF in Bacillus subtilis. J Bacteriol 188:5541–5550
15. Hoyland CN, Aldridge C, Cleverley RM, Duchêne MC, Minasov G, Onopriyenko O, Sidiq K, Stogios PJ, Anderson WF, Daniel RA, Savchenko A, Vollmer W, Lewis RJ (2014) Structure of the LdcB LD-carboxypeptidase reveals the molecular basis of peptidoglycan recognition. Structure 22:949–960
16. Todd JA, Bone EJ, Ellar DJ (1985) The sporulation-specific penicillin-binding protein 5a from *Bacillus subtilis* is a DD-carboxypeptidase in vitro. Biochem J 230:825–828

Chapter 8

Liquid Chromatography-Tandem Mass Spectrometry to Define Sortase Cleavage Products

Andrew Duong, Kalinka Koteva, Danielle L. Sexton, and Marie A. Elliot

Abstract

Sortase enzymes have specific endopeptidase activity, cleaving within a defined pentapeptide sequence at the C-terminal end of their protein substrates. Here, we describe how monitoring sortase cleavage activity can be achieved using peptide substrates. Peptide cleavage can be readily analyzed by liquid chromatography/tandem mass spectrometry (LC/MS/MS), which allows for the precise definition of cleavage sites. This technique could be used to analyze the peptidase activity of any enzyme, and identify sites of cleavage within any peptide.

Key words Sortase, Peptidase, Peptide cleavage, Liquid chromatography, Mass spectrometry

1 Introduction

Sortases are membrane-anchored, extra-cytoplasmic enzymes. They are largely confined to the Gram-positive bacteria (recently reviewed in [1–3]), although analogous systems have been identified for some Gram-negative bacteria (e.g., [4]) and archaea (e.g., [5]). Sortases have transpeptidase activity, cleaving first within a defined pentapeptide motif at the C-terminus of their target substrates [6, 7], before mediating covalent attachment of their substrates to either the peptide stem of the peptidoglycan (where peptidoglycan is a major structural component of bacterial cell walls), or to pilin subunits during pilus formation [8]. In recent years, there has been a move towards exploiting the specific and well-understood transpeptidase activity of sortases, for the purposes of protein tagging (e.g., [9, 10]), protein engineering [11–13], and general protein modification (e.g., [14–16]). This powerful technology is further benefiting from the development of sortase enzymes with altered target specificity [17].

There are a number of different ways in which sortase activity can be followed in vitro. For assessing the cleavage of full-length substrate proteins, this can be done using Edman degradation, or

Hee-Jeon Hong (ed.), *Bacterial Cell Wall Homeostasis: Methods and Protocols*, Methods in Molecular Biology, vol. 1440, DOI 10.1007/978-1-4939-3676-2_8,

SDS-PAGE coupled with Coomassie blue staining or immunoblotting (e.g., [18]). When using peptides as substrate, the most common means of defining sortase cleavage has typically involved the analysis of peptide cleavage products using liquid chromatography alone (e.g., [19]), or in conjunction with mass spectrometry (e.g., [20–23]). In particular, the latter combination has proven highly effective, allowing for precise determination of sortase cleavage sites through a two-step process of liquid chromatography-based product separation, and precise product identification using tandem mass spectrometry. Here, we describe the use of a coupled liquid chromatography/tandem mass spectrometry (LC/MS/MS) approach to analyze sortase cleavage products (an analogous approach could be used to follow the activity of any peptidase).

2 Materials

It is recommended that all solutions be prepared with ultrapure water, apart from the LC/MS/MS reagents, which must be prepared with HPLC-grade solutions. Solutions can be stored at room temperature, unless otherwise noted. Peptide and protein aliquots are to be stored at −80 °C. It is recommended that MSDS recommendations be followed when working with and disposing of solutions and other waste material.

2.1 Enzymes and Substrates

1. Purified, soluble sortase enzyme (*see* **Note 1**): Will need up to 50 μM for each reaction.
2. Purified, soluble active-site mutant (*see* **Note 2**): As for the wild-type enzyme, require up to 50 μM for each reaction.
3. Peptide substrate (*see* **Note 3**): To create a stock solution of the synthesized peptide substrate, the lyophilized peptide will be resuspended in sterile 100% dimethyl sulfoxide (DMSO) to a final concentration of 500 nM, before being dispensed in 50 μL aliquots and stored at −80 °C. Each aliquot can then be thawed once for use in the cleavage reactions, with any remaining peptide being discarded.

2.2 Enzyme Concentration Determination

Sortase concentration determination: Bradford protein assay [24] using BioRad dye reagents (*see* **Note 4**).

2.3 Cleavage Assay Components

1. Sortase storage buffer: 50 mM Tris–HCl pH 7.5, 150 mM NaCl, 1 mM DTT, 10% glycerol, final concentration.
   (a) Stock solutions.
   1.0 M Tris, pH 7.5: Dissolve 12.1 g of Tris base in 80 mL water. Adjust the pH to 7.5 using HCl. Make the final volume up to 100 mL with water, and sterilize by autoclaving.

5 M NaCl: Dissolve 29.2 g of NaCl in 80 mL water. Make volume up to 100 mL with water, and autoclave the resulting solution.

DTT: Dissolve 1.5 g DTT in 8 mL water, before making the volume up to 10 mL with water. Filter sterilize, dispense into 1 mL aliquots, and store frozen at −20 °C.

50 % Glycerol: Mix 50 mL 100 % glycerol with 50 mL water. Sterilize by autoclaving.

(b) Combine 500 μL 1.0 M Tris–HCl, pH 7.5; 300 μL 5 M NaCl; 10 μL 1 M DTT; and 2 mL 50 % glycerol in a sterile 15 mL Falcon tube. Make up to 10 mL with sterile water, and mix well. Keep solution on ice, or refrigerate at 4 °C.

2. Reaction buffer: 50 mM Tris–HCl, pH 7.5, 150 mM NaCl, 1 mM DTT. Prepare buffer as described for the sortase storage buffer, only omitting the glycerol (and replacing with water).
3. Calcium chloride: 0.5 M Stock solution (*optional—see* **Note 5**). Dissolve 5.6 g $CaCl_2 \cdot 6H_2O$ in 80 mL water. Make volume up to 100 mL with water. Sterilize by autoclaving.
4. 30 °C Incubator, temperature block, or water bath.
5. (*Optional*) If opting to follow peptide cleavage using a fluorophore/quencher system, this will require an instrument that can monitor changes in fluorescence (e.g., fluorimeter or plate reader with fluorescence detection capabilities). For example, the Abz/Dpn combination described in **Note 3** requires excitation at 320 nm and emission at 420 nm, while the DABCYL/EDANS combination used by others [25] requires excitation at 350 nm, and emission at 500 nm.

2.4 LC/MS Analysis

1. Liquid chromatography system (we use an Agilent 1100 series LC system, from Agilent Technologies, Canada): LC components include degasser (*see* **Note 6**); binary pump; autosampler (provides rapid sample injection); diode array detector; and a C18 column (3 μm, 120 Å, 4.6 × 100 mm; Dionex Corporation; *see* **Note 7**) equipped with a guard column.

(a) Solvent A: 0.05 % Formic acid (w/v) in HPLC-grade water (*see* **Note 8**): To 2 L of HPLC-grade water, add 819.6 μL concentrated formic acid and mix well. Filter sterilize the solution using a DURAPOLE membrane filter, 0.45 μm HW (Millipore), and degas for 5 min. To further minimize any microbial contamination, prepare fresh each time.

(b) Solvent B: 0.05 % Formic acid (w/v) in HPLC-grade acetonitrile (*see* **Note 9**). To 2 L of HPLC-grade acetonitrile, add 819.6 μL concentrated formic acid and mix well. As above, filter through a DURAPOLE membrane filter, 0.45 μm HW

(Millipore), to sterilize the solution, and degas for 5 min. To avoid microbial contamination, prepare fresh each time.

(c) Solvent line storage (*see* **Note 10**): 50% HPLC-grade acetonitrile, 50% HPLC-grade water. To 250 mL of HPLC-grade water, add 250 mL HPLC-grade acetonitrile.

2. QTRAP 2000 LC/MS/MS system from ABSciex (*see* **Note 11**).

3 Methods

3.1 Sortase Substrate Cleavage

1. Peptide substrate preparation: Commercially synthesized peptides are typically provided as lyophilized powders. Stock solutions can be created by resuspending the peptide powder in 100% DMSO to a concentration of 500 nM (taking into account the molecular weight of the peptide, and the mass of the peptide powder).
2. Determine the concentration of purified sortase/active site mutant proteins using the Bradford protein assay or another technique of choice (*see* **Note 4**). We use bovine serum albumin (BSA) as our protein standard, and dye reagents from BioRad.
3. Cleavage assay (Fig. 1): Mix together 10 or 50 μM purified sortase enzyme in sortase storage buffer (*see* **Note 12**), 2 μL of the 500 nM peptide stock solution, and 1 μL 0.5 M $CaCl_2$. Add reaction buffer to bring the volume up to 100 μL (*see* **Note 13**), and gently mix.
4. Negative controls for the assay include (1) reactions containing the peptide alone (no added sortase), to ensure that none of the reaction components are contaminated with a peptidase/protease, and (2) reactions using an active site mutant enzyme, which will test whether the purified protein preparations contain a contaminating peptidase from the *E. coli* overexpression host. If quantifying cleavage, the fluorescence values for the peptide-alone reaction can be subtracted from the overall fluorescence values for sortase enzyme/active site mutant reactions.

   To ensure that results are reproducible, reactions should be done in triplicate, and should include both technical and biological (independent enzyme preparations) replicates.
5. Incubate reactions for 18 h in a 30 °C incubator, temperature block, or water bath.
6. Cleavage products can be analyzed by LC/MS/MS immediately, or they can be frozen indefinitely at −80 °C.

3.2 Cleavage Product Analysis

1. When setting up the liquid chromatography aspect of the LC/MS/MS system, connect tube A to degassed solvent A and tube B to degassed solvent B. Flush tubing with respective solutions (*see* **Note 14**).

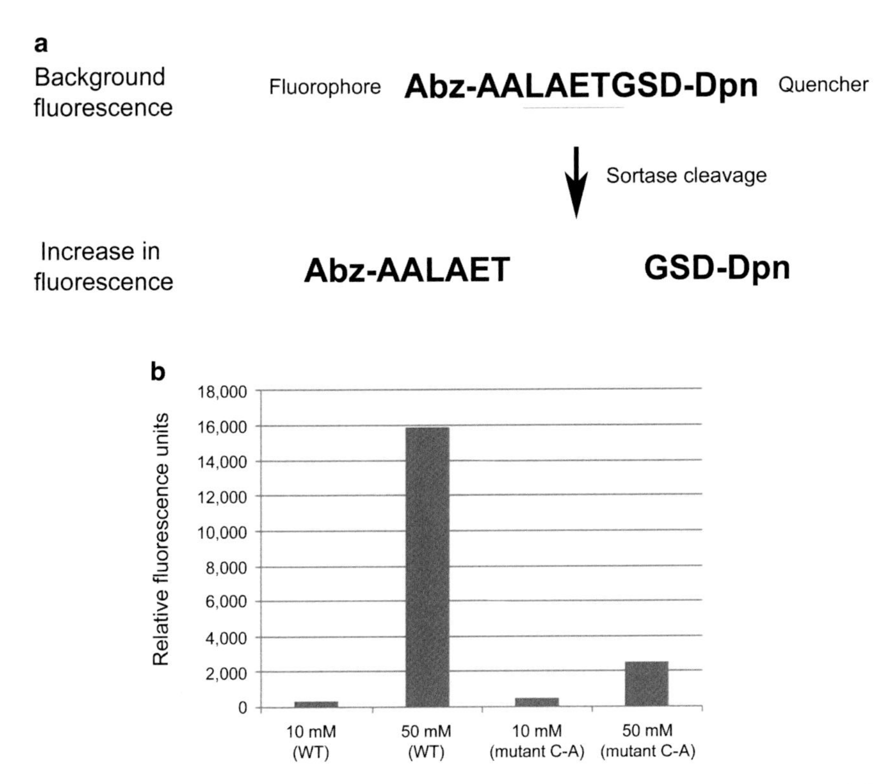

Fig. 1 Sortase cleavage assay. (**a**) Schematic diagram of peptide substrate, with an N-terminal fluorophore (Abz) and C-terminal quenching molecule (Dpn). Cleavage of the peptide relieves the inhibition mediated by the quencher, resulting in an increase in fluorescence. (**b**) Bar graph depicting relative sortase (SrtE1 from *Streptomyces coelicolor*) activity of wild-type (WT) and mutant (C-A) enzymes; data are shown as relative fluorescence units. Peptide-alone fluorescence has been subtracted from the graphed values

2. Affix C18 column (and associated guard column, *see* **Note 15**) to the LC system. Ensure that the column is firmly connected, but take care to avoid using excessive force when tightening the fittings. Wash the column with at least three column volumes of solvent A (maximal flow rate once the column is attached is 1 mL/min).
3. Set up LC/MS/MS protocol:
   (a) Flow rate: 1 min/mL, initially using 95 % solvent A, 5 % solvent B.
   (b) Cleavage product separation: 5 % solvent B for 5 min, followed by a linear gradient to 97 % solvent B over a 20-min period, and finally an isocratic elution with 97 % solvent B for 5 min.
   (c) Following LC separation, the flow was automatically split in a 9:1 ratio, with 1/10 of the flow volume (100 μL) being introduced into the connected mass spectrometer.

(d) QTRAP: operate in "enhance scan mode" (ESM), with switching polarity.

(e) Mass scan range: set to 100–1700 Da, at a scan rate of 4000 Da/s.

(f) Ion spray needle voltage: set to 4300 V for positive ion mode, and −4300 V for negative ion mode.

(g) Curtain gas: set to 12 psi.

(h) Collision (CAD) gas: set to high, at 20 psi.

(i) Source temperature: 100 °C.

(j) Interface heater: turned off.

4. Sortase enzyme removal from cleavage reaction: Mix the 100 μL cleavage reaction with an equal volume of methanol and place it at −20 °C for a minimum of 15 min to precipitate proteins and any additional impurities. Centrifuge the mixture for 10 min at 11,955 × *g* for 15 min, before transferring the supernatant to a fresh, sterile tube.
5. Using automatic sample injection (from a 10 × 10 vial sample tray), directly inject 100 μL of the enzyme-free peptide cleavage reaction onto the coupled LC/MS/MS system (*see* **Notes 16** and **17**). Separation of the different peptide fragments by LC is based on differences in polarity, while the mass spectrum for relevant fractions depends on the mass/charge ratio of the separated peptide fragments (Fig. 2).

4 Notes

1. It can be useful to remove the N-terminal membrane anchor when overexpressing and purifying sortase proteins from *Escherichia coli*. In our experience, and that of others (e.g., [7, 25]), N-terminally 6× His-tagged sortases are functional, and the tag does not need to be removed prior to use in the cleavage assay.
2. Sortases are cysteine transpeptidases, and consequently substituting the active site cysteine residue with an alanine residue can be an effective way of generating a nonfunctional enzyme (e.g., [23]).
3. If there is any interest in following sortase enzyme kinetics (in addition to elucidating the cleavage site), it can be useful to append fluorophore/quencher molecules to either end of the peptide substrate [e.g., 2-aminobenzoyl (Abz) fluorophore/2,4-dinitrophenyl (Dpn) quencher]. Such modified peptides can be synthesized in-house, or can be ordered from any number of companies that specialize in peptide synthesis. If ordering modified peptides, it is worth taking care when selecting a

a Total wavelength chromatogram (TWC) of diode array detector (DAD) spectral data

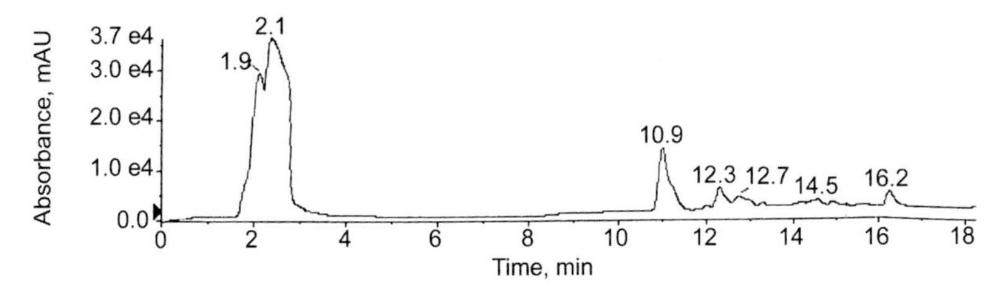

b Positive ion mode, sample taken at 14.6 min

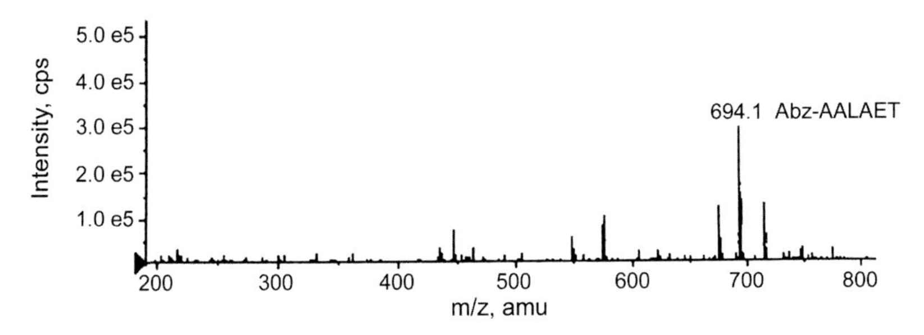

c Positive ion mode, sample taken at 12.3-12.4 min

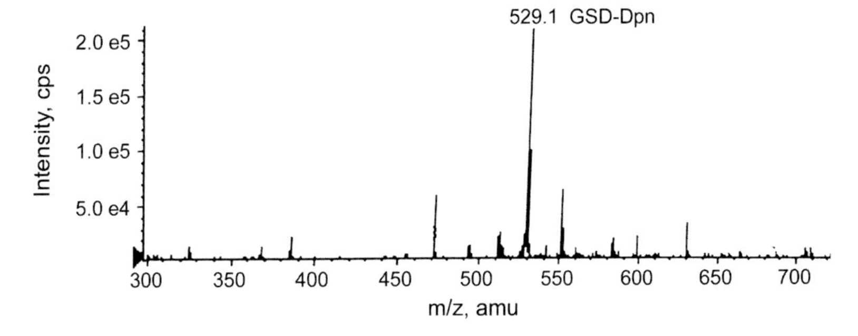

Fig. 2 Chromatogram and mass spectra of wild-type sortase (SrtE1 from *Streptomyces coelicolor*) cleavage products resulting from the reaction shown in Fig. 1. (**a**) Total wavelength chromatogram (TWC) of diode array detector (DAD) spectral data from the liquid chromatography of the cleavage products resulting from SrtE1 (purified as an N-terminally His-tagged protein from *Escherichia coli*) incubation with the Abz-AALAETGSD-Dpn peptide (sortase recognition sequence is underlined). (**b**, **c**) Mass/charge ratio of peptide fragments corresponding to the fractions separated in (**a**) at 14.6 min and 12.3–12.4 min, respectively. Both profiles are from the positive ion mode

company to work with; not all are able to successfully deliver, despite their claims to the contrary.

In addition to ordering/synthesizing peptides that contain the expected sortase cleavage site, it is recommended to include a negative control peptide that is not expected to be recognized/cleaved by the sortase enzyme of interest; this can be used to evaluate the substrate specificity of the sortase enzyme.

4. There are many ways to determine protein concentrations, including Lowry, Smith (or bicinchoninic acid—BCA), and

UV spectroscopic protein assays (reviewed by [26]). Any of these methods could be used here.

5. While the best studied *Staphylococcus aureus* SrtA enzyme requires Ca^{2+} for activity [27, 28], this is not a universal requirement, with the activity of SrtB from *S. aureus*, SrtA from group B *Streptococcus*, and SrtA, SrtB, and SrtC from *Bacillus anthracis* being unaffected by Ca^{2+} supplementation [29, 30].
6. There are a number of ways in which solutions can be degassed, including bubbling with helium, sonicating, or using vacuum filtrations. Here, we use an in-line vacuum degasser (likely the most common option available to most people). Degassing is important to remove the air from the LC solutions, as air bubble can interfere with pump function, impact flow rate (and consequently retention time), and contribute to noise during compound detection.
7. The C18 column that we used here is relatively long and thin, which allows for better resolution of sample components having similar polarity than shorter, wider columns.
8. Including low concentrations of formic acid helps to improve peak shapes during the separation of cleavage products by liquid chromatography. It also helps promote ionization, specifically for analysis in the positive ion mode, during mass spectrometry.
9. Acetonitrile is preferred over other solvents like methanol, as it typically yields less background noise, has higher elution strength, and applies less pressure to the column. However, more care is required when degassing acetonitrile-containing solutions, given its tendency to absorb heat, and later regenerate bubbles.
10. Storing solvent lines in 50% acetonitrile and 50% water helps to minimize microbial growth and prevent the accumulation of dust and other debris in the lines.
11. The liquid chromatography (LC) system used here (Agilent 1100 series; Agilent Technologies, Canada) was coupled to an electrospray mass spectrometry system (QTRAP 2000 LC/MS/MS; ABSciex), such that once the cleavage projects were injected onto the LC system, there was no manipulation required for the MS analysis.
12. It can be useful to initially perform the cleavage reactions with different concentrations of enzyme, to determine the optimal concentration (i.e., lowest enzyme concentration given maximal cleavage) for use in downstream LC/MS analyses.
13. This volume is optimized for use in 96-well plates, and can be adjusted, depending on tubes/plates used for the reactions and downstream fluorescence readings.

14. The initial wash of the system can be done using a high flow rate (e.g., 10 mL/min for 2–3 min) if the column is not attached. Once the column is attached, maximal flow rate is 1 mL/min.
15. Guard columns are designed to sit between the injector and an analytical column, and help to prevent contamination of the column by any impurities that were not effectively removed from the sample.
16. If using manual rather than automatic injection, it is important that an ordinary syringe is not used for sample injection. Liquid chromatography injectors are special syringes with a flat tip.
17. Prior to sample injection, it can be a good idea to do a "test run," in which only sample buffer is injected, to ensure that everything is behaving as expected, and that there are no residues left in the system from previous runs, as this could complicate downstream analyses.

Acknowledgements

This work was supported by a CIHR grant (MOP—137004).

References

1. Schneewind O, Missiakas D (2014) Sec-secretion and sortase-mediated anchoring of proteins in Gram-positive bacteria. Biochim Biophys Acta 1843:1687–1697
2. Schneewind O, Missiakas DM (2012) Protein secretion and surface display in Gram-positive bacteria. Phil Trans R Soc B Biol Sci 367:1123–1139
3. Spirig T, Weiner EM, Clubb RT (2011) Sortase enzymes in Gram-positive bacteria. Mol Microbiol 82:1044–1059
4. Haft DH, Varghese N (2011) GlyGly-CTERM and rhombosortase: a C-terminal protein processing signal in a many-to-one pairing with a rhomboid family intramembrane serine protease. PLoS One 6:e28886
5. Abdul Halim MF, Pfeiffer F, Zou J et al (2013) *Haloferax volcanii* archaeosortase is required for motility, mating, and C-terminal processing of the S-layer glycoprotein. Mol Microbiol 88:1164–1175
6. Mazmanian SK, Liu G, Ton-That H, Schneewind O (1999) *Staphylococcus aureus* sortase, an enzyme that anchors surface proteins to the cell wall. Science 285:760–763
7. Ton-That H, Liu G, Mazmanian SK et al (1999) Purification and characterization of sortase, the transpeptidase that cleaves surface proteins of *Staphylococcus aureus* at the LPXTG motif. Proc Natl Acad Sci U S A 96:12424–12429
8. Ton-That H, Schneewind O (2004) Assembly of pili in Gram-positive bacteria. Trends Microbiol 12:228–234
9. Ritzefeld M (2014) Sortagging: a robust and efficient chemoenzymatic ligation strategy. Chemistry 20:8516–8529
10. Theile CS, Witte MD, Blom AEM et al (2013) Site-specific N-terminal labeling of proteins using sortase-mediated reactions. Nat Protoc 8:1800–1807
11. Haridas V, Sadanandan S, Dheepthi NU (2014) Sortase-based bio-organic strategies for macromolecular synthesis. Chembiochem 15:1857–1867
12. Popp MW-L, Ploegh HL (2011) Making and breaking peptide bonds: protein engineering using sortase. Angew Chem Int Ed Engl 50:5024–5032
13. Tsukiji S, Nagamune T (2009) Sortase-mediated ligation: a gift from Gram-positive bacteria to protein engineering. Chembiochem 10:787–798
14. Williamson DJ, Webb ME, Turnbull WB (2014) Depsipeptide substrates for sortase-

mediated N-terminal protein ligation. Nat Protoc 9:253–262

15. Schmohl L, Schwarzer D (2014) Sortase-mediated ligations for the site-specific modification of proteins. Curr Opin Chem Biol 22:122–128
16. Proft T (2010) Sortase-mediated protein ligation: an emerging biotechnology tool for protein modification and immobilisation. Biotechnol Lett 32:1–10
17. Dorr BM, Ham HO, An C et al (2014) Reprogramming the specificity of sortase enzymes. Proc Natl Acad Sci U S A 111:13343–13348
18. Budzik JM, Marraffini LA, Souda P et al (2008) Amide bonds assemble pili on the surface of bacilli. Proc Natl Acad Sci U S A 105:10215–10220
19. Gaspar AH, Marraffini LA, Glass EM et al (2005) *Bacillus anthracis* sortase A (SrtA) anchors LPXTG motif-containing surface proteins to the cell wall envelope. J Bacteriol 187:4646–4655
20. Ton-That H, Mazmanian SK, Faull KF, Schneewind O (2000) Anchoring of surface proteins to the cell wall of *Staphylococcus aureus*. Sortase catalyzed in vitro transpeptidation reaction using LPXTG peptide and NH(2)-Gly(3) substrates. J Biol Chem 275:9876–9881
21. Kruger RG, Otvos B, Frankel BA et al (2004) Analysis of the substrate specificity of the *Staphylococcus aureus* sortase transpeptidase SrtA. Biochemistry 43:1541–1551
22. Maresso AW, Chapa TJ, Schneewind O (2006) Surface protein IsdC and sortase B are required for heme-iron scavenging of *Bacillus anthracis*. J Bacteriol 188:8145–8152
23. Duong A, Capstick DS, Di Berardo C et al (2012) Aerial development in *Streptomyces coelicolor* requires sortase activity. Mol Microbiol 83:992–1005
24. Bradford MM (1976) A rapid and sensitive method for the quantitation of microgram quantities of protein utilizing the principle of protein-dye binding. Anal Biochem 72:248–254
25. Mazmanian SK, Ton-That H, Su K, Schneewind O (2002) An iron-regulated sortase anchors a class of surface protein during *Staphylococcus aureus* pathogenesis. Proc Natl Acad Sci U S A 99:2293–2298
26. Olson BJSC, Markwell J (2007) Assays for determination of protein concentration. Curr Protoc Protein Sci Chapter 3: Unit 3.4–3.4.29. Wiley
27. Ilangovan U, Ton-That H, Iwahara J et al (2001) Structure of sortase, the transpeptidase that anchors proteins to the cell wall of *Staphylococcus aureus*. Proc Natl Acad Sci U S A 98:6056–6061
28. Naik MT, Suree N, Ilangovan U et al (2006) *Staphylococcus aureus* sortase A transpeptidase. Calcium promotes sorting signal binding by altering the mobility and structure of an active site loop. J Biol Chem 281:1817–1826
29. Necchi F, Nardi-Dei V, Biagini M et al (2011) Sortase A substrate specificity in GBS pilus 2a cell wall anchoring. PLoS One 6:e21317
30. Maresso AW, Schneewind O (2008) Sortase as a target of anti-infective therapy. Pharmacol Rev 60:128–141

Chapter 9

Genetics and Cell Morphology Analyses of the *Actinomyces oris srtA* Mutant

Chenggang Wu, Melissa Elizabeth Reardon-Robinson, and Hung Ton-That

Abstract

Sortase is a cysteine-transpeptidase that anchors LPXTG-containing proteins on the Gram-positive bacterial cell wall. Previously, sortase was considered to be an important factor for bacterial pathogenesis and fitness, but not cell growth. However, the *Actinomyces oris* sortase is essential for cell viability, due to its coupling to a glycosylation pathway. In this chapter, we describe the methods to generate conditional *srtA* deletion mutants and identify *srtA* suppressors by Tn5 transposon mutagenesis. We also provide procedures for analyzing cell morphology of this mutant by thin-section electron microscopy. These techniques can be applied for analyses of other essential genes in *A. oris*.

Key words *Actinomyces oris*, Glycosylation, Sortase, Allelic exchange, Tn5 transposon mutagenesis, Electron microscopy

1 Introduction

Actinomyces oris is a cariogenic bacterium that is important for the formation of oral biofilm, commonly known as dental plaque [1, 2]. The ability of *A. oris* to cultivate biofilm is dependent on the adhesive type 1 and 2 fimbriae, which, via a C-terminal cell-wall-sorting signal (CWSS), are assembled and anchored to the cell surface by cysteine transpeptidase sortase (Srt) enzymes. Type 1 fimbriae, composed of the fimbrial shaft FimP and tip fimbrillin FimQ, facilitate *A. oris* colonization by binding to proline-rich salivary deposits on the tooth surface [3]. Type 2 fimbriae, consisted of the fimbrial shaft FimA and tip fimbrillins FimB or CafA, promote biofilm development by mediating interactions with bacterial co-colonizers and host cells [4, 5]. Type 1 and 2 fimbriae are assembled by the pilus-specific sortases SrtC1 and SrtC2, respectively, but they are linked to the cell wall by a single housekeeping sortase (SrtA). In additional to targeting type 1 and 2 fimbriae, SrtA is also predicted to anchor 14 non-pilus proteins containing the CWSS [6].

Hee-Jeon Hong (ed.), *Bacterial Cell Wall Homeostasis: Methods and Protocols*, Methods in Molecular Biology, vol. 1440,
DOI 10.1007/978-1-4939-3676-2_9, © Springer Science+Business Media New York 2016

Housekeeping sortases are conserved in Gram-positive bacteria and serve as major virulence factors. The first housekeeping sortase was discovered in *Staphylococcus aureus* by Schneewind's group in 1999 [7]. Since then, many others have been characterized, but none were found to be required for viability [8, 9]. Recently, Wu and colleagues revealed that the *A. oris* housekeeping sortase (SrtA) is an exception [6]. Multiple attempts to delete the gene encoding this transpeptidase were fruitless, suggesting it is essential. However, further analysis of *srtA* was hampered by the lack genetic tools available for DNA manipulation in this organism. To overcome this, our lab has developed facile techniques for generating *A. oris* conditional mutants and studying the basis for gene essentiality. Using a combination of transcriptional and posttranscriptional mechanisms to control gene expression, we showed that depletion of *A. oris srtA* results in *severe* morphological abnormalities and aberrant division septa. To elucidate the pathway for essentiality, our lab then developed a Tn5 transposon system to identify suppressors of the conditional *srtA* deletion mutant. We discovered that the depletion of this gene resulted in a toxic accumulation of GspA, a glycosylated SrtA substrate, in the membrane. In this chapter, we describe the methods for gene deletions and Tn5-based mutagenesis in *A. oris*. Protocols for examining the cell morphologies of conditional deletion mutants are also provided.

1.1 Allelic Exchange in A. oris

To avoid potential polar effects by insertion mutations of genes, two methods were developed that generate markerless, in-frame deletion mutants in *A. oris*. In these methods, the suicide plasmids pCWU2 [10] and pCWU3 [11] were used that express GalK, which phosphorylates D-galactose to generate galactose-1-phosphate, and mCherry as counterselection markers, respectively. To create the deletion constructs, approximately 1 kB regions upstream and downstream of a target gene are PCR amplified, fused together, and then cloned into pCWU2 or pCWU3. The resulting plasmids are electroporated into *A. oris* and integrated into the chromosome via a single crossover event. To excise target genes from *A. oris,* a second homologous recombination event is induced by growing the co-integrant strains in the absence of selective antibiotics. When using pCWU2, excision of the plasmid is selected by growing bacteria with 2-deoxy-D-galactose (2-DG), which is converted into a toxic intermediate by GalK. For pCWU3, plasmid excision is selected by loss of cell fluorescence [11].

1.2 Generation of Conditional Deletion Mutants in A. oris

Approximately 20% of bacterial genes are required for growth and survival [12]. Deletion of these genes in many organisms is sometimes a daunting task. In *A. oris*, removal of an essential gene from the bacterial chromosome can be achieved by providing a copy of this gene ectopically. For example, to create a conditional *srtA* deletion mutant, a tetracycline-inducible expression system (P*tet*) was utilized [13].

However, the P*xyl/tetO* promoter of this system is leaky [14]. To provide a posttranscriptional level of gene regulation, we introduced a theophylline-responsive synthetic riboswitch element E* into the P*tet* system [15]. E* is composed of a small sequence whose secondary RNA structure inhibits protein translation by blocking ribosome binding. Inhibition of translation by E*, however, is relieved upon the addition of theophylline. This additional level of control was used for generation of the conditional *srtA* deletion mutant [6].

1.3 Random gene Disruptions by Tn5-Based Transposition

To screen for *srtA* depletion suppressors, we developed a highly efficient Tn5 system for *A. oris* [6, 16]. This system is based on EZ-Tn5, an in vitro transposon widely used for bacterial mutagenesis [16]. A kanamycin (Kan) resistance gene cassette derived from the *Actinomyces/E. coli* shuttle plasmid pJRD215 was cloned into pMOD-2<MCS>. The cassette is flanked by 19 bp mosaic ends (ME), which are recognized by EZ-Tn5 transposase for random insertion into the bacterial genome. The newly generated DNA fragment was then PCR amplified, combined with EZ-Tn5 transposase, and electroporated into *A. oris.*

2 Materials

2.1 Plasmids (Fig. 1)

1. pHTT177, a derivative of pUC19 with a Kan resistance gene derived from pJRD215 in place of the original ampicillin (Amp) resistance gene [4].
2. pCWU2, a derivative of pHTT177 containing *galK* under the control of the *rpsJ* promoter [10].
3. pCWU3, a derivative of pHTT177 containing *rfp* (mCherry) under the control of the *rpsJ* promoter [11].
4. pJRD215, an *Actinomyces/E. coli* shuttle vector containing Kan and streptomycin (Sm) resistance gene cassettes [17].
5. pJRD-Sm, a derivative of pJRD215 containing only the Sm resistance gene cassette.

2.2 Preparation of A. oris Competent Cells

1. *A. oris* MG-1 Δ*galK*.
2. Heart infusion agar (HIA) plates.
3. Heart infusion broth (HIB).
4. 15% glycine in HIB.
5. Sterile water.
6. Sterile 10% glycerol.
7. 37 °C water bath shaker.
8. 37 °C, 5% CO_2 incubator.
9. Disposable culture glass tubes.

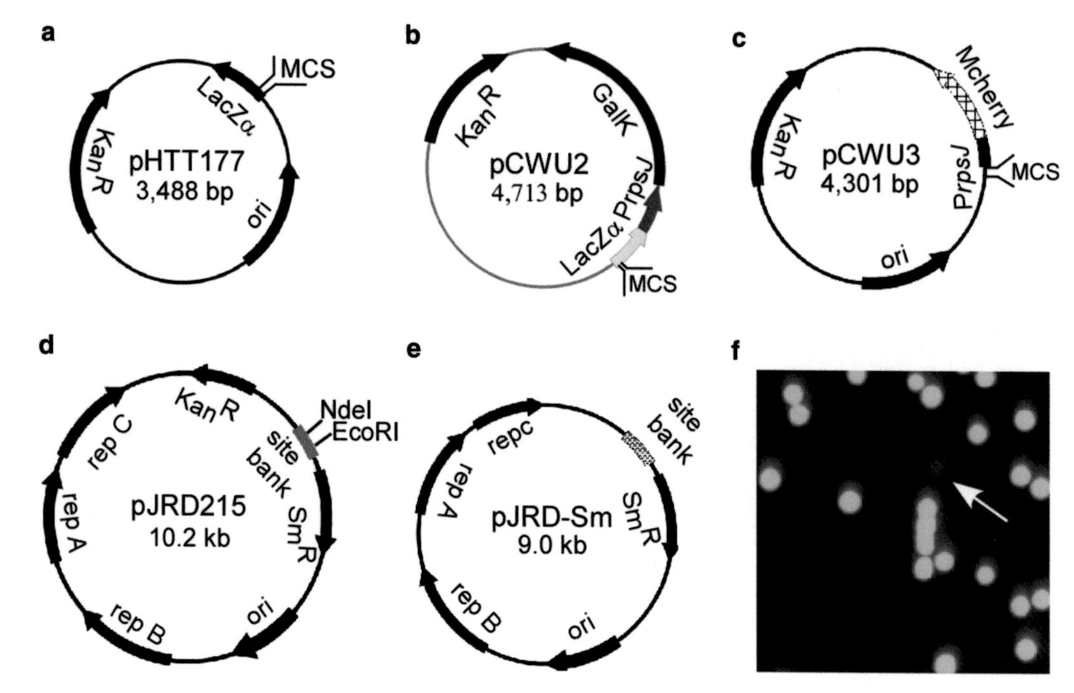

Fig. 1 Vectors used for genetic manipulation of *A. oris*—(**a–c**) Derived from pHTT177, pCWU2 and pCWU3 are two suicide plasmids in *A. oris.* (**d–e**) pJRD-Sm is a derivative of the *Actinomyces/E. coli* shuttle vector pJRD215. MCS indicates multiple cloning sites; Kan for kanamycin and Sm for streptomycin. (**f**) An unmarked, in-frame deletion mutant of *A. oris* was obtained using pCWU3 by screening for the loss of fluorescence; adapted from Wu and Ton-That [11]

10. 125 mL Erlenmeyer flask.
11. 50 mL centrifuge tubes with printed graduations.
12. Spectrophotometer and plastic cuvettes for measuring cell density at 600 nm (A_{600}).
13. Sterile 1.5 mL centrifuge tubes.
14. Dry ice–ethanol bath.

2.3 Electroporation

1. *A. oris* competent cells.
2. Plasmid or transposon DNA.
3. HIB.
4. HIA plates with 50 μg/mL Kan.
5. 0.2 cm electroporation cuvettes.
6. Electroporator.
7. Sterile 1.5 mL centrifuge tubes.
8. Sterile spreader.
9. 37 °C incubator with 5% CO_2 (*see* **Note 1**).

2.4 Allelic Replacement

1. Phusion® High-fidelity DNA polymerase.
2. DNA oligonucleotide Primers.
3. Agarose gel matrix.
4. DNA extraction kit (for gel purification) (*see* **Note 2**).
5. T4 DNA ligase and buffer.
6. Restriction enzymes.
7. *E. coli* DH5α competent cells.
8. Luria broth (LB) and agar.
9. 5 M 2-deoxy-D-galactose (2-DG) dissolved in HIB.
10. 42 °C water bath.
11. 37 °C incubators with and without 5 % CO_2.
12. Olympus X171 inverted microscope with TRITC filter FluorChem Q imaging system (Alpha Innotech).

2.5 Conditional srtA Deletion in A. oris

1. Anhydrotetracycline hydrochloride (AHT) suspended in methanol.
2. Theophylline dissolved in HIB.
3. HIA plates with 50 μg/mL Kan and 50 μg/mL Sm.
4. HIA plates with 100 ng/mL AHT and 2 mM theophylline.
5. Additional reagents and equipment listed above.

2.6 Tn5 Transposition

1. pMOD-2<MCS>® (Epicentre).
2. T4 polynucleotide kinase.
3. 100 mM Adenosine 5′-triphosphate (ATP).
4. Phusion® high fidelity DNA polymerase.
5. EZ-TN5™ Transposase from Epicentre.
6. Topo blunt-ending cloning kit.
7. 0.2 cm electroporation cuvettes.
8. Electroporator.

2.7 Thin Section and Electron Microscopy

1. Sterile 1.5 mL centrifuge tubes.
2. Sterilized water.
3. 0.1 mM NaCl.
4. Phosphate buffer saline (PBS).
5. 10 % formalin.
6. Glutaraldehyde.
7. Sodium borohydride.
8. Ethanol.
9. Millonig's buffer (EMS; Hatfield, PA).
10. LR (London Resin) white resin.

11. Oven.
12. Rotator.
13. BEEM® Capsules.
14. Diatome diamond knife.
15. Leica Ultracut microtome.
16. Formvar-carbon-coated 200-mesh nickel grids.
17. 1 % uranyl acetate (UA) for negative staining.
18. Filter paper.
19. Dumont tweezers.
20. Transmission Electron Microscope (TEM).

3 Methods

3.1 Construction of Deletion Plasmids

1. Design two sets of primers to amplify ~1 kb regions upstream and downstream of your target gene. As a reference, refer to Table 1 for the design of *srtA* deletion primers (*srtA* is annotated as *ana*_2245 at www.oralgen.org) (*see* **Note 3**). For a DNA template, isolate *A. oris* genomic DNA (*see* **Note 4**).
2. Purify the resulting PCR products using a DNA gel purification kit. Treat the PCR products with the appropriate restriction enzymes, and ligate the fragments into pCWU2 or pCWU3, precut with the appropriate restriction enzymes (*see* Table 1).
3. Transform *E. coli* DH5α with the resulting plasmid. Extract the plasmid from the positive clones and verify the insert.

3.2 Preparation of A. oris Competent Cells

1. Streak *A. oris* MG-1 or Δ*galK* from a frozen stock on a HIA plate, and incubate 2 days at 37 °C with 5 % CO_2 incubator to obtain single colonies (*see* **Note 3**).
2. Inoculate a colony of *A. oris* into 6 mL of HIB, and incubate overnight at 37 °C with minimal shaking. The next day, dilute 5 mL of the overnight culture into 65 mL of fresh HIB. Grow cells at 37 °C until the OD_{600} reaches approximately 0.6. Add 25 mL of prewarmed 15 % glycine in HIB to the culture, and incubate at 37 °C for 1 h.
3. Transfer the culture to centrifuge tubes and chill on ice for at least 10 min. Harvest the cell pellet by centrifugation at 4 °C, and discard the supernatant. Wash the cell pellet twice with 30 mL of prechilled 10 % glycerol. Harvest the cell pellet by centrifugation at 4 °C, resuspend the bacteria in 1 mL of 10 % cold glycerol, and then aliquot into prechilled 1.5 mL centrifuge tubes (~200 μL each).
4. Snap-freeze the samples using a dry ice–ethanol bath, and store at −80 °C.

Table 1
Primers used for constructing *srtA* conditional mutant

Primer	Sequence[a]	Used for
srtAupF	GGCGGAATTCATCGTCTCGGCGATCTACGC	pCWU2-Δ*srtA*
srtAupR	GGCGGGTACCTCGCACAAGACCTCCTCTAGTCA	pCWU2-Δ*srtA*
srtAdnF	GGCGGGTACCGGGGGTCAACTGATGTACGGCTTCA	pCWU2-Δ*srtA*
srtAdnR	GGCGTCTAGATAGGACTGGCGCAGCCACTTCT	pCWU2-Δ*srtA*
kpnI-tetR-R	GGCGGGTACCTTTAAGACCCACTTTCACATTTAAG	ptTetR-SrtA
EcoRI-tetR-F	GGCGGAATTCTCAAGCTTATTTTAATTATACTCTATC	ptTetR-SrtA
EcoRI-srtA-F(rbs)	GGCGGAATTCGAGGAGGGGCGATGACTAGAG	ptTetR-SrtA
HindIII-srtA-R	GGCGAAGCTTCCGGCAGGTGCCGCCAGATGAAG	ptTetR-SrtA
Ribo-srtA-F	ATGCCCTTGGCAGCACCCTGCTAAGGAGGCAACAAGATGACTAGAGGAGGTCTTGTGCGA	pTetR-R*-SrtA
Ribo-tet-R	CAAGACGATGCTGGTATCACCGGTACCTATAGTGAGTCGTATAGAATTGGACATCATCAGGCTAG	pTetR-R*-SrtA
PfimQ-tetR-R	GGCGGAATTCGCCGATGGATTCCGATCATGAG	pTetR-Ω-SrtA
PfimQ-tetR-F	GGCGAGATCTGATTCCTGCGCCCAGGAAAGTG	pTetR-Ω-SrtA

[a] *Underlined* are the restriction sites in the primers

3.3 Electroporation of A. oris with pCWU2 or pCWU3

1. Add at least 1 μg of plasmid, which harbors a deletion construct (Δ*geneX*), to an aliquot of *A. oris* competent cells thawed on ice, and then leave on ice for at least 10 min. Transfer the contents of the tube to a prechilled 0.2 mm cuvette.
2. Electroporate *A. oris* using the following conditions: Voltage = 2.5 kV, Resistance = 400 Ω, Capacity = 25 μF.
3. Following electroporation, immediately add 1 mL of prewarmed HIB to the cuvette, transfer the contents to a sterile 1.5 mL centrifuge tubes, and recover the cells for 2-3 h at 37 °C with minimal shaking.
4. Harvest the cell pellet by centrifugation, and remove 1 mL of the supernatant. Resuspend the cell pellet in the remaining HIB, and spread the cells onto HIA plates containing 50 μg/mL of Kan.
5. Incubate the plates at 37 °C with 5 % CO_2 for approximately 3 days. Colonies that appear after that time should contain a copy of pCWU2 or pCWU3 that is integrated into the chromosome.

3.4 Allelic Exchange Using galK (pCWU2) as a Counterselection Marker

1. To promote excision of pCWU2-Δ*geneX* by a double crossover event, dilute an overnight culture of the co-integrant 1:50 into HIB without Kan, and incubate at 37 °C with minimal shaking for 24 h.
2. The following day, dilute the culture 1:100, spread onto HIA containing 0.25 % 2-DG, and incubate the plates at 37 °C with 5 % CO_2 for approximately 3 days.
3. Patch at least 20 colonies onto HIA plates with and without Kan, and incubate overnight at 37 °C in a 5 % CO_2 incubator. Select at least ten colonies that are sensitive to Kan (i.e., have lost pCWU2), and screen for the loss of the target gene by PCR.

3.5 Allelic Exchange Using mCherry (pCWU3) as a Counterselection Marker

1. To promote excision of pCWU3-Δ*geneX* by a double crossover event, dilute an overnight culture of the co-integrant into HIB without Kan, and incubate at 37 °C overnight with minimal shaking. Repeat seven times (*see* **Note 5**).
2. Dilute the final culture 1/10,000, spread onto HIA plates, and incubate at 37 °C with 5 % CO_2 for approximately 3 days.
3. Colonies that have lost pCWU3 should no longer be fluorescent. Screen these cells using a FluorChem Q imaging system with a Cy3 filter.
4. Patch the nonfluorescent colonies onto HIA with and without Kan. Bacteria that have lost pCWU3 should be Kan-sensitive. Screen for the loss of the target gene by PCR.

3.6 Generation of the *A. oris* Conditional *srtA* Deletion Mutant

Described here is a general protocol for generating *srtA* conditional mutants in *A. oris*, starting with a merodiploid strain, in which the deletion vector pCWU2-Δ*srtA* has been integrated into the Δ*galK* chromosome. Also used is pTetR-Ω-SrtA, a derivative of pJRD-Sm, in which *srtA* expression is tightly regulated under the control of a tetracycline-inducible promoter and a theophylline-responsive riboswitch (*see* **Note 6**) [6]. In principle, this protocol is applicable for any essential gene.

1. Incubate approximately 0.1 μg of pTetR-Ω-SrtA with the competent merodiploid cells on ice for at least 10 min. Transfer the sample into a prechilled cuvette, and electroporate. Following electroporation, immediately add 1 mL of pre-warmed HIB without antibiotics, and transfer the cuvette contents into a 1.5 mL centrifuge tube. Incubate the cells for 2–3 h at 37 °C without shaking. Next, take a 100 μL aliquot of the culture, and spread it on HIA containing 50 μg/mL of Kan and 50 μg/mL Sm. Incubate the plates at 37 °C with 5% CO_2. Colonies would be visible after 2 days of growth.
2. Pick 2 isolated colonies and patch them on agar plates containing Kan and Sm (50 μg/mL). After 3 days of growth, inoculate cells from the two patches in 6 mL of HIB supplemented with 50 μg/mL Kan and Sm for overnight growth at 37 °C.
3. Harvest cells by centrifugation from 100 μL aliquots of the overnight cultures, and collect the cell pellet by centrifugation. Wash the cells in HIB and then resuspend them into 6 mL of HIB containing 50 μg/mL Sm, 100 ng/mL AHT (*see* **Note 7**), and 2 mM theophylline (*see* **Note 8**).
4. To select for clones with plasmid excision, plate 1/100 dilutions of the overnight cultures above onto HIA plates containing 0.25% 2-DG, 100 ng/mL AHT, and 2 mM theophylline. Incubate the plates at 37 °C with 5% CO_2 for approximately 3 days.
5. Double-patch at least 20 colonies onto HIA plates supplemented with 50 μg/mL Sm, 100 ng/mL AHT, and 2 mM theophylline, as well as HIA containing 50 μg/mL Kan, 50 μg/mL Sm, 100 ng/mL AHT, and 2 mM theophylline. Colonies with pCWU2 excision should be sensitive to Kan.
6. Screen at least ten colonies for the loss of the chromosomal *srtA* gene by PCR.
7. If deletion mutants are not obtained, repeat **steps 4–6**.

3.7 Construction of the Tn5 Transposon Plasmid pMOD-2/Kan215

In our previous study, Tn5 transposon mutagenesis was used to find suppressors for *srtA* lethality [6]. This transposon approach can also be employed in high throughput screens for virulence factors in *A. oris*.

1. PCR-amplify the Kan resistance gene cassette from the shuttle plasmid pJRD215 (Table 2), and clone it into the MCS of pMOD-2<MCS>, which carries 19 bp mosaic ends (ME).
2. To prepare the transposon, phosphorylate ME plus primers (ME plus 9-3 and ME plus 9-5′) with T4 DNA kinase (see Table 2).
3. PCR amplify the ME and Kan resistance gene fragment (*kanR*) from pMOD-2/Kan215 using the phosphorylated ME plus primers and Phusion® DNA polymerase. The recommended PCR program is provided: 98 °C for 5 min, 30 cycles of 98 °C for 10 s and 55 °C for 20 s and 72 °C for 1 min, 72 °C for 10 min.

3.8 Production of the Tn5 Transposome

1. Prepare the transposome: 2 μL ME-*kanR*-ME PCR product, 4 μL EZ-Tn5 transposase (Epicentre), 2 μL 100% glycerol. Incubate the reaction for 2 h at room temperature, and then store at −20 °C.
2. Add approximately 1.5 μL of transposome to the *A. oris* competent cells. Incubate on ice for at least 10 min. Transfer the mixture to a cuvette for electroporation. Following electroporation, immediately add 1 mL of prewarmed HIB, and recover the cells in a 1.5 mL centrifuge tube at 37 °C for 3 h without shaking. Spread 40 μL aliquots of transposon-treated cultures onto HIA plates (approximately 30) with 50 μg/mL Kan and 50 μg/mL Sm. As a control, spread another aliquot onto HIA containing both antibiotics and 100 ng/mL AHT and 2 mM theophylline (*see* **Note 9**).
3. Inoculate plates into 37 °C with 5% CO_2 for 3 days.

3.9 Characterization of Tn5 Insertion by TAIL-PCR

Thermal asymmetric interlaced PCR (TAIL-PCR) is a useful tool to map out Tn5 insertion sites [18]. TAIL-PCR is performed by alternate rounds of low stringency and high stringency PCR cycles using transposon-specific and degenerate primers. Below is a simplified TAIL-PCR protocol to determine Tn5 transposon insertion sites in *A. oris*.

1. First round of PCR (PCR-1): Tn5-1 and AD1 primers (Table 2), *A. oris* Tn5 suppressor chromosomal DNA as a template, PCR parameters (98 °C for 5 min, 30 cycles of 98 °C for 15 s and 45 °C for 20 s and 72 °C for 1 min, 72 °C for 10 min).
2. Second round of PCR (PCR-2) (*see* **Note 10**): Tn5-2 and AD2 primers (see Table 2), 1 μL of unpurified product from PCR-1 as a template, PCR parameters (98 °C for 5 min, 30 cycles of 98 °C for 15 s and 60 °C for 20 s and 72 °C for 1 min, 72 °C for 10 min).

Table 2
The primer used for Tn*5* transposition assay

Primer	Sequence[a]	Used for
ME plus 9-3′	CTGTCTCTTATACACATCTCAACCATCA	Transposon region
ME plus 9-5′	CTGTCTCTTATACACATCTCAACCCTGA	Transposon region
Kan-F	GGCGGGATCCCCAAGCTAGCTTCACGCTGCCGCAAG	Kan cassette
Kan-R	GGCGGGATCCGCTCAGAAGAACTCGTCAAG	Kan cassette
Tn5-1	CGAACTGTTCGCCAGGCTCAAG	TAIL-PCR
Tn5-2	CTGACCGCTTCCTCGTGCTTTA	TAIL-PCR
AD1	NGTCGASWGANAWGAA	TAIL -PCR
M13F	CCGAGCAGTCTCTGTCCTTC	Sequencing
M13R	CCCTCTCACTCCCTTCCTG	Sequencing

[a] *Underlined* are the restriction sites in the primers

3. Third round of PCR (PCR-3): Unpurified products from PCR-2 are used for PCR-3 reactions with conditions similar to PCR-2.
4. Clone the products into a TOPO blunt-end cloning vector, and sequence the vector containing PCR-3 product using primers M13F/M13R (see Table 2) to identify the site of Tn5 insertion (*see* **Note 11**).

3.10 Preparation of A. oris Thin Sections

For high-resolution studies of bacterial morphology, thin-section electron microscopy is an excellent choice of methodology. Described here is a protocol for preparing *A. oris* thin sections.

1. Scrap *A. oris* cells grown from HIA plates and suspend in 1 mL of PBS. Harvest cells by centrifugation and resuspend them in a fixative solution (3 % formalin, 0.15 % glutaraldehyde in Millonig's buffer, pH 7.4). Incubate for 4 days at 4 °C.
2. After harvesting cells centrifugation, wash them in fresh 0.1 % borohydride in Millonig's buffer for 10 min, followed by washing three times in Millonig's buffer alone for 5 min. Harvest the cells by centrifugation and wash them in water for 5 min.
3. Dehydrate the samples with graded concentrations of cold 50 % methanol for 5 min three times, followed by dehydration with cold 70 % methanol three times.
4. Infiltrate with 50 % LR White resin (EMS; Hatfield, PA)–50 % methanol mixtures at −20 °C for 2 h. Repeat this step once, followed by infiltration with 75 % LR white resin: 25 % methanol mixtures at −20 °C for 4 h. Finally, infiltrate with 100 % LR white resin on rotation at room temperature overnight.

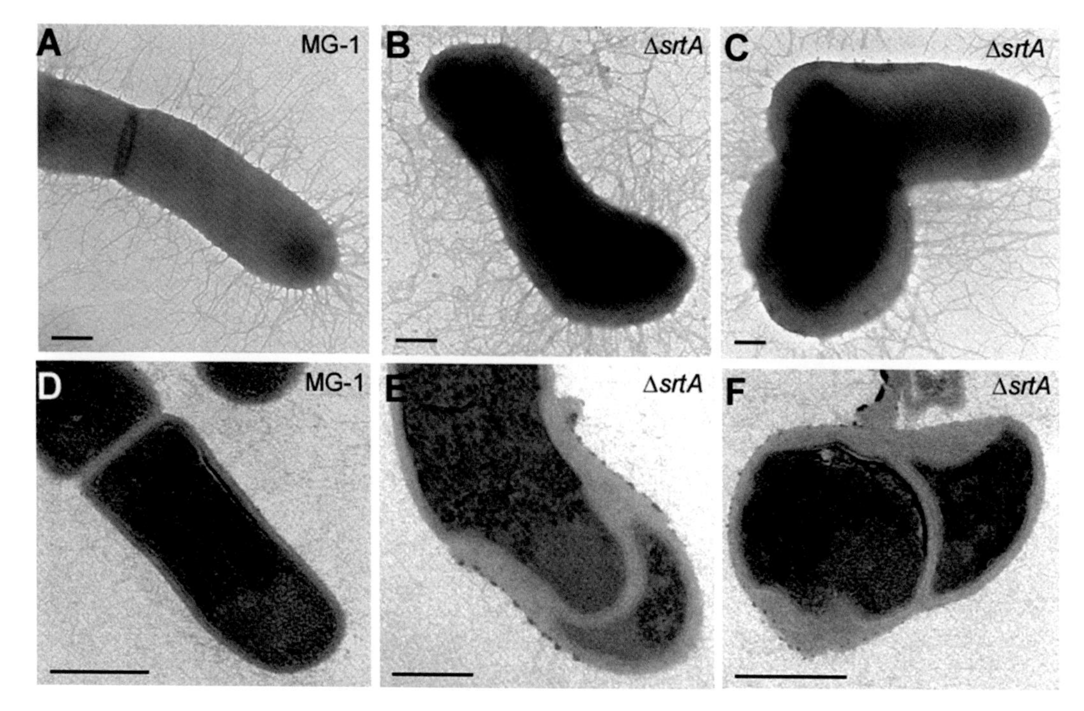

Fig. 2 Electron microscopic analysis of *A. oris*—Cells from the *A. oris* parental strain MG-1 (**a** and **d**) and *srtA*-depleted cells (**b**, **c**, **e**, and **f**) were examined by electron microscopy. Scale bars indicate 0.2 μm

5. Embed the pellets in BEEM capsules and polymerize in a 50 °C oven for 2 days.
6. Cut approximately 50 nm thin sections with a diamond knife using an ultramicrotome, and place onto Formvar-carbon-coated 200-mesh nickel grids.

3.11 Transmission Electron Microscopy (TEM)

1. Drop 10 μL of 0.25 % uranyl acetate on top of a grid containing the *A. oris* thin sections, and incubate at room temperature for 1 min (*see* **Note 12**).
2. Wick excess uranyl acetate using filter paper.
3. View the samples by TEM (Fig. 2).

4 Notes

1. If a CO_2 incubator is not available, wrap the HIA plates in Parafilm to limit their exposure to oxygen.
2. Qiagen kits are available for purchase.
3. Note that the upstream and downstream PCR products are fused by restrictions sites. This could also be achieved by overlapping PCR.

4. Genomic DNA can be isolated using phenol–chloroform extraction or a Promega Wizard® Genomic DNA purification kit.
5. The efficiency of *galk*-based counterselection is higher than the *mCherry*-based method. However, the mCherry counterselection does not require the *A. oris* Δ*galK* background.
6. The riboswitch element E* sequence (The RBS is highlighted in grey):
   *GGUACC*GGUGAUACCAGCAUCGUCUUGAUGCC CUUGGCAGCACCCUGCUAAGGAGGCAACAAGAUG
7. These are necessary to induce the plasmid-borne *srtA*. The absence of Kan will select for excision of the chromosomal copy of the gene of interest.
8. Anhydrotetracycline (AHT) is a non-bacteriostatic derivative of tetracycline.
9. Only *srtA* depletion suppressor mutants will grow in the absence of the inducers AHT and theophylline.
10. The annealing temperature for the second round of PCR can be adjusted from 55 to 65 °C only 1–2 major PCR fragments are obtained.
11. Sequences positive for Tn5 insertion should contain ME sequences.
12. Residual PBS will react with uranyl acetate resulting in poor visibility of the samples by TEM.

Acknowledgments

This work was supported by the National Institute of Dental and Craniofacial Research of the NIH under award numbers F31DE024004 (to M.E.R.-R.) and DE017382 and DE025015 (to H.T.-T.).

References

1. Jakubovics NS, Kolenbrander PE (2010) The road to ruin: the formation of disease-associated oral biofilms. Oral Dis 16(8):729–739
2. Chen L, Ma L, Park NH, Shi W (2001) Cariogenic actinomyces identified with a beta-glucosidase-dependent green color reaction to *Gardenia jasminoides* extract. J Clin Microbiol 39(8):3009–3012
3. Wu C, Mishra A, Yang J, Cisar JO, Das A, Ton-That H (2011) Dual function of a tip fimbrillin of *Actinomyces* in fimbrial assembly and receptor binding. J Bacteriol 193(13):3197–3206
4. Mishra A, Das A, Cisar JO, Ton-That H (2007) Sortase-catalyzed assembly of distinct heteromeric fimbriae in *Actinomyces naeslundii*. J Bacteriol 189(8):3156–3165
5. Reardon-Robinson ME, Wu C, Mishra A, Chang C, Bier N, Das A, Ton-That H (2014) Pilus hijacking by a bacterial coaggregation factor critical for oral biofilm development. Proc Natl Acad Sci U S A 111(10):3835–3840
6. Wu C, Huang IH, Chang C, Reardon-Robinson ME, Das A, Ton-That H (2014) Lethality of sortase depletion in *Actinomyces*

oris caused by excessive membrane accumulation of a surface glycoprotein. Mol Microbiol 94(6):1227–1241

7. Mazmanian SK, Ton-That H, Schneewind O (2001) Sortase-catalysed anchoring of surface proteins to the cell wall of *Staphylococcus aureus*. Mol Microbiol 40(5):1049–1057
8. Novick RP (2000) Sortase: the surface protein anchoring transpeptidase and the LPXTG motif. Trends Microbiol 8(4):148–151
9. Bradshaw WJ, Davies AH, Chambers CJ, Roberts AK, Shone CC, Acharya KR (2015) Molecular features of the sortase enzyme family. FEBS J 282(11):2097–2114
10. Mishra A, Wu C, Yang J, Cisar JO, Das A, Ton-That H (2010) The *Actinomyces oris* type 2 fimbrial shaft FimA mediates co-aggregation with oral streptococci, adherence to red blood cells and biofilm development. Mol Microbiol 77(4):841–854
11. Wu C, Ton-That H (2010) Allelic exchange in *Actinomyces oris* with mCherry fluorescence counterselection. Appl Environ Microbiol 76(17):5987–5989
12. Christen B, Abeliuk E, Collier JM, Kalogeraki VS, Passarelli B, Coller JA, Fero MJ, McAdams HH, Shapiro L (2011) The essential genome of a bacterium. Mol Syst Biol 7:528
13. Carroll P, Muttucumaru DG, Parish T (2005) Use of a tetracycline-inducible system for conditional expression in *Mycobacterium tuberculosis* and *Mycobacterium smegmatis*. Appl Environ Microbiol 71(6):3077–3084
14. Corrigan RM, Foster TJ (2009) An improved tetracycline-inducible expression vector for *Staphylococcus aureus*. Plasmid 61(2):126–129
15. Topp S, Reynoso CM, Seeliger JC, Goldlust IS, Desai SK, Murat D, Shen A, Puri AW, Komeili A, Bertozzi CR, Scott JR, Gallivan JP (2010) Synthetic riboswitches that induce gene expression in diverse bacterial species. Appl Environ Microbiol 76(23):7881–7884
16. Goryshin IY, Reznikoff WS (1998) Tn5 in vitro transposition. J Biol Chem 273(13):7367–7374
17. Yeung MK, Kozelsky CS (1994) Transformation of *Actinomyces* spp. by a gram-negative broad-host-range plasmid. J Bacteriol 176(13):4173–4176
18. Liu YG, Whittier RF (1995) Thermal asymmetric interlaced PCR: automatable amplification and sequencing of insert end fragments from P1 and YAC clones for chromosome walking. Genomics 25(3):674–681

Part IV

Reporter Assays for Cell Wall Stress

Chapter 10

Construction of a Bioassay System to Identify Extracellular Agents Targeting Bacterial Cell Envelope

Hee-Jeon Hong

Abstract

sigE in *Streptomyces coelicolor* encodes an extracytoplasmic function (ECF) sigma factor, σ^E, which is part of a signal transduction system that senses and responds to general cell wall stress in *S. coelicolor*. Expression of *sigE* is induced by a wide variety of agents that stress the cell wall under the control of two-component signal transduction system, CseB/CseC encoded in the same operon where *sigE* was identified from. Here we describe a method developing a bioassay system in *S. coelicolor* via a transcriptional fusion in which the promoter of *sigE* operon and a reporter gene (*neo*) conferring resistance to kanamycin were used. The effectiveness of the resulting bioassay system was determined by monitoring various agents that cause bacterial cell wall stress such as lysozyme or some antibiotics that target cell wall. In consequence, the result confirms that the bioassay system has a potential to be a simple but effective screening tool for identifying novel extracellular agents targeting bacterial cell wall.

Key words *Streptomyces coelicolor*, *sigE*, Bioassay system, Cell wall stress, Transcriptional fusion

1 Introduction

The bacterial cell envelope is crucial for bacterial cell growth because it provides a physical protective barrier between the cell and its environment, and giving cells their shape. It is also an important mediator of innate immune responses during bacterial infections. In particular, the bacterial cell wall peptidoglycan is a validated target for antibacterial chemotherapy such that antibiotics that inhibit cell wall peptidoglycan biosynthesis such as penicillin and vancomycin are clinically important in treatment of infectious disease. Bacteria must therefore have homeostatic mechanisms to monitor the integrity of their cell wall and respond accordingly; however, relatively little is known about the regulation of cell wall homeostasis in bacteria.

Actinomycete species live in soil which is an extremely complex and competitive habitat, and the ability of the bacteria to adapt and respond to changes in their cell wall is essential for survival.

Hee-Jeon Hong (ed.), *Bacterial Cell Wall Homeostasis: Methods and Protocols*, Methods in Molecular Biology, vol. 1440,
DOI 10.1007/978-1-4939-3676-2_10, © Springer Science+Business Media New York 2016

These changes may result from competition with other soil-dwelling microbes including other actinomycetes producing antibiotics that target the cell wall. To adapt to their environment, actinomycetes must possess diverse and efficient mechanisms for assessing the integrity of their cell wall, and signal transduction systems that respond by rapidly inducing appropriate sets of genes. The sensors for these systems are mostly found in the cell membrane and they transmit signals to the cytoplasm via regulatory proteins, resulting in changes in gene expression which protect the cell. One of the most profound systems defined to be involved in cell wall homeostasis in actinomycete species is *sigE* signal transduction system identified in a model actinomycetes, *Streptomyces coelicolor* [1]. The *sigE* system is composed of four proteins, encoded in an operon: σ^E itself; CseA, a negative regulator of undefined biochemical function; CseB, a response regulator; and CseC, a sensor histidine protein kinase with two-predicted transmembrane helices [2]. Expression of σ^E activity is governed at the level of *sigE* transcription by the CseB/CseC two-component signal transduction system. In response to signals that originate in the cell wall when it is under stress, the sensor kinases, CseC, becomes autophosphorylated at His-271, and, in accordance with the known mechanism for other two-component regulatory systems, this phosphate is then transferred to Asp-55 in the response regulator, CseB. Phospho-CseB activates the promoter of the *sigE* operon, and σ^E is recruited by core RNA polymerase to transcribe genes with cell-wall-related functions, including a putative operon of 12 genes likely to specify cell wall glycan synthesis [3]. *sigE* null mutants were extremely sensitive to cell wall hydrolytic enzymes and had an altered cell wall muropeptide profile, suggesting that *sigE* was required for normal cell wall integrity [2]. In addition, *sigE* null mutants required millimolar level of Mg^{2+} or Ca^{2+} for normal growth and sporulation. Divalent cations are known to interact with a number of different components of the cell envelop, including the membrane, accessory polymers and cell wall-associated proteins, thereby altering its structure [4]. Given that *sigE* mutants had an altered cell wall and that certain aspects of their phenotype were suppressed by Mg^{2+}, it was suggested that the CseB/CseC two-component system responded to changes in the cell envelope [5]. Analysis of the activity of the *sigE* promoter in different mutant backgrounds was also highly informative. The *sigE* promoter was inactive in a constructed *cseB* null mutant, such that *cseB* mutants lack σ^E, thus explaining why *cseB* and *sigE* mutants had the same phenotype [5]. In contrast, the *sigE* promoter was substantially upregulated in a *sigE* null mutant, suggesting that the cell envelope defect in *sigE* mutants is sensed by CseC, which responds by attempting to increase the expression of *sigE* [5]. Finally, most transcripts from the *sigE* promoter terminate immediately downstream of *sigE*, but about 10% read through into the downstream genes [5]. Thus, the

CseB/CseC signal transduction system is very likely to play a role in cell envelope homeostasis in the absence of exogenous agents that interfere with the integrity of the cell wall.

In attempt to better understand the nature of the signal sensed by CseC, a bioassay to test for compounds that induced the *sigE* promoter was developed. The *sigE* promoter was placed upstream from a plasmid-borne kanamycin resistance gene (*neo*) [6] to yield a construct that conferred a basal level of kanamycin resistance on the host. A selection of a wide range of compounds that cause bacterial cell wall stress would be then tested to see which increased kanamycin resistance above the basal level in a plate assay. The world is facing an urgent need to develop new antibacterial agents, for example, clinical isolates of vancomycin-resistant MRSA (VRSA) emerged in 2002 [7], and therefore the parallel development of techniques to screen a broad-range of new drugs is also necessary. The work has direct implications for medicine and for pharmaceutical companies, contributing both to their efforts to understand the molecular basis of defensive responses and resistance to antibiotics in bacteria, and to developing methods to discover new antibiotic activities.

2 Materials

2.1 Bacterial Strains and Culture (*See* Note 1)

1. Streptomyces strains: *Streptomyces lividans* 1326 (*S. lividans* 66 SLP2$^+$, SLP3$^+$) [8], *Streptomyces coelicolor* wild type M600 (SCP1$^-$, SCP2$^-$) [9].
2. YEME (yeast extract–malt extract) liquid medium: 0.3% (w/v) Difco Bacto-peptone, 0.5% (w/v) Difco yeast extract, 0.5% (w/v) Oxoid malt extract, 1% (w/v) glucose, 34% (w/v) sucrose. Dispense 100 mL aliquots and autoclave. At time of use, add 5 mM $MgCl_2 \cdot 6H_2O$ (*see* **Note 2**). For preparing protoplasts, also add 0.5% glycine (*see* **Note 3**).
3. TSB (tryptone soy broth): 3% Oxoid Tryptone Soya Broth powder (CM129). Dispense 100 mL aliquots and autoclave.
4. 0.05 M Tris[hydroxymethyl]methyl-2-aminoethanesulfonic acid (TES) buffer (pH 8). Dispense 100 mL aliquots and autoclave.
5. Double strength germination medium (2× GM) (*see* **Note 4**): 1% (w/v) Bacto yeast extract, 1% (w/v) casamino acids. Add 0.01 M $CaCl_2$ to 2× GM before use.
6. Mannitol soya flour (MS) agar medium (*see* **Note 5**): 2% (w/v) mannitol, 2% soya flour, 2% agar.
7. MMCGT (minimal medium supplemented by casamino acids, glucose, and tiger milk) agar medium (*see* **Note 6**): 0.05% (w/v) L-asparagine, 0.05% (w/v) K_4HPO_4, 0.02% (w/v)

$MgSO_4 \cdot 7H_2O$, 0.001 % (w/v) $FeSO_4 \cdot 7H_2O$, 0.6 % (w/v) Difco casamino acids, 0.5 % (w/v) glucose, 0.75 % (v/v) tiger milk (*see* **Note 7**), 1 % (w/v) agar.

2.2 Construction and Preparation of Plasmid DNA

1. pIJ5953 [5]: PvuII-SmaI fragment of 0.75 kb *sigE* promoter sequence ligated in pIJ2925 [10].
2. pIJ486 [6]: multicopy *Streptomyces* promoter-probe plasmid containing *neo* as reporter gene ($Thio^R$).
3. Dephosphorylation: Calf intestinal alkaline phosphatase (CIAP) and supplied 10× reaction buffer. Store at −20 °C.
4. Ligation: T4 DNA ligase and supplied 10× reaction buffer. Store at −20 °C.
5. TEG (Tris–HCl, EDTA (ethylenediamine tetraacetic acid) and glucose) buffer: 25 mM Tris–HCl, 10 mM EDTA, 50 mM glucose, pH 8.
6. RNaseA.
7. Lysozyme.
8. 0.2 M NaOH.
9. 11 % (w/v) SDS.
10. 3 M KCH_3COO (pH 4.8).
11. Isopropanol.
12. Phenol–chloroform (1:1, v/v).
13. 70 % (v/v) Ethanol.
14. QIAquick gel extraction kit.
15. Restriction enzymes: *Bgl*II and supplied 10× reaction buffer, *BamH*I and supplied 10× reaction buffer. Store at −20 °C.

2.3 Preparation of Streptomyces Protoplasts and DNA Transformation

1. 10.3 % (w/v) glucose.
2. P (protoplast) buffer: 10.3 % (w/v) sucrose, 0.025 % (w/v) K_2SO_4, 0.202 % (w/v) $MgCl_2 \cdot 6H_2O$, 0.2 % (v/v) trace element solution (*see* **Note 8**). Dispense 100 mL aliquots and autoclave.
3. 25 % PEG (polyethylene glycol) 6000. Dispense 5 mL aliquots and autoclave.
4. R5 agar medium (*see* **Note 9**): 10.3 % (w/v) sucrose, 0.025 % (w/v) K_2SO_4, 1.012 % (w/v) $MgCl_2 \cdot 6H_2O$, 1 % (w/v) glucose, 0.01 % (w/v) Difco casamino acids, 0.02 % (v/v) trace element solution, 0.5 % (w/v) Difco yeast extract, 0.573 % (w/v) TES buffer and supplements.

2.4 Agarose Gel Electrophoresis

1. TBE running buffer (10× stock solution): 10.8 % (w/v) Trizma base, 5.5 % (w/v) boric acid, 4 % (v/v) EDTA (0.5 M, pH 8). Dilute 1/10 in water to make 1× TBE for gel running.

2. Agarose gel: Melt 1% (w/v) agarose in 1× TBE using a microwave, and pour the hot gel solution into a gel casting tray to set, creating wells for sample loading using a suitable comb.
3. Gel electrophoresis unit with power supply.
4. Ethidium bromide (EtBr): DNA separations on agarose gels were visualized using a 0.5–1 μg/mL staining solution of EtBr in 1× TBE. EtBr is known to be a powerful mutagen and is moderately toxic. Take extra care working with EtBr and dispose of the waste as advised.
5. DNA loading dye: 0.25% (w/v) bromophenol blue, 0.25% (w/v) xylene cyanol FF, 30% (w/v) glycerol. Store in aliquots at −20 °C.
6. DNA size marker: 1 kb DNA ladder. Store at −20 °C.

2.5 Antibiotics

1. Antibiotic stock solutions: vancomycin (10 mg/mL), A47934 (10 mg/mL), ristocetin (10 mg/mL), teicoplanin (10 mg/mL), bacitracin (10 mg/mL), moenomycin A (10 mg/mL), ramoplanin (10 mg/mL), enduracidin (10 mg/mL), tunicamycin (10 mg/mL), D-cycloserine (30 mg/mL), polymyxin B (10 mg/lmL), phosphomycin (30 mg/mL), apramycin (50 mg/mL), thiostrepton (50 mg/mL), novobiocin (50 mg/mL), streptomycin (10 mg/mL). Thiostrepton is dissolved in DMSO and the other antibiotics in water. Filter-sterilize antibiotic solutions and store in aliquots at −20 °C. At time of use, each antibiotic stock solution was diluted in water and applied to 6 mm paper disk.
2. Antibiotic paper disks: penicillin G (30 μg), amoxicillin (25 μg), ticarcillin (75 μg), ampicillin (25 μg).

3 Methods

3.1 Preparation of Protoplasts

1. 25 mL YEME liquid medium was inoculated with 0.1 mL *Streptomyces* spore suspension in a baffled flask and grown for about 40 h at 30 °C in an orbital shaker. The culture was centrifuged for 7 min at 4000 × *g* and the mycelial pellet was washed twice with 15 mL 10.3% sucrose.
2. The mycelial pellet was resuspended in 4 mL lysozyme solution (1 mg/mL in P buffer) and incubated at 30 °C for 15–60 min, triturated three times with a 5 mL pipette and incubated for a further 15 min.
3. 5 mL P buffer was added, the trituration was repeated, and protoplasts were filtered through cotton wool, transferred to a plastic tube and centrifuged gently. The protoplast pellet was resuspended in the drop of buffer left after pouring off the supernatant by tapping the tube. 1 mL P buffer was added and the protoplast suspension was left on ice for immediate use.

4. To freeze for storage, protoplasts were aliquoted into small plastic tubes and placed in ice in a plastic beaker, which was placed at −70 °C overnight. Frozen tubes were freed from the ice and stored at −70 °C. Protoplasts were thawed by shaking the frozen tube under running warm water. Protoplasts were counted using a hemocytometer and samples of c. 4×10^9 protoplasts were distributed into 50 mL conical tubes for immediate use.

3.2 Construction and Preparation of sigEp-neo Reporter Plasmid

1. Digest pIJ5953 with *BglII* with *BamHI* at 37 °C for 2 h. Mix 2–5 μg of the digested DNA samples with 1/6 volume of 6× loading dye, and load into the sample wells of an agarose gel. For reference, the DNA size marker is also loaded alongside samples.
2. Mix 2–5 μg of the digested DNA samples with 1/6 volume of 6× loading dye, and load into the sample
3. Run the gel using a power supply set to provide a constant voltage of 100 V, until the fastest moving blue dye has migrated 90% down the gel (about 30–60 min).
4. Stain the gel for approximately 10 min in EtBr stain, and record the result by photographing while illuminating at a wavelength of 254 nm on the UVIdoc (or equivalent).
5. Cut approximately 750 bp size of DNA fragment (which carries *sigE* promoter (*sigEp*)) out from the agarose gel and purify the DNA using gel extraction kit according to the manufacturer's instruction.
6. Digest pIJ486 with *BamHI* and purify the DNA similarly as shown above. Dephosphorylate the linear pIJ486 cut with *BamHI* by adding 0.5 U CIAP and incubating for 30 min at 37 °C. Purify the enzyme-treated DNA by first extracting twice with double the volume of phenol–chloroform, then once with double the volume of chloroform (*see* **Note 10**). Ethanol-precipitate the DNA in the aqueous layer by adding 1/10th the volume of 3 M KCH_3COO (pH 4.8), then double the volume of isopropanol and incubating at −80 °C for 30 min. Collect the pellet by centrifugation, wash with 70% ethanol, and allow to dry by exposure to the air for 10 min (*see* **Note 11**). Dissolve the pellet in sterile water.
7. Ligate the dephosphorylated pIJ486 vector DNA to the 0.75 kb *sigEp* insert by incubating with T4 DNA ligase at 18 °C for 4–24 h (*see* **Note 12**).
8. Transform the ligated DNA mixture into protoplast of *S. lividans* 1326. Immediately before they are required, quickly thaw a frozen 50 μL aliquot of *S. lividans* 1326 protoplasts and maintain on ice. Immediately after DNA (in up to 20 μL) was added to 50 μL protoplast suspension, 0.5 mL 25% PEG made up in P buffer was added and mixed by pipetting up and

down once. 0.5 mL P buffer was added as soon as possible afterwards, and the mixture was plated on R5 medium and incubated at 30 °C (*see* **Note 13**).

9. After overnight culture, 20 μL of thiostrepton solution (50 mg/mL) was diluted in 1 mL sterile water and added to the plate. The plate was swirled vigorously to cover the entire surface, dried in a laminar flow hood and incubated at 30 °C. Colonies were visible after about 2 days.
10. A single colony isolated from the thiostrepton plate was then inoculated in a mixture of 5 mL YEME and 5 mL TSB with 50 μg/mL of thiostrepton and cultured for 2 days at 30 °C with shaking at 250 rpm. The cells were harvested by centrifugation (10 min at 3000 × *g*). The cells were resuspended in 200 mL of TEG buffer containing 100 mg of lysozyme and 20 mg of RNaseA. The suspension was incubated at 37 °C for 30 min (or until the cells lysed if this occurred first). 400 mL of 0.2 M NaOH, 11 % (w/v) SDS were added and the mixture was left at room temperature for 5 min. 300 mL of 3 M KCH_3COO (pH 4.8) were added, and the mixture was vortexed, left on ice for 10 min, and microcentrifuged for 10 min. The supernatant was extracted with an equal volume of phenol–chloroform (1:1, v/v), and the nucleic acid precipitated by adding an equal volume of isopropanol. The nucleic acid was harvested by centrifugation, washed in 70 % (v/v) ethanol, dried, and redissolved in sterile water.
11. To confirm the resulting reporter plasmid carries *sigEp-neo* (designated as pIJ6880) [3], the plasmid DNA was then digested with several restriction enzymes and verified its digestion pattern using agarose gel electrophoresis. Mix 2–5 μg of the digested DNA samples with 1/6 volume of 6× loading dye, and load into the sample wells of an agarose gel (prepared as detailed in Subheading 2.5). For reference, the DNA size marker is also loaded alongside samples. Run the gel using a power supply set to provide a constant voltage of 100 V, until the fastest moving blue dye has migrated 90 % down the gel (about 30–60 min). Stain the gel for approximately 10 min in EtBr stain, and record the result by photographing while illuminating at a wavelength of 254 nm on the UVIdoc (or equivalent).

3.3 Construction of sigEp-neo Bioassay System in S. coelicolor M600

1. Transform the unmethylated pIJ6880 DNA [3] into protoplasts of *S. coelicolor* M600 and isolate single colony of the resulting strain of *S. coelicolor* as described in **steps 8** and **9** in Subheading 3.2.
2. Prepare plasmid DNA and confirm it by restriction digestion and gel electrophoresis as described in **steps 10** and **11** in Subheading 3.2.

3. The isolated *S. coelicolor* single colony confirmed to harbor the resulting plasmid, pIJ6880, was then resuspended in 300 μL dH_2O and 100 μL was spread on an SFM plate to yield a confluent lawn. The plates were incubated at 30 °C for about 6 days or until confluent lawns of grey spores were visible. The plates were not left for more than 2 weeks to prevent significant loss of spore viability. The spores were harvested and stored at −20 °C. The viable spore concentration was determined by plating out a dilution series on SFM plates.
4. To access the basal resistance concentration to kanamycin of the resulting strain, confluent lawns of spores of *S. coelicolor* M600 carrying pIJ6880 were spread on plates carrying different concentrations of kanamycin. *S. coelicolor* M600 carrying pIJ6880 was completely killed by 100 μg/mL kanamycin but was resistant to 80 μg/mL kanamycin, resulting from the basal activity of the *sigEp*. A diagram of genetic map of pIJ6880 and introduction of pIJ6880 into *S. coelicolor* M600 is shown in Fig. 1.

3.4 Bioassay for Inducers of the sigEp-neo Reporter System

1. To see if the *sigEp* could be induced by antibiotics, chemicals or enzymes known to target the cell envelope, confluent lawns of spores (approximately 10^7 spores) of *S. coelicolor* M600 carrying pIJ6880 were spread on 144 cm^2 square plates of MMCGT agar medium contained a lethal concentration of kanamycin (100 μg/mL), and potential inducers were applied on 6-mm paper disks to the freshly spread plates. Plates were scored after incubation at 30 °C for 4 days.
2. As Fig. 2a demonstrates, inducers of *sigEp* raise the level of expression of the *neo* gene and hence induce a halo of kanamycin-resistant growth around the disk. A wide range of antibiotics that inhibit intermediate and late steps in peptidoglycan biosynthesis, including β-lactams (e.g., penicillin G, amoxycillin, ticarcillin, ampicillin), glycopeptides (e.g., A47934, ristocetin, teicoplanin, vancomycin), a peptide (bacitracin), a phosphoglycolipid (moenomycin A) and a cyclic depsipeptide (ramoplanin, enduracidin) induced a halo of kanamycin-resistant growth, whereas antibiotics that inhibit early steps in peptidoglycan biosynthesis such as tunicamycin, D-cycloserine, polymyxin B, and phosphomycin did not. Negative control antibiotics that target the ribosome (e.g., apramycin, thiostrepton, streptomycin) or DNA gyrase (novobiocin) also did not induce the *sigEp-neo* system. Given that a variety of structurally unrelated antibiotics with varied targets in the cell envelope induced the *sigE* signal transduction system, a cell wall hydrolytic enzyme such as lysozyme was also test for the bioassay to determine if it was also capable of acting as an inducer. A confluent lawn of spores of *S. coelicolor*

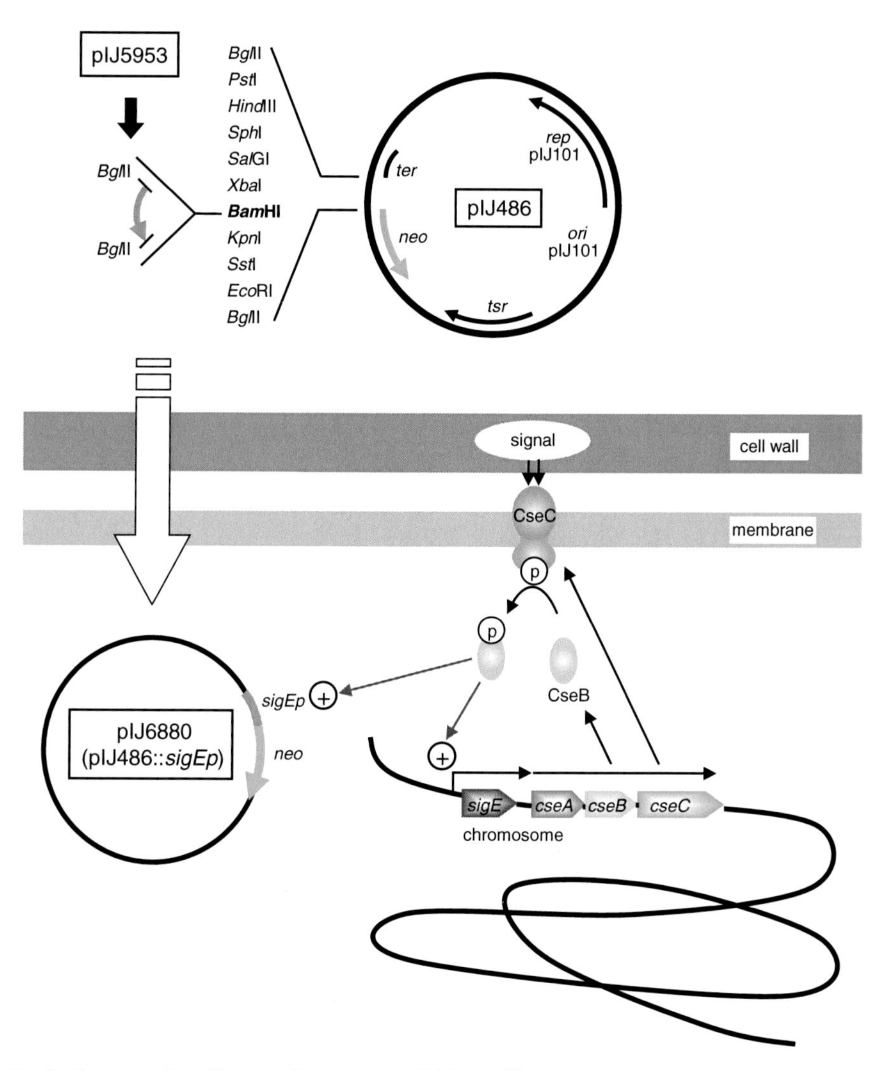

Fig. 1 An illustration describes the genetic map of pIJ6880 and introduction of pIJ6880 into *S. coelicolor* M600 results in *sigEp-neo* bioassay system

M600 carrying pIJ6880 was again spread on plates carrying a lethal concentration of kanamycin (100 μg/mL), and 5 μL samples of a twofold serial dilution series of lysozyme (from 1 mg/mL) in 10 mM Tris–HCl (pH 8) were spotted directly onto the plates. At the higher concentrations (1 mg/mL and 0.5 mg/mL), patches of lysozyme-induced kanamycin-resistant growth were observed as shown in Fig. 2b. To ensure that this induction was caused by the cell wall hydrolytic activity of lysozyme and not, for example, by a putative carbohydrate contaminant of the enzyme, the experiment was repeated after heat denaturation. No induction was seen with the heat-treated preparations (data not shown).

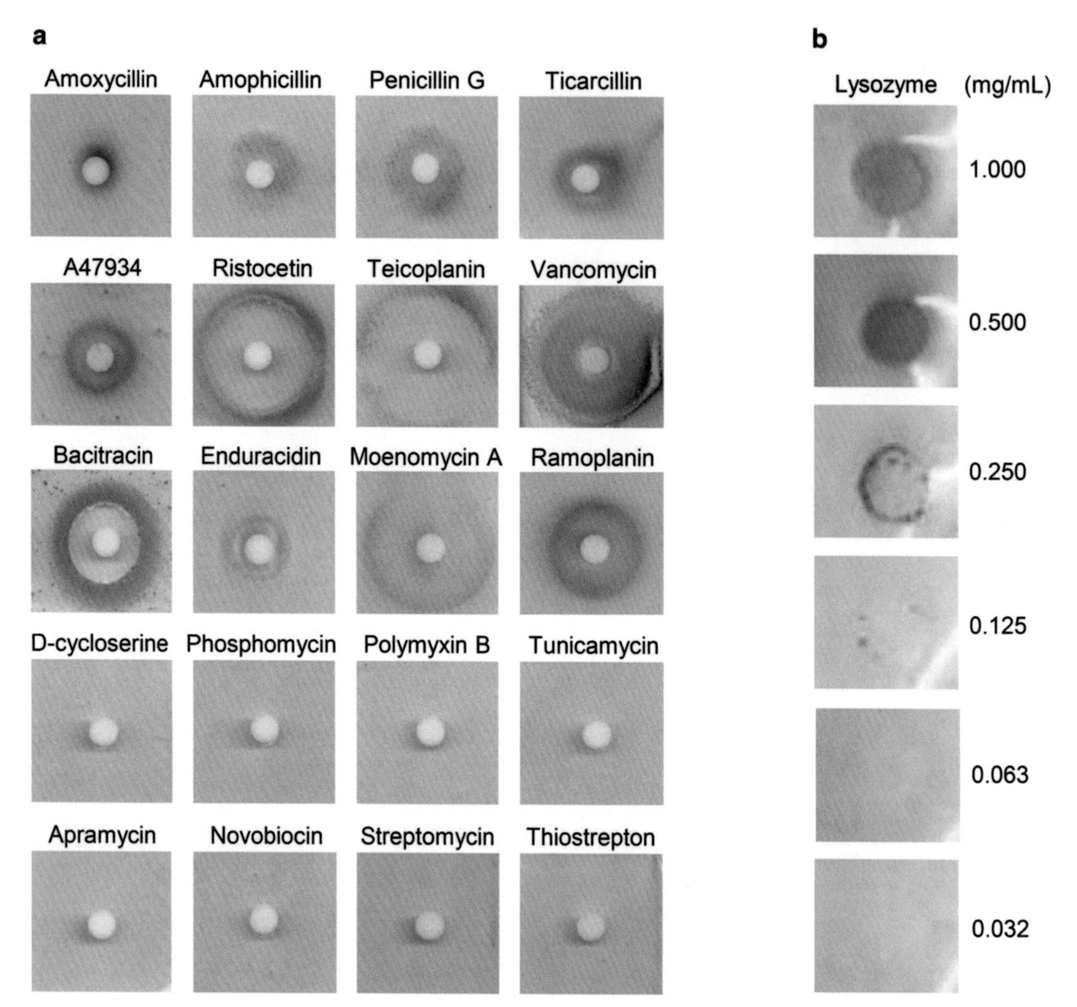

Fig. 2 Bioassay for inducers of *sigEp-neo* system. Approximately 10^7 spores of *S. coelicolor* M600 carrying pIJ6880 were spread on MMCGT agar medium plates containing 100 μg/mL kanamycin. Inducers were applied on 6-mm paper disks to the freshly spread plate. The plates were scored after 4–5 days of incubation at 30 °C. Inducers of the *sigEp* raise the level of expression of the *neo* gene and hence induce a halo of kanamycin-resistant growth around the disk. (**a**) Response of the *sigEp-neo* system to various antibiotics e.g., those target late steps cell wall peptidoglycan biosynthesis (amoxycillin, ampicillin, penicillin G, ticarcillin, A47934, ristocetin, teicoplanin vancomycin, bacitracin, enduracidin, moenomycin A, ramoplanin), those target early steps cell wall peptidoglycan biosynthesis (D-cycloserine, phosphomycin, polymyxin B, tunicamycin) and those target the ribosome or DNA gyrase (apramycin, thiostrepton, streptomycin, novobiocin). The amount of antibiotic in each disk is indicated. (**b**) Response of the *sigEp-neo* system to a cell wall hydrolytic enzyme, lysozyme. A confluent lawn of spores of *S. coelicolor* M600 carrying pIJ6880 was spread on MMCGT agar plates carrying a lethal concentration of kanamycin (100 μg/mL), and 5 μL samples of a twofold serial dilution series of egg white lysozyme (from 1 mg/mL, as indicated) in 10 mM Tris–HCl (pH 8) were spotted directly onto the plates immediately after plating

4 Notes

1. Due to the slow growth of *Streptomyces* strains, there is a greater risk of contamination than when using other, more rapidly growing bacteria such as *E. coli* or *Bacillus*. Manipulation of *Streptomyces* strains and cultures should therefore be done in a suitable laminar flow hood, and extra attention given to aseptic technique. All the solutions for the culture *Streptomyces* strains should be prepared as a small aliquots and autoclave at 115 °C for more than 15 min.
2. Prepare $MgCl_2{\cdot}6H_2O$ as 2.5 M stock solution and autoclave. To make the final concentration of 5 mM $MgCl_2{\cdot}6H_2O$ in YEME, add 2 mL of 2.5 M stock solution of $MgCl_2{\cdot}6H_2O$ in total 1 L of YEME.
3. Prepare glycine as 20 % stock solution and autoclave. To make the final concentration of 0.5 % glycine in YEME, add 25 mL of 20 % stock solution of glycine in total 1 L of YEME.
4. Prepare $CaCl_2$ as 1 M stock solution and autoclave. To make the final concentration of 0.01 M $CaCl_2$ in 2× GM, add 100 μL of 1 M $CaCl_2$ solution in total 10 mL of 2× GM.
5. This medium is used to prepare spores of *Streptomyces* strains. Autoclave twice at 115 °C and store at room temperature as 100 mL aliquots in 250 mL flasks. Re-melt the medium using a microwave prior to use.
6. This medium is used for the bioassay experiment. Dissolve L-asparagine, K_4HPO_4, $MgSO_4{\cdot}7H_2O$ and $FeSO_4{\cdot}7H_2O$ in the distilled water, adjust to pH 7.0–7.2 and dispense 200 mL into 250 mL Erlenmeyer flasks each containing 2 g agar. Close the flasks and autoclave. Prepare Difco casamino acids (30 % stock solution), glucose (50 % stock solution), and tiger milk separately and autoclave. At time of use, re-melt the agar medium and add 4 mL of Difco casamino acids (30 % stock solution), 2 mL of glucose (50 % stock solution), and 1.5 mL of tiger milk to each flask.
7. Tiger milk: 0.75 % (w/v) L-arginine, 0.75 % (w/v) L-cystine, 1 % (w/v) L-histidine, 0.75 % (w/v) DL-homoserine, 0.75 % (w/v) L-leucine, 0.75 % (w/v) L-phenylalanine, 0.75 % (w/v) L-proline, 0.15 % (w/v) adenine, 0.15 % (w/v) uracil, 0.01 % (w/v) nicotinamide. Dispense 100 mL aliquots and autoclave.
8. Trace element solution: 0.004 % (w/v) $ZnCl_2$, 0.02 % (w/v) $FeCl_3{\cdot}6H_2O$, 0.001 % (w/v) $CuCl_2{\cdot}2H_2O$, 0.001 % (w/v) $Na_2B_4O_7{\cdot}10H_2O$, 0.001 % (w/v) $(NH_4)_6Mo_7O_{24}{\cdot}4H_2O$. Dispense 100 mL aliquots and autoclave. Make a fresh solution every 2–4 weeks and store at 4 °C.

9. This medium is used for vigorous regeneration of *Streptomyces* protoplast. Pour 100 mL of solution into 250 mL Erlenmeyer flasks each containing 2.2 g agar and autoclaved. At time of use, re-melt and add following supplements to each flask in the order listed: 1 mL of KH_2PO_4 (0.5 % w/v), 0.4 mL of $CaCl_2{\cdot}2H_2O$ (3.68 % w/v), 1.5 mL of L-proline (20 % w/v), NaOH (1 M), 0.75 mL of any required growth factors for auxotroph. Each stock solution should be prepared and autoclaved (except NaOH) before use.
10. When extracting DNA-containing solutions with phenol–chloroform it is important that the phenol–chloroform has been equilibrated to the correct pH. Above pH 8, extraction will remove contaminating protein into the lower organic layer, leaving DNA in the upper aqueous phase. Acidic phenol–chloroform, however, will extract both DNA and protein into the organic layer. A chloroform extraction step is included after extracting with phenol containing solutions to help remove all traces of the phenol that would inhibit enzyme activity in any subsequent use of the DNA, e.g., ligation, PCR. Ethanol precipitation of the DNA also assists in this.
11. The precipitated DNA pellet may be very difficult to see by eye, and care should be taken not to dislodge and lose the pellet while removing the supernatant, or when washing with 70 % ethanol. The 70 % ethanol wash is important for removing salt from the DNA preparation that may interfere with the efficiency of the subsequent ligation reactions.
12. The optimal ratio of vector–insert DNA used for cloning can be determined experimentally, but a ratio between 3:1 and 1:3 is usually effective. When using linearized vector DNA that can potentially self-ligate, as is the case here, it is preferable to provide an excess of insert DNA, e.g., by using a 1:3 ratio. Ligation temperature is a compromise between the optimal conditions for the enzyme (usually 25 °C) and the temperature required to ensure annealing of the DNA ends (which can vary with the length and base composition of any overhanging DNA sequences). It is good practice to include a positive control for the ligation, which is usually self-ligation of cut vector DNA that has not been dephosphorylated, and also a negative control containing the cut and dephosphorylated vector but no insert.
13. *S. coelicolor* contains at least four methyl-specific restriction endonucleases that restrict (reduce or prevent) the introduction of methylated DNA, e.g., from Dam^+ Dcm^+ Hsd^+ *E. coli* K-12 strains. Therefore, DNA is generally passaged through a non-methylating Dam^- Dcm^- Hsd^- *E. coli* host before introduction into *S. coelicolor*. Alternatively, the less restricting

S. lividans 66 has been used as a recipient for methylated DNA (*see* Ref. *8*). In this study, we used *S. lividans* 1326 as a non-methylation host to bypass the methyl-specific restriction system of *S. coelicolor*.

Acknowledgments

This work was supported by funding from the Medical Research council, UK (G0700141) and the Royal Society, UK (516002. K5877/ROG). The author would like to thank Andy Hesketh (University of Cambridge, Systems Biology Centre) for comments on the manuscript.

References

1. Lonetto MA, Gribskov M, Gross CA (1992) The σ^{70} family: sequence conservation and evolutionary relationships. J Bacteriol 174: 3843–38491
2. Paget MSB, Chamberlin L, Atrih A, Foster SJ, Buttner MJ (1999) Evidence that the extracytoplasmic function sigma factor, σ^E, is required for normal cell wall structure in *Streptomyces coelicolor* A3(2). J Bacteriol 181:204–211
3. Hong H-J, Paget MS, Buttner MJ (2002) A signal transduction system in *Streptomyces coelicolor* that activates the expression of a putative cell wall glycan operon in response to vancomycin and other cell wall-specific antibiotics. Mol Microbiol 44:1199–1211
4. Doyle RJ (1989) How cell walls of Gram-positive bacteria interact with metal ions. In: Beveridge TJ, Doyle RJ (eds) Metal ions and bacteria. John Wiley & Sons, New York, pp 275–293
5. Paget MSB, Leibovitz E, Buttner MJ (1999) A putative two-component signal transduction system regulates σ^E, a sigma factor required for normal cell wall integrity in *Streptomyces coelicolor* A3(2). Mol Microbiol 33:97–107
6. Ward JM, Janssen GR, Kieser T, Bibb MJ, Buttner MJ, Bibb MJ (1986) Construction and characterisation of a series of multi-copy promoter-probe plasmid vectors for *Streptomyces* using the aminoglycoside phosphotransferase gene from Tn5 as indicator. Mol Gen Genet 203:468–478
7. Pearson H (2002) 'Superbug' hurdles key drug barrier. Nature 418:469–470
8. Bibb MJ, Cohen SN (1982) Gene expression in *Streptomyces*: construction and application of promoter-probe plasmid vectors in *Streptomyces lividans*. Mol Gen Genet 187: 265–277
9. Chakraburtty R, Bibb MJ (1997) The ppGpp synthetase gene (*relA*) of *Streptomyces coelicolor* A3(2) plays a conditional role in antibiotic production and morphological differentiation. J Bacteriol 179:5854–5861
10. Janssen GR, Bibb MJ (1993) Derivatives of pUC18 that have *Bgl*II sites flanking a multiple cloning site and that retain ability to identify recombinant clones by visual screening of *Escherichia coli* colonies. Gene 124:133–134

Chapter 11

Luciferase Reporter Gene System to Detect Cell Wall Stress Stimulon Induction in *Staphylococcus aureus*

Vanina Dengler and Nadine McCallum

Abstract

Luciferase reporter gene fusions provide an extremely rapid and sensitive tool for measuring the induction or repression of stress responses in bacteria. *Staphylococcus aureus* activates the expression of a cell wall stress stimulon (CWSS) in response to the inhibition or disruption of cell wall synthesis. The highly sensitive promoter–reporter gene fusion construct p*sas016*$_p$-*luc+* can be used to quantify and compare any changes in CWSS expression levels and induction kinetics. Potential uses of this system include identifying and characterizing novel cell wall-targeting antibacterial agents, identifying genomic loci influencing cell envelope synthesis and detecting changes in CWSS expression that could be linked to decreased antibiotic susceptibility profiles in clinical isolates.

Key words Cell wall stress stimulon, VraTSR, *Staphylococcus aureus*, Luciferase reporter gene fusion, Cell wall stress

1 Introduction

The bacterial cell envelope is one of the most important antibiotic targets and there are several different classes of antibacterial agents that act by directly or indirectly blocking or disrupting envelope-biosynthetic pathways [1]. When exposed to such cell wall stress, most gram-positive bacteria mount protective stress responses [2]. In *Staphylococcus aureus*, exposure to cell wall-active antibiotics or depletion of essential cell wall synthesis enzymes triggers the activation of the VraTSR three-component sensor-transducer system which controls the transcription of up to 50 genes, collectively known as the cell wall stress stimulon (CWSS) [3–8] (Fig. 1).

Point mutations leading to constitutive upregulation of the CWSS have been linked to increased levels of glycopeptide, beta-lactam and daptomycin resistance in several clinical isolates [9–12], while experimental deletion of the VraTSR system has been shown to decrease resistance levels to most CWSS-inducing agents [3, 6]. The CWSS is also induced by the experimental depletion of genes

Hee-Jeon Hong (ed.), *Bacterial Cell Wall Homeostasis: Methods and Protocols*, Methods in Molecular Biology, vol. 1440, DOI 10.1007/978-1-4939-3676-2_11, © Springer Science+Business Media New York 2016

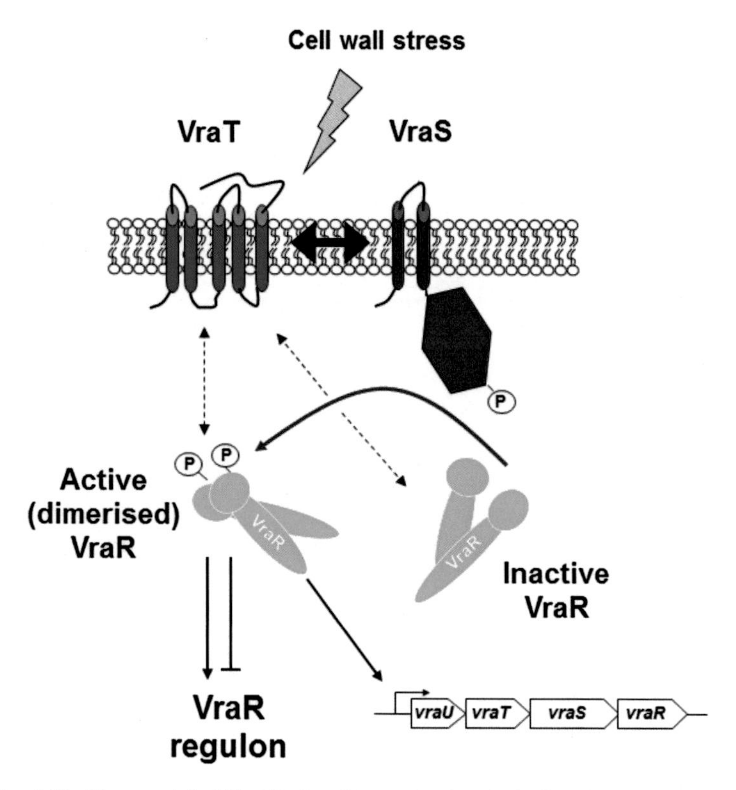

Fig. 1 Working model of VraSR signal transduction, including factor VraT. *Black arrows* and the *black blocked arrow* indicate transcriptional regulation, the *red arrow* indicates phosphorylation of VraR by VraS, *double-pointed arrows* indicate suggested interaction; *dashed double-pointed arrows* indicate possible additional interactions or signal transduction pathways. Adapted from [6, 23]

essential for cell wall biosynthesis including MurA, MurZ, MurB, MurF, and PBP2 [4, 13, 14].

CWSS induction is predicted to protect cells against envelope damage as several CWSS-encoded proteins are directly linked to cell envelope biosynthesis, such as MurZ [13, 15]; PBP2 [15, 16]; SgtB [15, 17]; FmtA [18]; the predicted autolysins Atl and SA0424 [19]; and the predicted wall teichoic acid ligases MsrR, SA0908, and SA2103 [20, 21]. There are also several genes of unknown or poorly characterized functions in the VraTSR regulon, including the gene with the highest induction level, *sas016* (also called *vraX*) [4, 8, 22].

Induction of VraTSR can be measured by directly quantifying CWSS transcription products using methods such as Northern blotting [22, 23], quantitative real-time reverse-transcription PCR [14, 23], or microarray [5, 8, 23]. However, these methods are generally time consuming, difficult to standardize and expensive. The luciferase reporter gene construct p*sas016*$_p$-*luc+* is an *Escherichia coli–S. aureus* shuttle plasmid containing the promoter of *sas016* fused to the firefly luciferase gene (*luc+*), which was developed as a highly sensitive, rapid, and inexpensive tool for quantifying relative levels of CWSS expression [3, 6, 24, 25]. This construct has been used to compare levels of CWSS expression in different strain backgrounds [6, 24, 25] and to compare the induction kinetics and peak induction levels of the CWSS in response to varying concentrations of different cell wall-active compounds [3]. These studies demonstrated the extremely high sensitivity of this system, which was able to accurately measure differences in expression ranging from as low as 1.5-fold up to a magnitude of over 4-log fold. Such high levels of sensitivity make this construct an ideal tool for identifying or characterizing novel antimicrobial agents that target the cell envelope or specific genetic loci that influence cell envelope homeostasis [24–26].

2 Materials

Prepare all reagents using ultrapure water and sterilize by autoclaving or filtration. Prepare and store all reagents at room temperature (optimum temperature range: 20–25 °C) unless directed otherwise. Perform all steps involving bacterial cultures under PC2 laboratory conditions.

2.1 Luciferase Fusion Construct Components

1. pSP-luc+NF fusion vector (Promega, Cat. No. E4471).
2. *E. coli–Staphylococcus aureus* shuttle vector (e.g., pBUS1 [27]).
3. *S. aureus* genomic DNA, prepared using standard protocols (*see* **Note 1**), from a strain with a published genome sequence (e.g., *S. aureus* strain COL; accession number: CP000046.1).
4. Thermocycler (any standard PCR machine).
5. PCR enzymes and reagents, e.g., Expand High Fidelity PCR system (Roche; Product No. 11732641001); PCR Nucleotide Mix (Roche; Product No. 11581295001).
6. Primers for amplification of the *sas016* promoter region: SAS016.lucF (AATTAGGTACCTGGATCACGGTGCATACAAC) and SAS016.lucR (AATTACCATGGCCTATATTACCTCCTTTGCT), promoters from other VraTSR-regulated genes can also be used to construct promoter–luciferase fusion constructs (*see* **Note 2**).
7. Gel electrophoresis system.
8. PCR-product purification kit, e.g., QIAquick PCR Purification Kit (Qiagen; Cat. No. 28104).
9. Agarose gel extraction kit, e.g., QIAquick Gel Extraction Kit (Qiagen; Cat. No. 28704).
10. Restriction enzymes Asp718 (Roche; Product No. 10814245001), EcoRI (Roche; Product No. 10703737001), and NcoI (Roche; Product No. 10835315001).
11. Water bath or thermomixer.

12. Alkaline phosphatase, e.g., Alkaline Phosphatase (AP) (Roche; Product No. 10108138001).
13. T4 DNA ligase, e.g., T4 DNA Ligase (Roche; Product. No. 10481220001).
14. *E. coli* competent cells for plasmid cloning, e.g., *E. coli* strain DH5α (Life Technologies; Cat. No. 18263012) (*see* **Note 3**).
15. Luria-Bertani (LB) liquid broth and LB agar plates: 10 g/L tryptone, 5 g/L yeast extract, and 10 g/L NaCL. Dissolve reagents in distilled water and sterilize by autoclaving for 15 min. For LB agar, add 15 g/L of powdered agar before autoclaving.
16. Antibiotics for the selection of bacterial subclones containing fusion plasmid constructs. When appropriate, medium should be supplemented with 100 μg/mL of ampicillin (Sigma; Cat. No. A9518-5G) or 10 μg/mL of tetracycline (Sigma; Cat. No. T7660-5G).
17. X-Gal (5-Bromo-4-chloro-3-indolyl β-D-galactopyranoside) (GoldBio; Cat. No. X4281C) stock solution: 20 mg/mL of X-Gal dissolved in dimethylformamide (DMF) and IPTG (GoldBio; Cat. No. 12481C) stock solution: IPTG (100 mM) dissolved in distilled H_2O.
18. Plasmid DNA extraction kit, e.g., QIAprep Spin Miniprep Kit (Qiagen; Cat. No. 27104).
19. Competent cells of restriction negative *S. aureus* strain RN4220 (*see* **Note 4**).
20. Lysostaphin, e.g., AMBICIN L (AMBI Products; Cat. No. LSPN-50). Prepare stock solution of 10 mg/mL in H_2O and store at −20 °C.
21. Electroporation system, e.g., Bio-Rad Gene Pulser or MicroPulser and electroporation cuvettes with a gap width of 0.1 cm.
22. Brain Heart Infusion (BHI) broth (BD; Cat. No. 211059) and BHI agar. For BHI agar, prepare BHI broth according to manufacturer's instructions and add 15 g/L of powdered agar prior to autoclaving.
23. Competent cells of *S. aureus* recipient strain(s) to be analyzed for basal CWSS expression levels or induction studies (*see* **Note 4**).
24. Incubator for growing bacterial cultures on agar plates at 37 °C and orbital-shaking incubator for growing bacterial cultures at 37 °C with shaking at 180 rpm.

2.2 Luciferase Assay Components

1. LB broth and orbital-shaking incubator for growing bacterial cultures (as described above).
2. Spectrophotometer for measuring the optical density (OD) of bacterial cultures at 600 nm (all OD measurements specified in this study are taken at OD 600 nm).

3. Antibiotic stock solutions required for measuring induction.
4. Phosphate-buffered saline (PBS): 137 mM NaCl, 2.7 mM KCl, 4.3 mM $Na2HPO_4$, and 1.47 mM KH_2PO_4; adjust final pH to 7.4.
5. Luciferase assay system (Promega; Cat. No. E1500): Dissolve lyophilized luciferase assay substrate in provided volume of luciferase assay buffer, aliquot into amber colored (light protective) Eppendorf tubes (Eppendorf; Cat. No. 22363221) and store at −70 °C.
6. Luminometer, e.g., Turner Designs TD-20/20 luminometer (Promega) or GloMax-Multi Jr Single Tube Multimode luminometer (Promega), fitted with a 0.5 mL PCR tube sample adapter.

3 Methods

3.1 Construction of Promoter–Luciferase Reporter Gene Fusion Plasmids

1. Amplify the *sas016* promoter using the primers listed above and suitable *S. aureus* genomic DNA as a template (fusions can also be constructed using promoters from additional CWSS genes; *see* **Note 2** for a list of potential promoters and corresponding primers). Prepare PCR reactions according to the enzyme manufacturer's instructions (*see* **Note 5**).
2. Electrophorese a small volume (2–5 μL) of the amplified product on a 1 % agarose gel to confirm the amplification of a single 585-bp fragment (for the *sas016* promoter). If a single clear band is visible purify the remainder of the amplification product using a PCR-product purification kit (if multiple bands are visible, *see* **Note 6**).
3. Digest the purified PCR product and the vector pSP-luc+ using the restriction enzymes EcoRI and NcoI according to the instructions from the enzyme manufacturer(s).
4. Purify the digested insert using standard procedures, e.g., a PCR-product purification kit.
5. Dephosphorylate the digested vector fragment using alkaline phosphatase and then inactivate the alkaline phosphatase, in accordance with the manufacturer's instructions, then purify the vector DNA (*see* **Note 7**).
6. Ligate the *sas016* promoter immediately upstream of the promoterless luciferase (*luc+*) gene in pSP-luc+ (Promega) using T4 DNA ligase by combining the following reaction components in a microcentrifuge tube on ice: T4 DNA ligase buffer (final conc. 1×), prepared vector DNA (approx. 0.02 pmol), prepared insert DNA (approx. 0.6 pmol), T4 DNA ligase (1 U), ultrapure water. Follow manufacturer's instructions for optimal ligation conditions and necessary purification steps (*see* **Note 8**).

7. Transform the ligation into competent *E. coli* cells (*see* **Note 3**) and plate aliquots of transformation onto LB agar containing ampicillin, X-Gal and IPTG.
8. Screen white colonies to obtain a plasmid containing the correct insert (*see* **Note 9**) and prepare plasmid DNA of the construct using a plasmid DNA extraction kit.
9. Excise the plasmid region containing the *sas016* promoter-*luc+* gene fusion by digesting plasmid DNA with the restriction enzymes EcoRI and Asp718. Electrophoreses the digested fragments through a 1 % agarose gel and excise and purify the correct band (2256 bp), using a gel extraction kit.
10. Digest an *E. coli–S. aureus* shuttle vector (e.g., pBUS1) with EcoRI and Asp718, then dephosphorylate and purify the vector as described above.
11. Ligate the *sas016* promoter-*luc+* fusion fragment into the shuttle vector using the ligation procedures described above.
12. Transform the ligation into competent *E. coli* cells and select transformants on LB agar containing tetracycline.
13. Screen transformants by preparing plasmid miniprep DNA and digesting with EcoRI and Asp718 to identify plasmids with correct vector and insert sizes.
14. Transform the resulting fusion plasmid, p*sas016p-luc+*, into a restriction negative *S. aureus* strain (e.g., RN4220) by electroporation (*see* **Note 10**) and select transformants on LB agar containing tetracycline.
15. Extract plasmid DNA from RN4220 using the QIAprep Spin Miniprep Kit (Qiagen; Cat. No. 27104), following the manufacturer's instructions with minor modifications (*see* **Note 11**).
16. Electroporate the plasmid isolated from RN4220 into competent cells of *S. aureus* strains to be analyzed (*see* **Note 12**).

3.2 Luciferase Assay

1. Prepare an overnight culture of an *S. aureus* strain carrying p*sas016*$_{p}$-luc+, by inoculating a single colony into a flask containing LB broth supplemented with tetracycline and incubating for approx. 16 h at 37 °C with shaking at 180 rpm (*see* **Note 13**).
2. Dilute the overnight culture to an OD of 0.05 by adding the culture to flasks containing fresh LB broth supplemented with tetracycline (pre-warmed to 37 °C) (*see* **Note 14**) and grow culture at 37 °C with shaking at 180 rpm.
3. To compare relative CWSS expression levels in different *S. aureus* strains containing p*sas016*$_{p}$-*luc+*, collect samples of cell cultures at various time points, e.g., after 1.5, 3, 4.5, 6, 7.5, and 9 h of growth, or collect samples when cells reach specific OD values, e.g., OD 0.25, 0.5, 1, 2, and 4. Collect and store samples as described below in **step 5**.

4. To measure reporter gene induction by an antibiotic, grow the culture from **step 2** until it reaches OD 0.5 and then split the culture evenly into smaller pre-warmed flasks (the number of flasks will depend upon the number of different antibiotic concentrations to be tested). Leave one flask as an uninduced control culture and induce the remaining flasks with predetermined concentrations of antibiotic (*see* **Note 15**).
5. Collect samples from all flasks at selected post-induction time points, e.g., after 5, 10, 20, 30, 60, and 120 min. To collect samples, centrifuge 1 mL of culture for 2 min at 15,000 × *g*, remove all supernatant and freeze and store cell pellets at −20 °C or immediately proceed to **step 6**. Measure and record the OD of all cultures at each sampling point.
6. To measure luciferase activity, first resuspend cell pellets in PBS to an OD of 10 (*see* **Note 16**) and allow suspensions to equilibrate to room temperature.
7. Transfer 10 μL of cell suspension into a clear 0.5 mL PCR tube.
8. Thaw luciferase substrate stock solution and let it equilibrate to room temperature (*see* **Note 17**). Add 10 μL of luciferase substrate to the 10 μL of cell suspension from **step** 7 and mix by pipetting up and down ten times before measuring relative light units (RLU) for 15 s after a delay of 3 s (*see* **Note 18**). If values approach or exceed the maximum limit of detection for the luminometer (*see* **Note 19**), prepare an appropriate dilution of the cell suspension and repeat **steps** 7 and **8**.
9. Record and compare RLU values to compare relative CWSS expression and/or induction levels (Fig. 2).

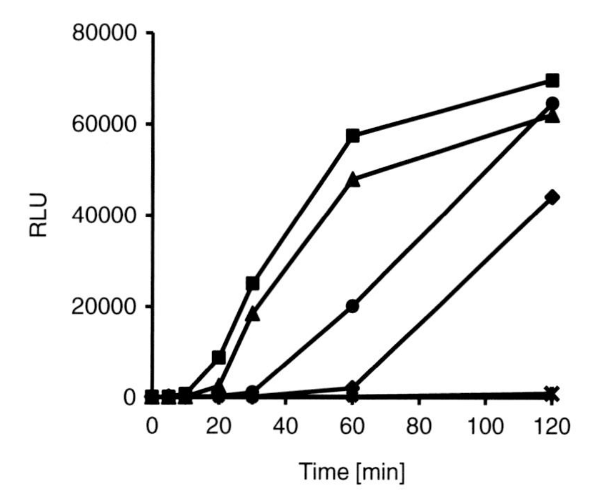

Fig. 2 Example of induction curves showing RLU detected from the construct p*sas016*$_p$-*luc+* in *S. aureus* cultures exposed to different oxacillin concentrations. Oxacillin was added to mid exponential cell cultures and relative light units (RLU) were measured at time points up to 2 h post addition. Concentrations were chosen in relation to the oxacillin minimum inhibitory concentration (MIC) of the strain. *Squares*—5× MIC, *triangles*—2× MIC, *circles*—1× MIC, *diamonds*—0.5× MIC, and *asterisks*—no oxacillin

4 Notes

1. Genomic DNA of S. *aureus* can be prepared using several well-established methods, e.g., using the DNeasy Blood and Tissue Kit (Qiagen; Cat. No. 69504), according to the manufacturer's protocol for Gram-positive bacteria. This protocol can be modified to increase DNA yield, by adding lysostaphin to the lysis buffer (to a final concentration of 100 μg/mL).
2. Table 1 shows VraR-regulated promoters and corresponding primers that have previously been used to construct luciferase reporter gene fusions.
3. Competent *E. coli* cells can be prepared using several well-established methods, e.g., the rubidium chloride protocol from Promega (*Protocols and Applications Guide* (3rd edition), p. 45–46).
4. Competent cells of *S. aureus* can be prepared using the protocol established by Katayama et al. [28].
5. PCR amplification using the Expand High Fidelity PCR System (Roche), prepare reactions containing: 1× Expand High Fidelity buffer with $MgCl_2$, 200 μM dNTPs, 0.2 μM of each primer, approx. 500 ng of template DNA, and 1.75 U Expand High Fidelity enzyme mix. For amplifying the *sas016* promoter the following cycling conditions can be used: denaturation at 94 °C for 2 min, followed by 30 cycles of denaturation at 94 °C for 15 s, annealing at 48 °C for 30 s, and elongation at 72 °C for 1 min, followed by a final elongation step at 72 °C for 5 min.
6. If multiple bands are present, purify the fragment of the correct size by gel extraction.
7. To inactivate the enzyme alkaline phosphatase (AP) (Roche; Product No. 10108138001) and purify the vector backbone

Table 1
Primers used to amplify VraR-regulated promoter regions

Gene/operon promoter	Primers[a]
tcaA	tcaA.lucF (TAATGGTACCAGTATTAGAAGTCATCAATCA) tcaA.lucR (TAAT CCATGGTTTCACCTCAATTCTGTTCCT)
sa0908	sa0908.lucF (AATTAGGTACCATAA TAGTACACACGCATGT) sa0908.lucR (TTAATCCATGGTTGATGCTCCTA TATTAAATT)
vraUTSR operon	vra.lucF (AATTTGGTACCGCACATGTACTTAATTACTT) vra.lucR (ATTAACCATGGCTATCACCTTTTATAATAAGT)

[a]Restriction sites are *underlined*

fragment, the vector can be can be electrophoresed into a 1 % agarose gel, then excised and purified using an agarose gel extraction kit.

8. Ligations using T4 DNA ligase (Roche; Product. No. 10481220001) can be performed overnight at 18 °C. Ligation products can then be purified using a PCR-product purification kit or by ethanol precipitation.
9. White colonies can be screened by extracting plasmid DNA using a standard plasmid miniprep procedure, e.g., QIAprep Spin Miniprep Kit (Qiagen; Cat. No. 27104). Plasmids can then be digested using restriction enzymes EcoRI and NcoI to confirm that both vector and insert fragments are the correct size. Alternatively, plasmids can be sequenced using standard Sanger sequencing protocols and the SAS016.lucF primer to ensure that they contain the correct promoter–luciferase fusion sequence.
10. The following electroporation procedure can be used to transform competent *S. aureus* cells prepared using the protocol suggested in **Note 4**: Add approx. 500 ng of plasmid DNA to 50 μL of competent cells and incubate on ice for 10 min. Transfer mixture to a 0.1 cm gap electroporation cuvette (pre-chilled on ice) and electroporate (25 μF, 100 Ω and 2.5 kV). Immediately resuspend cells in 0.5 mL of pre-warmed BHI containing 1.1 M sucrose and incubate for 2 h at 37 °C, before plating aliquots on BHI agar containing the appropriate selective antibiotic.
11. The QIAprep Spin Miniprep Kit (Qiagen; Cat. No. 27104) protocol can be modified to enhance lysis of *S. aureus*. During **step 1** of the handbook protocol, 2 μL of lysostaphin (10 mg/mL) can be added to the cell suspension prepared using buffer P1 and the suspension incubated at 37 °C for 30 min before proceeding to **step 2**.
12. The plasmid can also be transduced from RN4220 into different *S. aureus* strains using phage 80α, following standard phage-transduction protocols.
13. The ratio of LB medium to air in culture flasks should be approx. 1:5.
14. To obtain the correct cell dilution, dilute the overnight culture 1/10 in H_2O and measure the OD. Use this value to calculate a dilution factor for diluting the overnight culture to an OD of 0.05, e.g., If the OD of the 1/10 dilution is 0.7, the overnight culture will have an OD of approx. 7.0 and a dilution factor of 140 will be required. The OD can be checked after dilution and adjusted if required, by adding additional overnight culture or pre-warmed LB containing tetracycline.

15. It is recommended to first determine the minimum inhibitory concentration (MIC) of the antibiotic against the strain(s) being testing. Appropriate antibiotic concentrations can then be selected, e.g., 0.5×, 1×, 2×, and 10× the MIC, to compare induction kinetics at different antibiotic concentrations.
16. The volume of PBS required to resuspend cell pellets to an OD of 10 can be calculated using the following formula: volume of PBS (μL) = OD of suspension at time of sampling × 100. Therefore, if the OD of the suspension sampled was 0.2 add 20 μL of PBS and if the OD was 2.0 add 200 μL of PBS.
17. The optimum temperature range for all assay reagents is 20–25 °C.
18. Reproducibility of experiments is greatly increased if you try to normalize the time between adding the luciferase substrate and measuring RLU, e.g., after adding and mixing the luciferase assay substrate with the cell suspension wait for exactly 20 s before initiating RLU measurement on the luminometer.
19. Protocols for optimizing light detection and determining the linear range of light detection for your luminometer, to avoid signal saturation, can be found in the manufacturer's instructions for the Luciferase assay system (Promega; Cat. No. E1500).

Acknowledgements

This work was supported by the Swiss National Science Foundation Fellowship No. P2ZHP3_151582 to V.D., and by funding from the Centre for Infectious Diseases and Microbiology—Public Health, ICPMR, Westmead Hospital to N.M.

References

1. Bugg TDH (1999) Bacterial peptidoglycan biosynthesis and its inhibition. In: Pinto M (ed) Comprehensive natural products chemistry. Elsevier, Oxford, pp 241–294
2. Jordan S, Hutchings MI, Mascher T (2008) Cell envelope stress response in Gram-positive bacteria. FEMS Microbiol Rev 32(1):107–146
3. Dengler V, Stutzmann Meier P, Heusser R, Berger-Bächi B, McCallum N (2011) Induction kinetics of the *Staphylococcus aureus* cell wall stress stimulon in response to different cell wall active antibiotics. BMC Microbiol 11:16
4. Gardete S, Wu SW, Gill S, Tomasz A (2006) Role of VraSR in antibiotic resistance and antibiotic-induced stress response in *Staphylococcus aureus*. Antimicrob Agents Chemother 50(10):3424–3434
5. Kuroda M, Kuroda H, Oshima T, Takeuchi F, Mori H, Hiramatsu K (2003) Two-component system VraSR positively modulates the regulation of cell-wall biosynthesis pathway in *Staphylococcus aureus*. Mol Microbiol 49(3): 807–821
6. McCallum N, Stutzmann Meier P, Heusser R, Berger-Bächi B (2011) Mutational analyses of open reading frames within the *vraSR* operon and their roles in the cell wall stress response of *Staphylococcus aureus*. Antimicrob Agents Chemother 55(4):1391–1402
7. Sobral RG, Ludovice AM, de Lencastre H, Tomasz A (2006) Role of murF in cell wall biosynthesis: isolation and characterization of a murF conditional mutant of *Staphylococcus aureus*. J Bacteriol 188(7):2543–2553

8. Utaida S, Dunman PM, Macapagal D, Murphy E, Projan SJ, Singh VK, Jayaswal RK, Wilkinson BJ (2003) Genome-wide transcriptional profiling of the response of *Staphylococcus aureus* to cell-wall-active antibiotics reveals a cell wall stress stimulon. Microbiology 149(Pt 10):2719–2732
9. Cui L, Neoh HM, Shoji M, Hiramatsu K (2009) Contribution of vraSR and graSR point mutations to vancomycin resistance in vancomycin-intermediate *Staphylococcus aureus*. Antimicrob Agents Chemother 53(3): 1231–1234
10. Kato Y, Suzuki T, Ida T, Maebashi K (2010) Genetic changes associated with glycopeptide resistance in *Staphylococcus aureus*: predominance of amino acid substitutions in YvqF/VraSR. J Antimicrob Chemother 65(1):37–45
11. Mehta S, Cuirolo AX, Plata KB, Riosa S, Silverman JA, Rubio A, Rosato RR, Rosato AE (2011) VraSR two-component regulatory system contributes to *mprF*-mediated decreased susceptibility to daptomycin in vivo-selected clinical strains of methicillin-resistant *Staphylococcus aureus*. Antimicrob Agents Chemother 56(1):92–102
12. Yoo JI, Kim JW, Kang GS, Kim HS, Yoo JS, Lee YS (2013) Prevalence of amino acid changes in the *yvqF*, *vraSR*, *graSR*, and *tcaRAB* genes from vancomycin intermediate resistant *Staphylococcus aureus*. J Microbiol 51(2):160–165. doi:10.1007/s12275-013-3088-7
13. Blake KL, O'Neill AJ, Mengin-Lecreulx D, Henderson PJ, Bostock JM, Dunsmore CJ, Simmons KJ, Fishwick CW, Leeds JA, Chopra I (2009) The nature of *Staphylococcus aureus* MurA and MurZ and approaches for detection of peptidoglycan biosynthesis inhibitors. Mol Microbiol 72(2):335–343
14. Sobral RG, Jones AE, Des Etages SG, Dougherty TJ, Peitzsch RM, Gaasterland T, Ludovice AM, de Lencastre H, Tomasz A (2007) Extensive and genome-wide changes in the transcription profile of *Staphylococcus aureus* induced by modulating the transcription of the cell wall synthesis gene *murF*. J Bacteriol 189(6):2376–2391
15. Sengupta M, Jain V, Wilkinson BJ, Jayaswal RK (2012) Chromatin immunoprecipitation identifies genes under direct VraSR regulation in *Staphylococcus aureus*. Can J Microbiol 58(6):703–708. doi:10.1139/w2012-043
16. Pinho MG, de Lencastre H, Tomasz A (2001) An acquired and a native penicillin-binding protein cooperate in building the cell wall of drug-resistant staphylococci. Proc Natl Acad Sci U S A 98(19):10886–10891
17. Wang QM, Peery RB, Johnson RB, Alborn WE, Yeh WK, Skatrud PL (2001) Identification and characterization of a monofunctional glycosyltransferase from *Staphylococcus aureus*. J Bacteriol 183(16):4779–4785. doi:10.1128/JB.183.16.4779-4785.2001
18. Fan X, Liu Y, Smith D, Konermann L, Siu KW, Golemi-Kotra D (2007) Diversity of penicillin-binding proteins. Resistance factor FmtA of *Staphylococcus aureus*. J Biol Chem 282(48): 35143–35152
19. Oshida T, Sugai M, Komatsuzawa H, Hong YM, Suginaka H, Tomasz A (1995) A *Staphylococcus aureus* autolysin that has an *N*-acetylmuramoyl-L-alanine amidase domain and an endo-beta-*N*-acetylglucosaminidase domain: cloning, sequence analysis, and characterization. Proc Natl Acad Sci U S A 92(1):285–289
20. Kawai Y, Marles-Wright J, Cleverley RM, Emmins R, Ishikawa S, Kuwano M, Heinz N, Bui NK, Hoyland CN, Ogasawara N, Lewis RJ, Vollmer W, Daniel RA, Errington J (2011) A widespread family of bacterial cell wall assembly proteins. EMBO J 30(24):4931–4941
21. Over B, Heusser R, McCallum N, Schulthess B, Kupferschmied P, Gaiani JM, Sifri CD, Berger-Bächi B, Stutzmann Meier P (2011) LytR-CpsA-Psr proteins in *Staphylococcus aureus* display partial functional redundancy and the deletion of all three severely impairs septum placement and cell separation. FEMS Microbiol Lett 320(2):142–151
22. McCallum N, Spehar G, Bischoff M, Berger-Bächi B (2006) Strain dependence of the cell wall-damage induced stimulon in Staphylococcus aureus. Biochim Biophys Acta 1760(10):1475–1481
23. Boyle-Vavra S, Yin S, Jo DS, Montgomery CP, Daum RS (2013) VraT/YvqF is required for methicillin resistance and activation of the VraSR regulon in *Staphylococcus aureus*. Antimicrob Agents Chemother 57(1):83–95. doi:10.1128/AAC.01651-12
24. Dengler V, McCallum N, Kiefer P, Christen P, Patrignani A, Vorholt JA, Berger-Bächi B, Senn MM (2013) Mutation in the c-di-AMP cyclase *dacA* affects fitness and resistance of methicillin resistant *Staphylococcus aureus*. PLoS One 8(8):e73512. doi:10.1371/journal.pone.0073512
25. Dengler V, Stutzmann Meier P, Heusser R, Kupferschmied P, Fazekas J, Friebe S, Burger Staufer S, Majcherczyk PA, Moreillon P, Berger-Bächi B, McCallum N (2012) Deletion of hypothetical wall teichoic acid ligases in *Staphylococcus aureus* activates the cell wall stress response. FEMS Microbiol Lett 333(2):109–120

26. Campbell J, Singh AK, Swoboda JG, Gilmore MS, Wilkinson BJ, Walker S (2012) An antibiotic that inhibits a late step in wall teichoic acid biosynthesis induces the cell wall stress stimulon in *Staphylococcus aureus*. Antimicrob Agents Chemother 56(4):1810–1820
27. Rossi J, Bischoff M, Wada A, Berger-Bächi B (2003) MsrR, a putative cell envelope-associated element involved in Staphylococcus aureus sarA attenuation. Antimicrob Agents Chemother 47(8):2558–2564
28. Katayama Y, Zhang HZ, Chambers HF (2003) Effect of disruption of Staphylococcus aureus PBP4 gene on resistance to beta-lactam antibiotics. Microb Drug Resist 9(4): 329–336

Part V

Analysis of the Non-Protein Components of the Cell Wall

Chapter 12

Extraction and Analysis of Peptidoglycan Cell Wall Precursors

Elisa Binda, Lùcia Carrano, Giorgia Letizia Marcone, and Flavia Marinelli

Abstract

Extraction and analysis by LC-MS of peptidoglycan precursors represent a valuable method to study antibiotic mode of action and resistance in bacteria. Here, we describe how to apply this method for: (1) testing the action of different classes of antibiotics inhibiting cell wall biosynthesis in *Bacillus megaterium*; (2) studying the mechanism of self-resistance in mycelial actinomycetes producing glycopeptide antibiotics.

Key words Peptidoglycan precursors, Antibiotics, Glycopeptides, Lantibiotics, Ramoplanin, Bacitracin, LC-MS, *Bacillus megaterium*, *Actinoplanes teichomyceticus*, *Nonomuraea* sp. ATCC 39727

1 Introduction

The peptidoglycan (PG) cell wall is a unique macromolecule responsible for both shape determination and cellular integrity under osmotic stress in virtually all bacteria [1]. Notwithstanding the great diversity of shapes and sizes across the bacterial domain, PG is universally composed of sugar strands (glycans) cross-linked by short peptides. PG network consists in linear glycan chains alternating β-(1-4)-linked units of *N*-acetylglucosamine (GlcNAc) and *N*-acetylmuramic acid (MurNAc)-residues, which are cross-linked by a peptide bond between peptide subunits (Fig. 1). PG structure is highly conserved in bacteria, but some modifications occur in its chemical composition. Most of variations are related to the amino acid composition of the peptide stem and the method by which the glycan strands are linked [2, 3]. For example, in gram-negative bacteria and in the two groups of gram-positives described in this chapter (bacilli and actinomycetes), a *meso*-diaminopimelic acid (*m*DAP), instead of lysine, participates to the cross-linking between glycan strands. In some species, the peptide cross-link is direct, whereas in other there is an additional cross-bridge as a penta-glycine in *Staphylococcus aureus* [4] and a mono glycine in *Streptomyces* [5]. Modification such as *N*-deacetylation

Hee-Jeon Hong (ed.), *Bacterial Cell Wall Homeostasis: Methods and Protocols*, Methods in Molecular Biology, vol. 1440,
DOI 10.1007/978-1-4939-3676-2_12,

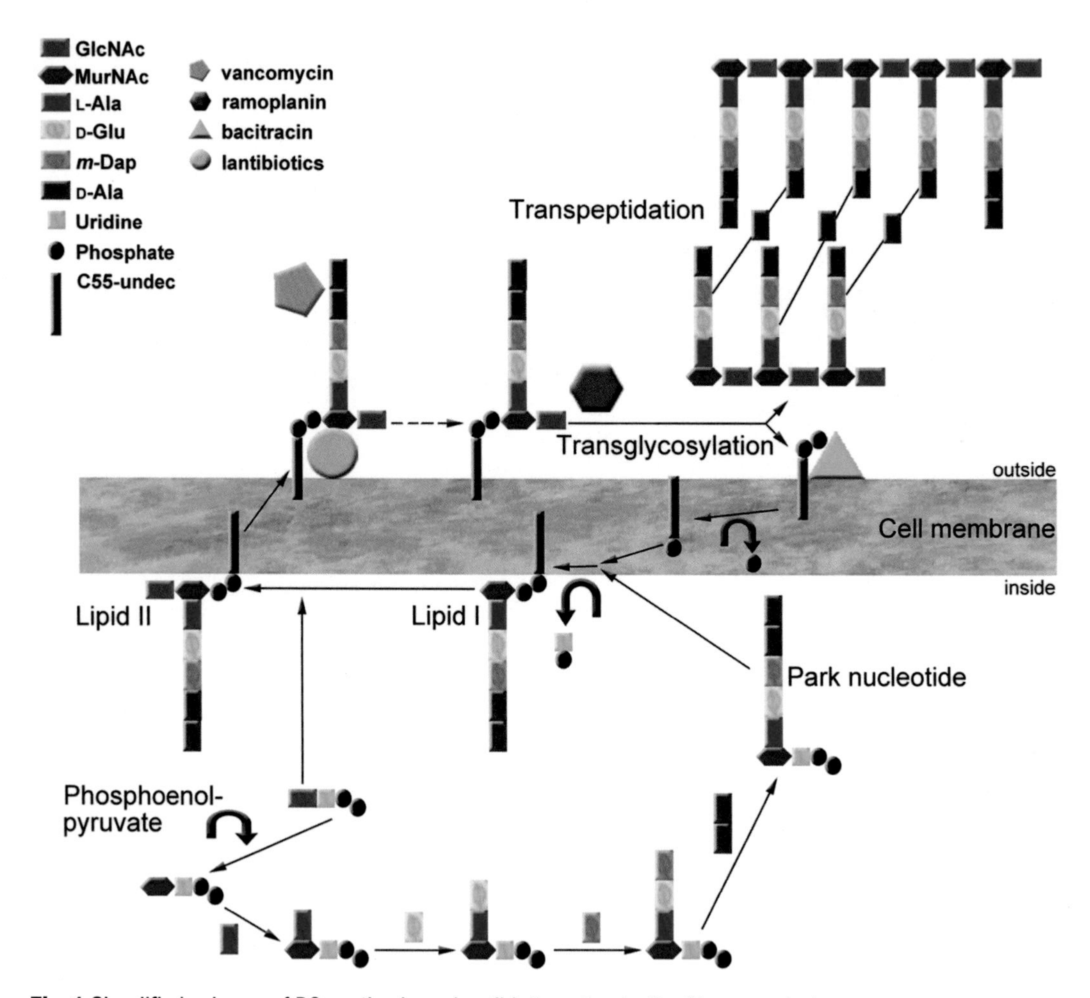

Fig. 1 Simplified scheme of PG synthesis and antibiotic action in *Bacillus megaterium*

and *O*-acetylation of the glycan strands in some bacterial species are reported to render them more resistant to the hydrolytic activity of lysozymes [6].

The biosynthesis of PG is a multistep process, which requires numerous enzymatic reactions, occurring in three compartments of a bacterial cell: the cytoplasm (synthesis of uridine diphosphate (UDP)-bond precursors, also named Park nucleotides), the inner face of the membrane (synthesis of the cell wall building block lipid II and lipid II modifications), and the outer face of the membrane (polymerization of lipid II into the growing PG). In detail, biosynthesis starts in the cytoplasm (Fig. 1), where the MurA-F ligases catalyze the formation of the ultimate soluble cell wall precursor UDP-MurNAc-pentapeptide [7]. In the following membrane associated step, UDP-MurNAc-pentapeptide is linked to the membrane carrier undecaprenol-phosphate (C_{55}-P) by the translocase MraY, resulting in the formation of lipid I (undecaprenylphosphate-MurNAc-pentapeptide). Then, the membrane-associated transferase

MurG links UDP-GlcNAc to the MurNAc moiety of lipid I, forming the lipid II (undecaprenylphosphate-GlcNAc-MurNAc-pentapeptide), that represents the central wall building block, which can be further modified strain dependently and in response to environmental conditions [8]. Lipid II is then transferred across the cytoplasmic membrane and incorporated into the growing PG network through the activity of the penicillin-binding proteins (PBPs) by transglycosylation and transpeptidase reactions, realizing C_{55}-PP, which after dephosphorylation enters a new synthesis cycle [9]. The standard substrate of PBPs is the terminal D-Ala-D-Ala residues of the peptide stems. The structural similarity between the D-Ala-D-Ala residues and the β-lactam antibiotics facilitate binding of these antibiotics to the active site of PBPs, with the β-lactam's nucleus of the molecule irreversibly binding to the catalytic PBP site [10]. This covalent binding typically leads to loss of cell shape and integrity, often causing cell death.

Several antibiotic chemical classes other than β-lactams, exert their action by inhibiting specific steps of the bacterial cell wall synthesis (Fig. 1) [8, 11]. Some of them are in clinical use for a long time, such as the glycopeptides vancomycin and teicoplanin [12], that block cell wall biosynthesis in gram-positive bacteria by binding to the D-Ala-D-Ala dipeptide terminus of the PG precursors, when they are translocated onto the external face of cytoplasmic membrane, thereby sequestering the substrate for the PBPs. The specific interaction between vancomycin and the C-terminal D-Ala-D-Ala dipeptide of the PG precursor is mediated by five hydrogen bonds and additional hydrophobic interactions [13].

As reported in this chapter, a method for extracting and analyzing PG precursors may be useful for studying the mode of action of such antibiotic classes that, affecting the late stages of cell wall biosynthesis, bring to the accumulation of the cytoplasmic PG precursors (Fig. 1). In the following paragraphs we use: (1) vancomycin, (2) ramoplanin, which is a lipodepsipeptide antibiotic in Phase II of clinical development [14] whose mechanism of action is not yet completely clarified [15]; (3) bacitracin, a small circular metallopeptide that blocks C_{55}-P dephosphorylation, and thus hampers the recycling of the membrane carrier [16]; and two more recently discovered lantibiotics (microbisporicin and planosporicin), that are active on vancomycin resistant gram-positives and whose mechanism of action is currently intensively investigated [17–21].

The same method can be applied for investigating how bacteria evolve resistance to those antibiotic classes that inhibit cell wall biosynthesis. Resistance to glycopeptides is greatly studied in enterococci and staphylococci, where the expression of genes (named *van*) encoding proteins reprogram cell wall biosynthesis and thus evade the action of the antibiotics. The detailed mechanism of *van* gene-mediated glycopeptide resistance in enterococci was elucidated by Courvalin, Walsh, and their coworkers in the 1990s [22–24]. In the two most prominent manifestations of resistance (VanA and VanB phenotypes), the PG precursor is remodeled to the terminal D-Ala-D-Lac, incorporating an ester linkage in place of the amide of the D-Ala-D-Ala. The replacement of a dipeptide with a depsipeptide removes one of the hydrogen bonding interactions and leads to lone pair–lone pair repulsion, reducing of 1000-fold the affinity of GPAs to their molecular target, and resulting in a corresponding 1000-fold loss in antimicrobial activity [23, 25]. In these organisms, Arthur et al. [26] reported the accumulation of PG precursors ending in D-Ala-D-Lac. Moreover, an increased transcription of the *van* genes was associated with increased incorporation of D-Ala-D-Lac into peptidoglycan precursors to the detriment of D-Ala-D-Ala, and with a gradual increase in the vancomycin-resistance levels. Hong et al. [5] adapted a similar method to the nonpathogenic, non-glycopeptide-producing actinomycete *Streptomyces coelicolor*, where Van enzymes reprogram the cell wall such that precursors terminate in D-Ala-D-Lac rather than D-Ala-D-Ala, thus conferring resistance to vancomycin.

As explained below, we have modified our method used for detecting PG precursors in *Bacillus megaterium*, adapting it to those uncommon actinomycetes producing glycopeptides. In detail we report on *A. teichomyceticus* (*Micromonosporaceae* family) that produces teicoplanin [27] and on the *Nonomuraea* sp. ATCC 39727 (*Streptosporangiaceae* family), the producer of A40926 [28] which is the precursor of the second generation glycopeptide Dalvance (approved by Food and Drug Administration on May 2014) [29]. The main technical issue is due to the complex cell life-cycle of these microorganisms, in fact they form spores, vegetative and aerial mycelium and especially aggregate in pellets when cultivate in liquid cultures. The onset of morphological differentiation generally coincides with the production of glycopeptides [30]. Antibiotic-producing actinomycetes possess mechanisms to avoid suicide by their own toxic products by generating modified PG precursors [12, 31, 32]. The investigation of these mechanisms in the producing strains could improve the knowledge about the novel strategies of resistance that might emerge in environmental and pathogenic bacteria, due to the exposure to old and new glycopeptides, and provide new insights into the evolution of resistant determinants. It may thus contribute to an early warning system for emerging resistance mechanisms [33].

2 Materials

When ultrapure water is used, it is prepared from deionized water to attain a sensitivity of 18 MΩ cm at 25 °C. Prepare and store all the reagents at room temperature unless indicated otherwise.

2.1 Strains

Bacillus megaterium ATCC 13632, *Nonomuraea* sp. ATCC 39727, and *Actinoplanes teichomyceticus* ATCC 31121 are from American Type Culture Collection (ATCC). Strains are conserved in Working Cell Banks (WCBs), i.e., 2 mL cryo-vials at −80 °C containing 1 mL cultures growing exponentially in the media described below.

2.2 Cultivation Media

1. For *B. megaterium*, prepare Difco Mueller Hinton Broth (MHB) according to the supplier indications.
2. For *A. teichomyceticus*, prepare MS medium [31] following this protocol: weight 10 g glucose, 4 g Bacto peptone, 4 g Bacto yeast extract, 0.5 g $MgSO_4{\cdot}7H_2O$, 2 g KH_2PO_4, 4 g K_2HPO_4 (*see* **Note 1**). Mix all the components and add deionized water up to 1 L; adjust pH to 7.4 with NaOH and sterilize in autoclave for 20 min at 120 °C.
3. For *Nonomuraea* sp., prepare MVSP medium [34] following this protocol: weight 24 g soluble starch, 1 g dextrose, 3 g meat extract, 5 g yeast extract, 5 g tryptose, 0.5 g L-proline, 50 g sucrose (*see* **Note 1**). Mix all the components and add deionized water up to 1 L; adjust pH to 7.4 with NaOH and sterilize in autoclave for 20 min at 120 °C.

2.3 Antibiotics and Solution Preparation

1. Weigh vancomycin, ramoplanin, bacitracin, planosporicin [17] and microbisporicin [18, 19] and dissolve the powders in ultrapure water at the concentration of 1–5 mg/mL. Filter these solutions with a syringe and a 0.2 μm (7 bar maximum of pressure) filter to guarantee the sterility of the solutions. Store antibiotic solutions at −20 °C.
2. Samples of UDP-MurNAc-L-Ala-D-Glu-*meso*-Dap-D-Ala-D-Ala and UDP-MurNAc-L-Ala-D-Glu-*meso*-Dap-D-Ala-D-Lac are obtained from the UK-BaCWAN PG precursor synthesis facility (University of Warwick) [35] and used as standards. Weigh powders and dissolve them in ultrapure water at the final concentration of 10 mg/mL. Store standard solutions at −20 °C.

2.4 Instruments

1. Sonics VibraCell VCX 130 (Sonics and Materials) is used to sonicate mycelium (*see* **Note 2**).
2. Zeiss Primo Star Optical Microscope (Zeiss), lens 40× and optical magnification 400× is used for checking cell and mycelium morphology.
3. LC-MS and MS/MS experiments are conducted using a ThermoQuest Finnigan LCQ Deca mass detector equipped with an ESI interface and a Thermo Finnigan Surveyor MS pump, PDA detector (UV6000, Thermo Finnigan).

3 Methods

Carry out all the procedures at room temperature unless otherwise specified.

3.1 Cultivation of Microbes and Extraction of PG Precursors

3.1.1 Cultivation of B. megaterium and Extraction of PG Precursors

1. Thaw cryo-vials from the Working Cell Bank (WCB) at room temperature, and use 0.5 mL to inoculate 100 mL MHB medium in 500 mL Erlenmeyer baffled flasks.
2. Incubate flasks at 37 °C with shaking at 200 rotations per min (r.p.m.).
3. Sample (1 mL) the culture each 30 min, centrifuge at 6000 × *g* for 10 min, collect the supernatant, and measure the optical density (O.D.) at 540 nm using a spectrophotometer.
4. When the $O.D._{540\ nm}$ reaches ca. 0.7, add 10 or 20 μg/mL (final concentration) of each antibiotic solution (planosporicin, microbisporicin, ramoplanin, vancomycin) (*see* **Note 3**). Control cells are incubated with an equivalent volume of water.
5. Incubate flasks for further 60 min at 37 °C and 200 r.p.m. (*see* **Note 3**).
6. Harvest cells by centrifugation at 12,000 × *g* for 10 min and collect them in ice.
7. Suspend collected cells (0.1 g fresh weight per mL) in deionized water and boil for 20 min.
8. After cooling first at room temperature and then in ice, centrifuge the suspension at 39,000 × *g* for 30 min. Completely lyophilize the supernatant and dissolve the powder in 0.1 volumes of water adjusted to pH 3 with formic acid (*see* **Note 4**).

3.1.2 Cultivation of A. teichomyceticus and Nonomuraea sp. and Extraction of PG Precursors

1. Thaw one cryo-vial from the WCB at room temperature, and use 0.5 mL to inoculate 10 mL of the cultivation medium in 100 mL Erlenmeyer baffled flasks.
2. After 72 h incubation at 28 °C, use 5 mL to inoculate each 300 mL Erlenmeyer flask containing 50 mL cultivation medium (*see* **Note 5**).
3. Incubate flasks at 28 °C, with shaking at 200 r.p.m. for 72 h (*see* **Note 5**).
4. Add bacitracin or ramoplanin at 100 or 150 μg/mL (final concentration) respectively, and incubate the cultures for further 90 min at 28 °C and 200 r.p.m. (*see* **Note 6**). Control cells are incubated with an equivalent volume of water.
5. Harvest mycelium by centrifugation 10,000 × *g* for 10 min, washed the pellet twice with deionized water and resuspend it in polypropylene tubes.
6. Sonicate (*see* **Note 7**) the washed mycelium in the following conditions: Power 130 W, 230 VV, 50–60 Hz—Frequency

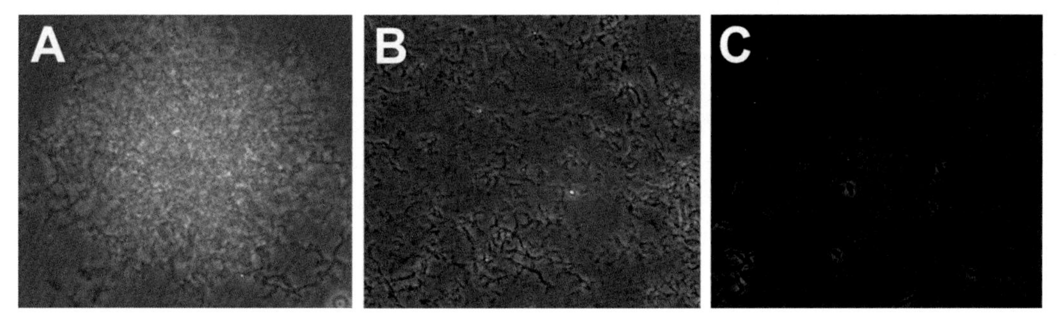

Fig. 2 (**a**) Mycelium of *Nonomuraea* sp. ATCC 39727 growing in MV liquid media. (**b**) Mycelium of *Nonomuraea* sp. ATCC 39727 after few cycles of sonication: most of the mycelium is still aggregated even if in a more dispersed mode. (**c**) Completely sonicated mycelium, no aggregates are visible and the preparation is ready for PG precursor extraction. Pictures by the optical microscope Zeiss Primo Star lens 40× and optical magnification 400×

20 Hz—Amplitude 90% (90% of 60 Hz) for a minimum of 5 min. During the sonication, keep the samples in ice, and perform sonication cycles of 30 s, followed by 10 s of interval. Check the population morphology by microscope analyses (*see* **Note** 7 and Fig. 2).

7. Collect the sonicated mycelium by centrifugation 12,000 × *g* for 20 min and resuspend it in deionized water (0.2 g of wet weight per mL). Boil for 20 min.
8. Centrifuge the suspension at 39,000 × *g* for 60 min. Completely lyophilize the supernatant and dissolve the powder in 0.1 volumes of water adjusted to pH 3 with formic acid.

3.2 LC and MS-MS Analysis

1. Analyze the antibiotic-treated samples containing the PG precursors in parallel with the control samples by Liquid Chromatography–Mass Spectrometry (LC-MS) using a C18 250 × 4.6 mm column (Phenomenex Luna, Torrance, CA) 5-μm particle size, eluted at a flow rate of 1 mL/min with 2 min with 100% phase A and then a 50 min linear gradient to 100% phase B. Recommended phase A is composed of 2% CH_3CN, 97.9% H_2O, 0.1% HCOOH (v/v/v); and phase B is composed of 95% CH_3CN, 4.915% H_2O, and 0.085% HCOOH (v/v/v). The recommended column temperature is 22 °C.
2. Register the chromatographic UV absorption profile by a photo diode array detector (PDA) and in the meantime split one-fifth of the detector elution flow into the mass spectrometer. The effluent from the column is splitted in a ratio 5:95, the majority (ca. 950 μL/min) diverted to PDA detector, the remaining 50 μL/min are diverted into the mass spectrometer (*see* **Note 8**).
3. Achieve the MS spectra by electrospray ionization (both in positive and negative mode) of the peaks present in the antibiotic treated samples but not in the control ones, under the following conditions: (1) Sample Inlet Conditions, capillary temperature 250 °C; sheath gas (N_2), 80 LCQ arbitrary units; auxiliary

gas (N_2), 20 LCQ arbitrary units; (2) Sample Inlet Voltage Settings: positive polarity 4.5 kV, negative polarity 2.8 kV; capillary voltage 4 V; and tube lens offset 30 V. Nitrogen is used as sheath gas and auxiliary gas. Helium is used as the buffer and collision gas (*see* **Note 8**).

4. Perform MS/MS analyses of the selected peaks at collision energy levels ranging from 25 to 50 kV and acquire all the spectra in the 150–2000 mass unit range (*see* **Note 9**).

3.3 Data Analysis

3.3.1 Effect of Different Antibiotics on the PG Precursor Synthesis in B. megaterium

1. Figure 3 shows MS traces (A, B and C) of the PG precursor pool chromatographic profile in the planosporicin-(A) and vancomycin-(B) treated, or in untreated-(C) control cells of *B. megaterium*. A major peak is detectable in planosporicin- and vancomycin-treated cells at the retention time of 13.4 min (indicated by the arrow) (*see* **Note 10**).

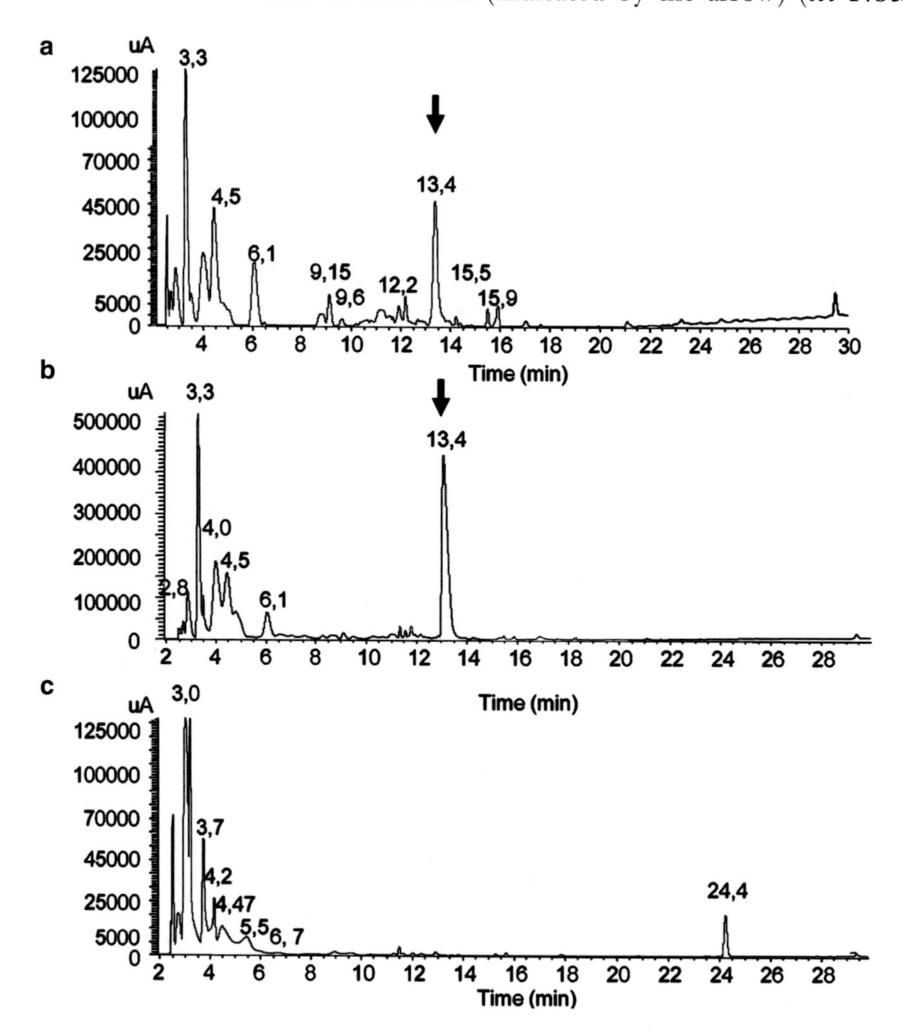

Fig. 3 LC-MS traces (**a**, **c**, **d**) from the analyses of the PG precursor pool in the planosporicin-(**a**) and vancomycin-(**b**) treated, or untreated-(**c**) cells of *B. megaterium* ATCC 13632. Adapted with permission from Ref. [17]. Copyright (2015) American Chemical Society

The other peaks present on the HPLC plots show MS profiles not related to soluble PG precursors.

2. Figure 4 shows the full mass spectrum in negative and positive mode of the peak eluted at 13.4 min. Molecular ions $[M\text{-}H]^{+}$ at *m/z* 1194.2 and at $[M\text{-}H]^{-}$ at *m/z* 1192.1 correspond to the UDP-MurNAc-L-Ala-D-Glu-*m*Dap-D-Ala-D-Ala, double-charged $[M\text{-}H]^{2-}$ at *m/z* 595.7. The ion at *m/z* 403.1 corresponds to UDP formed directly in the MS source (*see* **Note 9**).
3. The data reported in Figs. 3 and 4 indicate that the recently discovered lantibiotic planosporicin [17] as well as the glycopeptide vancomycin cause accumulation of UDP-MurNAc-pentapeptide precursors in growing cells of *B. megaterium*. Upon treatment with planosporicin or with vancomycin, a marked increase in a LC-MS peak is revealed, that cannot be detected in untreated control cells (Fig. 3). The corresponding

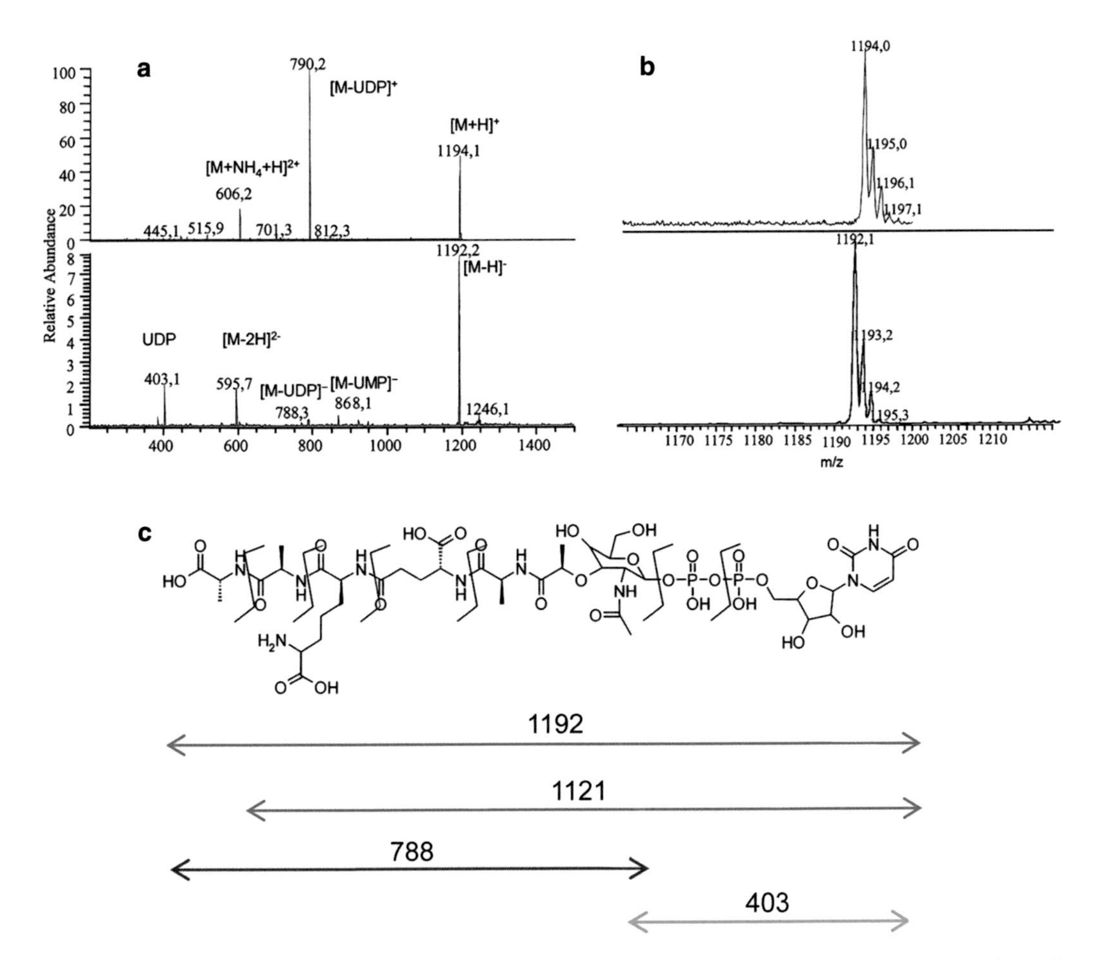

Fig. 4 (**a**) Full-scan mass spectrum in negative and positive mode of the peak eluted at 13.4 min shown in Fig. 3; (**b**) Zoom scan of the quasi-molecular ions in positive and negative ion mode; (**c**) chemical structure of the UDP-MurNAc-L-Ala-D-Glu-*m*Dap-D-Ala-D-Ala and structural assignments for the main fragment ions. Adapted with permission from Ref. [17]. Copyright (2015) American Chemical Society

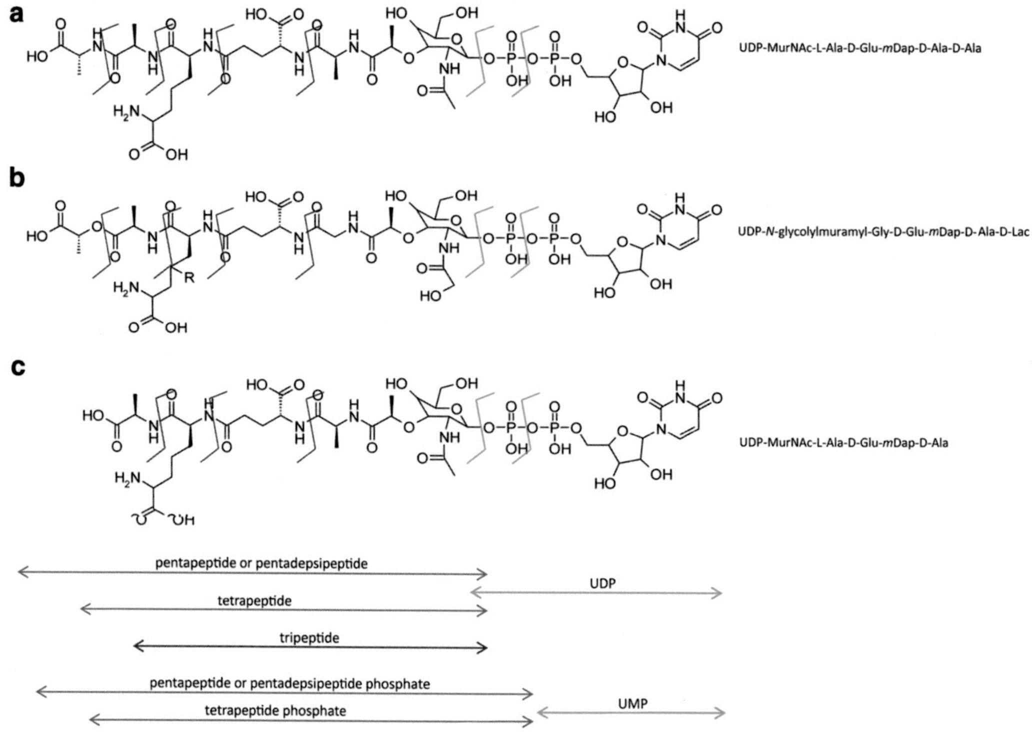

Fig. 5 Chemical structures of PG precursors *in B. megaterium* ATCC 13632 (**a**), *A. teichomyceticus* ATCC 31121 (**b**) and in *Nonomuraea* sp. ATCC 39727 (**c**). In (**b**), chemical structure of the UDP-*N*-glycolylmuramyl-Gly-D-Glu-*m*Dap-D-Ala-D-Lac (R=H) or of the UDP-*N*-glycomuramyl-Gly-D-Glu-*m*-3-hydroxy-Dap-D-Ala-D-Lac (R=OH) and structural assignments for the main fragment ions. In (**c**), chemical structure of the UDP-*N*-acetyl-muramyl-L-Ala-D-Glu-LL-Dap-D-Ala corresponding to the tetrapeptide stem-PG precursor and structural assignments for the main fragment ions

mass spectrum is determined in positive and negative ion mode, as shown in Fig. 4 (*see* **Note 8**): the quasi-molecular ionic area is expanded. The ions at *m/z* 1194.1 and 790.2 are attributed to $[M+H]^+$ and $[M\text{-}UDP]^+$, the ion at *m/z* 606.2 is attributed to the double charged species $[M+NH_4+H]^+$, those at *m/z* 1192.2, 595.7, and 403.1 to $[M\text{-}H]^-$, $[M\text{-}2H]^{--}$ and $[UDP]^-$, respectively, indicating the presence of a compound with a monoisotopic molecular mass of 1193 and containing the UDP moiety. This is consistent with the UDP-MurNAc-L-Ala-D-Glu-*m*Dap-D-Ala-D-Ala structure of *B. megaterium* PG precursor [17] shown in Figs. 4c and 5a. Further MS/MS studies confirm the fragmentation indicated in Fig. 4c (*see* **Note 9**). As reported in Table 1, the fragment ion at *m/z* 1121, in negative mode, corresponds to the loss of the alanine residue (71 uma) and those at *m/z* 868 and 788 to the loss of UMP and UDP, respectively, while the ion at *m/z* 403 corresponds to $[UDP]^-$ (*see* **Notes 9**).

Table 1
Signal detected in the MS/MS fragmentation pattern of the PG precursor in *B. megaterium*

$[M+H]^-$	$[M+H]^+$	Sequence of the peptidoglycan precursor
1192	1194	D-Ala-D-Ala-*m*Dap-L-Glu-L-Ala-*N*-acetylmuramyl-UDP
1121	1123	D-Ala-*m*Dap-L-Glu-L-Ala-*N*-acetylmuramyl-UDP
868		D-Ala-D-Ala-*m*Dap-L-Glu-L-Ala-*N*-acetylmuramyl-P
788	790	D-Ala-D-Ala-*m*Dap-L-Glu-L-Ala-*N*-acetylmuramyl
403		UDP

MS/MS analysis are performed on the molecular ion at the ionization energy of 30 eV

By using ramoplanin [17, 36] or another potent recently discovered lantibiotic named microbisporicin [19], similar results can be obtained (*see* **Note 10**). The accumulation of the UDP-MurNAc-L-Ala-D-Glu-*m*Dap-D-Ala-D-Ala in *B. megaterium* cells, treated with the use of antibiotics is consistent with the data reported by other authors in a variety of bacteria treated with PG biosynthesis inhibitors, acting at steps subsequent to the synthesis of UDP-MurNAc pentapeptide in the cytoplasm, such as ramoplanin, vancomycin, teicoplanin, bacitracin [26, 31, 34, 36], mannopeptimycins [37], and type B lantibiotics mersacidin and actagardine [38, 39].

3.3.2 PG Precursor Modification in Glycopeptide Producing Actinomycetes

The identification of the modified PG precursors in glycopeptide producing actinomycetes is achieved following the same data analysis procedure described above and reconstructing the chemical structures by comparing the LC-MS data with those reported for the PG precursor in *B. megaterium* (Fig. 5). LC-MS analysis of the PG precursors in ramoplanin-treated mycelium of *A. teichomyceticus* shows a major peak eluting at 17.8 min [31] (*see* **Note 11**). As untreated cells do not exhibit this peak, it thus appears related to a PG precursor that accumulates as the result of the effect of ramoplanin. The positive ion ESI-MS spectrum corresponding to this peak reveals the presence of two quasi-molecular ions $[M-H]^+$, at *m/z* 1196.9 and 1212.9. The negative ion spectrum shows quasi-molecular ions $[M-H]^-$, at *m/z* 1195.2 and 1211.0 and the double charged ions $[M-2H]^2$, at *m/z* of 597.1 and 605.1, in agreement with the presence of two molecular species with the molecular formulas $C_{40}H_{62}N_8O_{30}P_2$ and $C_{40}H_{62}N_8O_{31}P_2$ and corresponding to a calculated monoisotopic mass of 1196.30 and 1212.30 Da, respectively [31] (*see* **Note 9**). The UDP-*N*-glycolylmuramyl pentadepsipeptide structure UDP-*N*-glycolylmuramyl-Gly-D-Glu-*m*Dap-D-Ala-D-Lac is attributed to the species showing quasi-molecular ion, $[M-H]^-$ at *m/z* 1195.2 (Fig. 5b). This depsipeptide contains the D-Ala-D-Lac terminus characteristic of the vancomycin resistance phenotype described in enterococci and streptococci [22–25] and a remaining tripeptide part linked to a modified muramyl moiety as reported in literature [40, 41]. In fact, in *Actinoplanes* species and in some other actinomycetes, the glycolyl group is present instead of acetyl one in muramic acid and the composition of the peptide unit is glycine, glutamic acid, *meso*-diaminopimelic acid (or *meso*-3-hydroxy-diaminopimelic acid), and alanine [40] (*see* **Note 12**). The difference of 16 mass units between the molecular mass of the two species which co-elute by LC/MS at 17.8 min [31] suggests the presence of a hydroxy group on the UDP-*N*-glycolylmuramyl-Gly-D-Glu-*m*Dap-D-Ala-D-Lac, probably on the *meso*-diaminopimelic acid (indicated by R in Fig. 5b), as also reported for *Actinoplanes* species and in other actinomycetes [40, 41]. Additional evidence for the UDP-*N*-glycolylmuramyl-Gly-D-Glu-*m*Dap-D-Ala-D-Lac structure is provided by MS/MS analysis in the negative-ion mode of the quasi-molecular ion $[M-H]^-$ corresponding to the depsipeptide terminating PG precursor mentioned above (*see* **Note 9**). Significant fragments are highlighted in Fig. 5b and Table 2. Fragmentation of the molecular ion at *m/z* 1195.2 gives a product ion at *m/z* 1123.1 corresponding to the loss of 72 mass unit equivalent to a lactate (Lac) fragment and thus corresponding to the sequence UDP-*N*-glycolylmuramyl-Gly-D-Glu-*m*Dap-D-Ala. The fragmentation ions at *m/z* 871.2 and 791.1 correspond to the loss of UMP or UDP from the UDP-muropentadepsipeptide, giving P-*N*-glycolylmuramyl-Gly-D-Glu-*m*Dap-D-Ala-D-Lac and *N*-glycolylmuramyl-Gly-D-Glu-*m*Dap-D-Ala-D-Lac, respectively. The ion at *m/z* 403.0 corresponds to UDP, while the ions at *m/z* 1105.2 and 1087.3 are attributed to losses of lactic acid and lactic acid plus water, respectively, from the quasi-molecular ion. Among the ions with relative abundance of less than 30%, it is possible to assign those at *m/z* 951.2 and 799.3,

Table 2
Signal detected in the MS/MS fragmentation pattern of the PG precursors in *A. teichomyceticus*

$[M-H]^-$	Sequence of the peptidoglycan precursor
1195	D-Lac-D-Ala-*m*Dap-L-Glu-Gly-glycolylmuramyl-UDP
1123	D-Ala-*m*Dap-L-Glu-Gly-glycolylmuramyl-UDP
951	D-Lac-D-Ala-*m*Dap-L-Glu-Gly-glycolylmuramyl-PP
871	D-Lac-D-Ala-*m*Dap-L-Glu-Gly-glycolylmuramyl-P
799	D-Ala-*m*Dap-L-Glu-Gly-glycolylmuramyl-PP
791	D-Lac-D-Ala-*m*Dap-L-Glu-Gly-glycolylmuramyl
403	UDP

MS/MS analysis are performed on the molecular ion in negative mode, the ionization energy is 30 eV

which are attributed to the loss of uridine and UMP plus lactate, respectively. No traces of D-Ala-D-Ala UDP-*N*-glycolylmuramylpentapeptide (expected $[M\text{-}H]^-$ at *m/z* 1194) are detectable, indicating that *A. teichomyceticus* produces PG precursors resistant to the action of glycopeptide antibiotics (*see* **Note 13**). A glycopeptide-resistant cell wall biosynthesis is playing a crucial role in the teicoplanin producer *A. teichomyceticus,* as well as in other glycopeptide producers [32, 42, 43], contributing to avoiding self-inhibition during antibiotic production [12].

LC-MS analysis of PG precursors in bacitracin-treated cells of *Nonomuraea* sp. ATCC 39727 reveals two molecular species [32, 34, 43] (*see* **Note 14**). As untreated cells do not exhibit these peaks, they are related to the PG precursors that accumulate as the result of the effect of bacitracin. The comparison of LC–MS analyses between untreated and bacitracin-treated cells show in the latter case the occurrence of peaks eluting at 12.56 and 13.40 min (*see* **Note 15**). In correspondence of the peak eluting at 13.40 min, full scan mass spectrum in negative ion current shows quasi-molecular ions $[M\text{-}H]^-$ and $[M\text{-}2H]^{-2}$ at *m/z* 1192.3 and 595.7, respectively, corresponding to a calculated monoisotopic mass of 1193 (*see* **Note 8**). This monoisotopic mass is consistent with the presence of a UDP-MurNAc-pentapeptide precursor with structure UDP-MurNAc-L-Ala-D-Glu-*m*Dap-D-Ala-D-Ala (Fig. 5a), identical to the precursor described above for *B. megaterium.* MS–MS analysis confirm this identity (*see* **Note 9**) [32, 34, 43]. Interestingly, full scan mass spectrum in negative and positive ion current of the peak eluting at 12.56 min shows quasi-molecular ions $[M\text{-}H]^-$, $[M\text{-}H]^+$, and $[M\text{-}2H]^{-2}$ at *m/z* 1121.1, 1123.0, and 560.3, respectively, corresponding to a calculated monoisotopic mass of 1122. The structure attributable to the accumulated precursor is UDP-MurNAc-L-Ala-D-Glu-*m*Dap-D-Ala and corresponds to the UDP-MurNAc-L-Ala-D-Glu-*m*Dap-D-Ala-D-Ala depleted of the terminal D-Ala (UDP-MurNAc-tetrapeptide peptidoglycan precursor) (Fig. 5c) [32, 34]. Additional evidence for the structure UDP-MurNAc-L-Ala-D-Glu-*m*Dap-D-Ala is directly provided by the formation in the MS source of ions at *m/z* 719.2 (positive mode) corresponding to the loss of UDP and yielding MurNAc-L-Ala-D-Glu-*m*Dap-D-Ala, and 403.0 (negative mode) corresponding to UDP. The detection of the UDP-MurNAc tetrapeptide PG precursors in *Nonomuraea* sp. ATCC 39727 indicates the involvement of a novel mechanism of glycopeptide self-resistance in this actinomycete, described by Marcone et al. [32, 43], that is based on the conversion of the pentapeptide to tetrapeptide stem in the PG precursor, which has a lower affinity for glycopeptides [44]. No UDP-linked depsipeptide containing a D-Lac in the terminal position of the UDP precursor is detectable in *Nonomuraea* sp. extracts, indicating that the PG precursors in this actinomycete (producing the teicoplanin-like A40926) are different from those described above in the teicoplanin producer *A. teichomyceticus.*

4 Notes[1]

1. Actinomycetes as *Nonomuraea* sp. and *A. teichomyceticus* are usually cultivated in rich and complex industrial media containing insoluble component such flours or very viscous ones such as molasses and syrups. For the extractions of PG precursors, the use of limpid media is recommended to avoid interference, especially during the step of sample boiling. Please consider that the addition of sucrose and proline to the *Nonomuraea* sp. medium is not mandatory, but it favors mycelium dispersion. The same medium without the addition of sucrose and proline is named MV [32, 45].
2. As reported below in the method section, sonication is needed only for those microbes (filamentous actinomycetes) that form compact mycelium aggregates in liquid cultures.
3. Time of antibiotic addition corresponds to the early exponential growth phase. The rationale behind is that cells should be in the phase of actively synthesizing cell wall. Antibiotic concentration and time of incubation are defined by measuring the reduction of growth after antibiotic exposure. PG precursors are accumulated into cells retaining some degree of vitality/integrity. Approximately a tenfold reduction of the colony forming units (CFUs) and/or one third reduction of the O.D.$_{540\ nm}$ is recommended.
4. The protocol of PG precursor extraction is adapted from the one indicated by Kohlrausch and Höltje [46].
5. Filamentous actinomycetes grow slower than bacilli, thus two steps of inoculation and longer times of flask incubation are needed to reach the exponential growth phase.
6. As reported in **Note 3**, time of antibiotic addition corresponds to the early exponential growth phase. In the case of filamentous growing actinomycetes, microbial population growth is not measurable by counting CFUs or detecting O.D.$_{540\ nm}$. For establishing the correct antibiotic dose and time of incubation, 2 mL of the treated population are inoculated into 50 mL fresh medium without antibiotics: typical mycelium growth (checked at the optical microscope) should be restored within 24 h.
7. Filamentous actinomycetes form hyphae that aggregate in pellets in liquid cultures (Fig. 2a), whose density, size, and morphology vary depending both on the strain and on the cultivation conditions. Agglomerates should be dispersed before extraction. Sonication protocol should be adapted

[1]Dalmastri C, Gastaldo L, Marcone GL, Binda E, Congiu T, Marinelli F (2016) Classification of Nonomuraea sp. ATCC 39727, an actinomycete that produces the glycopeptide antibiotic A40926, as Nonomuraea gerenzanensis sp. nov. Int J Syst Evol Microbiol 66(2):912–921. doi:10.1099/ijsem.0.000810

accordingly, avoiding excessive lysis and protein denaturation. Checking the progressive disaggregation of pellets by optical microscopy is recommended (*see* Fig. 2).

8. Chromatographic profile is firstly obtained by recording UV or also UV–Visible traces. If a PDA detector is not available, following the absorbance at λ 260 or λ 210 nm is recommended: the uracil moiety of PG precursor absorbs at λ 260 nm, whereas the peptide moiety absorbs at 210 nm. Anyhow, the detection by UV of PG precursors is indicative but not diagnostic, and too many peaks are detectable in these chromatographic profiles due to the complexity of the extract mixture. On a contrary, MS traces, recorded both in negative and positive mode, better identify compounds by the molecular weight.
9. Identification of PG precursor is possible by MS/MS analysis performed on the precursor molecular ions identified in the MS traces. Particularly important is to annotate the fragmentation of the molecular ion: diagnostic signals are those corresponding to the UDP phosphate and to the peptide moiety without the UDP.
10. Results similar to those illustrated with planosporicin and vancomycin are achieved treating *B. megaterium* cells with ramoplanin or microbisporicin at the antibiotic concentrations reported in Subheading 3.1.1. The retention time of the peak corresponding to the UDP-MurNAc-L-Ala-D-Glu-*m*Dap-D-Ala-D-Ala may vary, according to the analytical conditions. The area of the peak corresponding to the UDP-MurNAc-L-Ala-D-Glu-*m*Dap-D-Ala-D-Ala is variable and it might depend on the used antibiotic class, concentration and time of exposure. Lantibiotics, glycopeptides, and ramoplanin in fact inhibit different steps of cell wall biosynthesis [8]. We have observed that the area of the peak is greater if the cells are treated with ramoplanin and vancomycin rather than planosporicin and microbisporicin [17, 19].
11. Since *A. teichomyceticus* produces the glycopeptide teicoplanin, it is preferable using an antibiotic with a mechanism of action differing from glycopeptides. Lantibiotics are not further used since they are apparently less efficacious in causing PG precursor accumulation (*see* **Note 10**); in addition, they are not commercially available and should be prepared by fermentation and extraction of active molecules [17–19]. Indeed, in *A. teichomyceticus*, ramoplanin can be replaced by bacitracin, which blocks dephosphorylation of undecaprenyl-phosphate resulting in the accumulation of PG precursors [16]. We have introduced the use of bacitracin following the protocol reported by Schäberle et al. [42]. In addition, bacitracin is less expensive than ramoplanin.

12. It is worthy of noting that the chemical composition of the cell wall is a diagnostic feature for the polyphasic taxonomy of actinomycetes [40, 41, 47].
13. Glycopeptide producing actinomycetes possess mechanisms to avoid suicide by their own toxic products [31, 42–44]. Several lines of evidences indicate that the producing actinomycetes may represent the evolutionary source of the resistance genes emerging in pathogens under the selective pressure due to the increase use of antibiotics [12, 33, 48].
14. Similar PG precursor accumulation is detectable following exposure to ramoplanin, instead of bacitracin. Bacitracin use is indeed preferred for the reason stated in **Note 8**.
15. Measuring the area of the peaks eluting at 12.56 and 13.40 min, it is possible to calculate the ratio between the UDP-MurNAc-tetrapeptide and UDP-MurNAc-pentapeptide precursors. In MV and MVSP media, *Nonomuraea* sp. produces mostly UDP-MurNAc-tetrapeptide (88 %), with UDP-MurNAc-pentapeptide present in much lower amounts (12 %). The ratio of the tetrapeptide to pentapeptide stem of the PG precursor might be changed by manipulating the culture conditions and introducing specific mutations affecting cell wall biosynthesis and the resistance phenotype [32, 43].

Acknowledgement

We thanks Fabrizio Beltrametti and Enrico Selva for their contribution in developing the protocols described in this chapter.

References

1. Höltje JV (1998) Growth of the stress-bearing and shape-maintaining murein sacculus of *Escherichia coli*. Microbiol Mol Biol Rev 62:181–203
2. Desmarais SM, De Pedro MA, Cava F, Huang KC (2013) Peptidoglycan at its peaks: how chromatographic analyses can reveal bacterial cell wall structure and assembly. Mol Microbiol 89:1–13
3. Vollmer W, Seligman SJ (2010) Architecture of peptidoglycan: more data and more models. Trends Microbiol 18:59–66
4. Koyama N, Tokura Y, Münch D, Sahl HG, Schneider T, Shibagaki Y, Ikeda H, Tomoda H (2012) The nonantibiotic small molecule cyslabdan enhances the potency of β-lactams against MRSA by inhibiting pentaglycine interpeptide bridge synthesis. PLoS One 7:e48981
5. Hong HJ, Hutchings MI, Hill LM, Buttner MJ (2005) The role of the novel Fem protein VanK in vancomycin resistance in *Streptomyces coelicolor*. J Biol Chem 280:13055–13061
6. Vollmer W (2008) Structural variation in the glycan strands of bacterial peptidoglycan. FEMS Microbiol Rev 32:287–306
7. Bouhss A, Trunkfield AE, Bugg TD, Mengin-Lecreulx D (2008) The biosynthesis of peptidoglycan lipid-linked intermediates. FEMS Microbio Rev 32:208–233
8. Münch D, Sahl HG (2015) Structural variations of the cell wall precursor lipid II in Gram-positive bacteria—impact on binding and efficacy of antimicrobial peptides. Biochim Biophys Acta S0005-2736(15)00137-6
9. Van Heijenoort J (2007) Lipid intermediates in the biosynthesis of bacterial peptidoglycan. Microbiol Mol Biol Rev 71:620–635

10. Sauvage E, Kerff F, Terrak M, Ayala JA, Charlier P (2008) The penicillin-binding proteins: structure and role in peptidoglycan biosynthesis. FEMS Microbiol Rev 32:234–258
11. Jovetic S, Zhu Y, Marcone GL, Marinelli F, Tramper J (2010) β-Lactam and glycopeptide antibiotics: first and last line of defense? Trends Biotechnol 28:596–604
12. Binda E, Marinelli F, Marcone GL (2014) Old and new glycopeptide antibiotics: action and resistance. Antibiotics 3:572–574
13. Cooper MA, Williams DH (1999) Binding of glycopeptide antibiotics to a model of a vancomycin-resistant bacterium. Chem Biol 6:891–899
14. http://www.nanotherapeutics.com/ramoplanin/
15. Bionda N, Pitteloud JP, Cudic P (2013) Cyclic lipodepsipeptides: a new class of antibacterial agents in the battle against resistant bacteria. Future Med Chem 5:1311–1330
16. Siewert G, Strominger JL (1967) Bacitracin: an inhibitor of the dephosphorylation of lipid pyrophosphate, an intermediate in the biosynthesis of the peptidoglycan of bacterial cell wall. Proc Natl Acad Sci U S A 57:767–773
17. Castiglione F, Cavaletti L, Losi D, Lazzarini A, Carrano L, Feroggio M, Ciciliato I, Corti E, Candiani G, Marinelli F, Selva E (2007) A novel lantibiotic acting on bacterial cell wall synthesis produced by the uncommon actinomycete *Planomonospora* sp. Biochemistry 46:5884–5895
18. Carrano L, Abbondi M, Turconi P, Candiani G, Marinelli F (2015) A novel microbisporicin producer identified by early dereplication during lantibiotic screening. BioMed Research Int 2015:419383
19. Castiglione F, Lazzarini A, Carrano L, Corti E, Ciciliato I, Gastaldo L, Candiani P, Losi D, Marinelli F, Selva E, Parenti F (2008) Determining the structure and mode of action of microbisporicin, a potent lantibiotic active against multiresistant pathogens. Chem Biol 15:22–31
20. Münch D, Müller A, Schneider T, Kohl B, Wenzel M, Bandow JE, Maffioli S, Sosio M, Donadio S, Wimmer R, Sahl HG (2014) The lantibiotic NAI-107 binds to bactoprenol-bound cell wall precursors and impairs membrane functions. J Biol Chem 289:12063–12076
21. Pozzi R, Coles M, Linke D, Kulik A, Nega M, Wohlleben W, Stegmann E (2015) Distinct mechanisms contribute to immunity in the lantibiotic NAI-107 producer strain *Microbispora* ATCC PTA-5024. Environ Microbiol. doi:10.1111/1462-2920.12892
22. Courvalin P (2006) Vancomycin resistance in gram-positive cocci. Clin Infect Dis 42(Suppl 1):S25–S34
23. Bugg TD, Wright GD, Dutka-Malen S, Arthur M, Courvalin P, Walsh CT (1991) Molecular basis for vancomycin resistance in *Enterococcus faecium* BM4147: Biosynthesis of a depsipeptide peptidoglycan precursor by vancomycin resistance proteins VanH and VanA. Biochemistry 30:10408–10415
24. Arthur M, Reynolds P, Courvalin P (1996) Glycopeptide resistance in enterococci. Trends Microbiol 4:401–407
25. Lessard IA, Healy VL, Park IS, Walsh CT (1999) Determinants for differential effects on D-Ala-D-lactate vs D-Ala-D-Ala formation by the VanA ligase from vancomycin-resistant enterococci. Biochemistry 38:14006–14022
26. Arthur M, Depardieu F, Reynolds P, Courvalin P (1996) Quantitative analysis of the metabolism of soluble cytoplasmic peptidoglycan precursors of glycopeptide-resistant enterococci. Mol Microbiol 21:33–44
27. Somma S, Gastaldo L, Corti A (1984) Teicoplanin, a new antibiotic from *Actinoplanes teichomyceticus* nov. sp. Antimicrob Agents Chemother 26:917–923
28. Goldstein BP, Selva E, Gastaldo L, Berti M, Pallanza R, Ripamonti F, Ferrari P, Denaro M, Arioli V, Cassani G (1987) A40926, a new glycopeptide antibiotic with anti-*Neisseria* activity. Antimicrob Agents Chemother 31:1961–1966
29. Traynor K (2014) Dalbavancin approved for acute skin infections. Am J Health Syst Pharm 71:1062
30. Marcone GL, Marinelli F (2013) Glycopeptides: an old but up to date successful antibiotic class. In: Marinelli F, Genilloud O (eds) Antimicrobials, vol 5. Springer, New York, pp 85–107
31. Beltrametti F, Consolandi A, Carrano L, Bagatin F, Rossi R, Leoni L, Zennaro E, Selva E, Marinelli F (2007) Resistance to glycopeptide antibiotics in the teicoplanin producer is mediated by *van* gene homologue expression directing the synthesis of a modified cell wall peptidoglycan. Antimicrob Agents Chemother 51:1135–1141
32. Marcone GL, Beltrametti F, Binda E, Carrano L, Foulston L, Hesketh A, Bibb M, Marinelli F (2010) Novel mechanism of glycopeptide resistance in the A40926 producer *Nonomuraea* sp. ATCC 39727. Antimicrob Agents Chemother 54:2465–2472
33. D'Costa VM, King CE, Kalan L, Morar M, Sung WW, Schwarz C, Froese D, Zazula G, Calmels F, Debruyne R, Golding GB, Poinar HN, Wright GD (2011) Antibiotic resistance is ancient. Nature 477:457–461
34. Marcone GL, Carrano L, Marinelli F, Beltrametti F (2010) Protoplast preparation and reversion to the normal filamentous

growth in antibiotic-producing uncommon actinomycetes. J Antibiot 63:83–88

35. http://www2.warwick.ac.uk/
36. Billot-Klein D, Shlaes D, Bryant D, Bell D, Legrand R, Gutmann L, van Heijenoort J (1997) Presence of UDP-*N*-acetylmuramyl-hexapeptides and -heptapeptides in enterococci and staphylococci after treatment with ramoplanin, tunicamycin, or vancomycin. J Bacteriol 179:4684–4688
37. Ruzin A, Singh G, Severin A, Yang Y, Dushin RG, Sutherland AG, Minnick A, Greenstein M, May MK, Shlaes DM, Bradford PA (2004) Mechanism of action of the mannopeptimycins, a novel class of glycopeptide antibiotics active against vancomycin-resistant gram-positive bacteria. Antimicrob Agents Chemother 48:728–738
38. Brötz H, Bierbaum G, Reynolds PE, Sahl HG (1997) The lantibiotic mersacidin inhibits peptidoglycan biosynthesis at the level of transglycosylation. Eur J Biochem 246: 193–199
39. Brötz H, Bierbaum G, Leopold K, Reynolds PE, Sahl HG (1998) The lantibiotic mersacidin inhibits peptidoglycan synthesis by targeting lipid II. Antimicrob Agents Chemother 42:154–160
40. Kawamoto I, Oka T, Nara T (1981) Cell wall composition of *Micromonospora olivoasterospora*, *Micromonospora sagamiensis*, and related organisms. J Bacteriol 146:527–534
41. Parenti F, Coronelli C (1979) Members of the genus *Actinoplanes* and their antibiotics. Annu Rev Microbiol 33:389–411
42. Schäberle TF, Vollmer W, Frasch HJ, Hüttel S, Kulik A, Röttgen M, von Thaler AK, Wohlleben W, Stegmann E (2011) Self-resistance and cell wall composition in the glycopeptide producer *Amycolatopsis balhimycina*. Antimicrob Agents Chemother 55:4283–4289
43. Marcone GL, Binda E, Carrano L, Bibb M, Marinelli F (2014) Relationship between glycopeptide production and resistance in the actinomycete *Nonomuraea* sp. ATCC 39727. Antimicrob Agents Chemother 58:5191–5201
44. Binda E, Marcone GL, Pollegioni L, Marinelli F (2012) Characterization of VanYn, a novel D, D-peptidase/D, D-carboxypeptidase involved in glycopeptide antibiotic resistance in *Nonomuraea* sp. ATCC 39727. FEBS J 279: 3203–3213
45. Marcone GL, Foulston L, Binda E, Marinelli F, Bibb M, Beltrametti F (2010) Methods for the genetic manipulation of *Nonomuraea* sp. ATCC 39727. J Ind Microbiol Biotechnol 37:1097–1103
46. Kohlrausch U, Höltje JV (1991) One-step purification procedure for UDP-*N*-acetylmuramyl-peptide murein precursors from *Bacillus cereus*. FEMS Microbiol Lett 62:253–257
47. Vandamme P, Pot B, Gillis M, de Vos P, Kersters K, Swings J (1996) Polyphasic taxonomy, a consensus approach to bacterial systematics. Microbiol Rev 60:407–438
48. Marshall CG, Lessard IA, Park I, Wright GD (1998) Glycopeptide antibiotic resistance genes in glycopeptide-producing organisms. Antimicrob Agents Chemother 42:2215–2220

Chapter 13

Continuous Fluorescence Assay for Peptidoglycan Glycosyltransferases

Alexander J.F. Egan and Waldemar Vollmer

Abstract

Bacterial cell wall peptidoglycan is synthesized from its precursor lipid II by two enzymatic reactions. First, glycosyltransferases polymerize the glycan strands and second, DD-transpeptidases form cross-links between peptides of neighboring strands. Most bacteria possess bifunctional peptidoglycan synthesis enzymes capable of catalyzing both reactions. Here, we describe a continuous fluorescence glycosyltransferase assay using Dansyl-labeled lipid II as substrate. Progression of the reaction is monitored by the reduction in fluorescence over time. The assay is suitable to investigate the effect of protein interaction partners on the glycan strand synthesis activity of peptidoglycan polymerases.

Key words Peptidoglycan, Glycosyltransferase, Peptidoglycan synthesis, Synthase, Continuous fluorescence assay

1 Introduction

The majority of bacteria surround their cytoplasmic membrane with a peptidoglycan sacculus, a continuous layer that is required to maintain cell shape and osmotic stability [1]. The basic chemical structure of peptidoglycan is well known; glycan strands consisting of alternating *N*-acetylglucosamine (Glc*N*Ac) and *N*-acetylmuramic acid (Mur*N*Ac) residues connected by short stem peptides protruding from Mur*N*Ac. Peptides of neighboring glycan strands may be connected (i.e., cross-linked) forming a net-like layer [2, 3]. Peptidoglycan is synthesized at the outer leaflet of the cytoplasmic membrane from its substrate lipid II, undecaprenolpyrophosphoryl-Mur*N*Ac(L-Ala-D-iGlu-*meso*-Dap-D-Ala-D-Ala)-Glc*N*Ac (Gram-negative version with *meso*-Dap, *meso*-diaminopimelic acid), by two enzymatic reactions. First, the disaccharide subunits are polymerized to glycan strands by glycosyltransferases and second, peptide cross-links are formed by DD-transpeptidases [4]. DD-transpeptidases covalently bind β-lactam antibiotics such as penicillin, and are hence named

Hee-Jeon Hong (ed.), *Bacterial Cell Wall Homeostasis: Methods and Protocols*, Methods in Molecular Biology, vol. 1440, DOI 10.1007/978-1-4939-3676-2_13,

penicillin-binding proteins (PBPs). Most bacteria possess several peptidoglycan synthases capable of catalyzing the glycosyltransferase (GTase) and/or transpeptidase (TPase) reactions. The Gram-negative model organism *Escherichia coli* has three enzymes capable of performing both reactions, so-called bifunctional synthases; PBP1A, PBP1B, and PBP1C, two monofunctional transpeptidases, PBP2 and PBP3, and the monofunctional glycosyltransferase MtgA [4]. The main peptidoglycan synthesis activity in the cell is provided by the semi-redundant PBP1A and PBP1B in consort with the transpeptidases PBP2 and PBP3; the latter have essential roles in cell elongation and division, respectively [4–6]. The functions of PBP1C and MtgA are not known, as both are dispensable for growth.

The enzymology of PBPs has been studied since the discovery of the mode of action of β-lactams [7, 8]. The molecular details about how these enzymes facilitate the insertion of new material into an existing peptidoglycan sacculus remain to be determined. PBPs are challenging to work with for several reasons. They are integral membrane proteins with a single membrane-spanning region near the N-terminus and additional interactions with the cytoplasmic membrane via hydrophobic surface residues in the GTase domain [9, 10], therefore requiring detergents for solubilization. Once solubilized, some PBPs are difficult to purify and are also prone to denaturation and aggregation. Regardless, we have been gaining significant insight into their activities and regulation in recent years thanks to improvements in the available enzymatic activity assays, for example the in vitro peptidoglycan synthesis assay with lipid II substrate, which is the subject of a previous article in Methods in Molecular Biology [11]. This assay uses natural lipid II substrate which is radiolabeled to allow the detection of the product. This substrate is incubated with the peptidoglycan synthase at buffer conditions which allow its utilization in producing a cross-linked peptidoglycan product. The product is digested with a muramidase which hydrolyses the glycan strands between the disaccharide units, leaving peptide cross-links intact. The resulting subunits (muropeptides) can be resolved and quantified by reversed phase high-pressure liquid chromatography (HPLC), demonstrating both enzymatic activities and giving a measure of the TPase activity of the synthase (i.e., the percentage of peptides in cross-links). This assay was crucial to discover key differences in the in vitro activities of the two major bifunctional peptidoglycan synthases in *E. coli*, PBP1A and PBP1B [12, 13] and to identify the first activator of a synthase, the cell division protein FtsN, which stimulates PBP1B [14]. Furthermore, the assay has been used to demonstrate the stimulatory effects of recently identified outer membrane anchored lipoproteins LpoA and LpoB on the transpeptidase activity of PBP1A and PBP1B, respectively [15, 16]. However, because of the discontinuous end-point measurement using HPLC time-course experiments are tedious making it unpractical to determine transpeptidation reaction rates with this assay. Moreover, although the assay is capable of quantifying the muropeptides residing at the Mur*N*Ac-terminus of the glycan chain ends, allowing the calculation of the average chain length, it does not easily allow determination of the rate of glycan chain polymerization.

In this paper we describe a currently used continuous assay for peptidoglycan glycosyltransferases which was originally developed by Schwartz et al. [17] and modified by Zapun et al. for multiwell microplate format [18]. In this assay, polymerization of Dansyl-lipid II to glycan strands followed by their digestion with a muramidase (such as mutanolysin or cellosyl) results in the formation of dansylated muropeptide that shows a lower fluorescence than the lipid II substrate (Fig. 1). Hence, the progression of the GTase reaction can be followed by the reduction in fluorescence over time. The dansyl group is attached to the ε-amino group of the lysine at position three of the stem peptide in lipid II. Thus, the transpeptidation reaction is prevented and the assay measures only glycosyltransferase activity. We have optimized the reaction conditions to omit the solvent DMSO and reduced the detergent concentration allowing us to better observe glycosyltransferase activity and, crucially, determine the effect of protein interaction partners on the reaction rates of *E. coli* PBP1A and PBP1B. We have reported the stimulation of the glycosyltransferase activity of PBP1B by LpoB and the effect of a novel PBP1B-interaction partner, TolA, on this stimulation [15, 16, 19]. We have also reported the stimulation of glycosyltransferase activity of PBP1A by PBP2 [5]. The particular advantages of this continuous glycosyltransferase assay are that it is quick to set up and execute and that it allows sensitive monitoring of glycan strand synthesis activity, providing deeper insights into the regulation of this key reaction in bacterial cell wall growth.

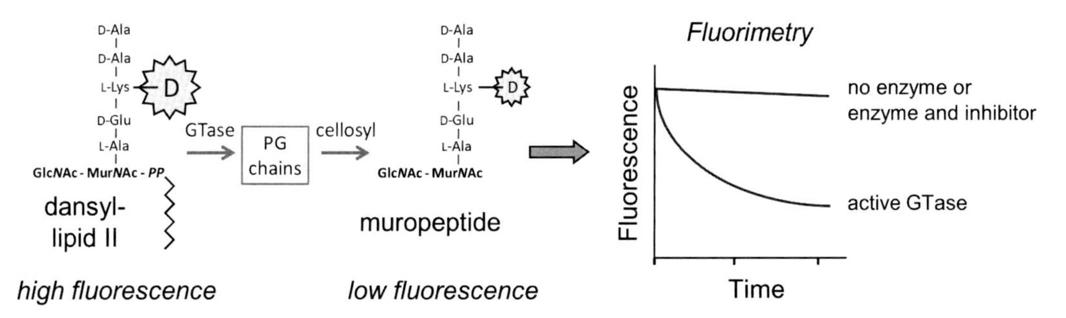

Fig. 1 Principle of the glycosyltransferase assay with dansylated lipid II substrate. In proximity to the undecaprenyl moiety of lipid II the dansyl fluorophore has high fluorescence. The polymerization of lipid II into glycan chains and their subsequent hydrolysis by a muramidase leads to the loss of the lipid moiety and a decrease in fluorescence, which can be observed in real-time using fluorimetry

2 Materials

All solutions should be prepared using ultrapure water (through purification of deionized water to 18 MΩ cm at 25 °C) such as that produced by a MilliQ water system. In this work we use a PureLab Flex system (Elga-Veola, Paris, France). In the following sections this water is referred to as ddH_2O. All reagents are prepared and stored at room temperature unless otherwise stated.

2.1 Purification of the Peptidoglycan Synthase PBP1B

The materials required for the purification of the peptidoglycan synthase PBP1B of *E. coli* have been previously published [11] and are given below.

2.1.1 Overproduction of PBP1B and Its Solubilization from Cell Membranes

1. *Escherichia coli* BL21 pDML924 producing His-tagged form of PBP1B [20].
2. Stock solution of 50 mg/mL kanamycin. Store at −20 °C.
3. LB medium: 10 g NaCl, 5 g yeast extract, and 10 g tryptone per L, pH 7.2.
4. Orbital shaker.
5. 250 mL and 2 L flasks.
6. Spectrophotometer.
7. Centrifuge and associated rotor.
8. 0.5 M IPTG (isopropyl 1-thio-β-D-galactopyranoside) in water. Prepare fresh each time.
9. PBP1B basic buffer: 25 mM Tris–HCl pH 7.5, 10 mM $MgCl_2$, 1 M NaCl, 1 mM EGTA.
10. DNAse (freeze-dried).
11. 100 mM PMSF (phenylmethanesulfonylfluoride) in ethanol.
12. 5 mg/mL protease inhibitor cocktail in DMSO.
13. French press or probe sonicator.
14. Potter S Homogenizer 30 mL.
15. Light microscope.
16. Ultracentrifuge and associated rotor.
17. Triton X-100, purified for membrane research. Store at 4 °C (*see* **Note 1**).
18. PBP1B extraction buffer: 25 mM Tris–HCl pH 7.5, 10 mM $MgCl_2$, 1 M NaCl, 20 % glycerol, 2 % Triton X-100.

2.1.2 Ni-NTA Affinity Chromatography

1. Ni-NTA-beads.
2. PBP1B extraction buffer: *see* Subheading 2.1.1, **item 18**.
3. Rotary wheel.
4. D-Tube Dialyzer Maxi; MWCO 6–8 kDa (Merck Millipore, Billerica, USA).

5. SDS-PAGE apparatus and buffers.
6. Coomassie Blue.
7. 1.8 U/μL thrombin (restriction grade).
8. 5 M imidazole, pH 7.5.
9. Disposable gravity columns (10 mL).
10. PBP1B dialysis buffer 1: 25 mM Tris–HCl pH 7.5, 10 mM $MgCl_2$, 1 M NaCl, 0.5 mM EGTA, 20% glycerol.

2.1.3 Ion Exchange Chromatography

1. A FPLC (Fast Protein Liquid Chromatography) system. In this work we use an ÄKTA Prime+ system from GE Healthcare (Little Chalfont, UK).
2. PBP1B dialysis buffer 2: 10 mM sodium acetate pH 5.0, 10 mM $MgCl_2$, 0.02% NaN_3, 10% glycerol, 1 M NaCl.
3. PBP1B dialysis buffer 3: 10 mM sodium acetate pH 5.0, 10 mM $MgCl_2$, 0.02% NaN_3, 10% glycerol, 300 mM NaCl.
4. PBP1B dialysis buffer 4: 10 mM sodium acetate pH 5.0, 10 mM $MgCl_2$, 0.02% NaN_3, 10% glycerol, 100 mM NaCl.
5. PBP1B dialysis buffer 5: 10 mM sodium acetate pH 5.0, 10 mM $MgCl_2$, 0.02% NaN_3, 10% glycerol, 500 mM NaCl, 0.2% Triton X-100.
6. HiTrap SP HP 5 mL ion exchange column.
7. PBP1B FPLC buffer A: 10 mM sodium acetate pH 5.0, 10 mM $MgCl_2$, 0.02% NaN_3, 10% glycerol, 100 mM NaCl, 0.2% Triton X-100.
8. PBP1B FPLC buffer B: 10 mM sodium acetate pH 5.0, 10 mM $MgCl_2$, 0.02% NaN_3, 10% glycerol, 2 M NaCl, 0.2% Triton X-100.
9. Protein concentration determination assay kit. In this work we use the colorimetric BCA protein assay (ThermoFisher Scientific, Waltham, USA).

2.2 Glycosyltransferase Reaction

1. In this work we use Dansyl-lipid II from the laboratory of Dr. Eefjan Breukink (Department of Biochemistry of Membranes, University of Utrecht, The Netherlands) in which the fluorophore is attached via a lysine residue at position three of the peptide. Lipid II is dissolved in 1:1 mix of chloroform–methanol and stored at −20 °C, our stock is 0.44 mM. Various dansylated versions of lipid II (with *meso*-Dap or lysine, and different lipid length) can be purchased from the synthetic facility of the UK Bacterial Cell Wall Biosynthesis Network (UK-BaCWAN), University of Warwick, UK (http://www2.warwick.ac.uk/fac/sci/lifesci/people/droper/bacwan/). Additional labeled and unlabeled versions of lipid II have been produced in the laboratories of Dr. Susanne Walker

(Harvard University, Massachusetts, USA) and Dr. Shahriar Mobashery (University of Notre-Dame, Indiana, USA).

2. FLUOTRAC™200 96-well plate (Greiner Bio-One, Freickenhausen, Germany).
3. Plate reader capable of fluorescence measurement, temperature control and orbital shaking and with an excitation filter of 340 nm and an emission filter of 540 nm. In this work we use a FLUOstar OPTIMA microplate reader (BMG Labtech, Offenberg, Germany).
4. 1 M HEPES/NaOH pH 7.5.
5. 1 M $MgCl_2$.
6. 2 M NaCl.
7. 0.2% solution of Triton X-100, purified for membrane research. Store at 4 °C (*see* **Note 1**). In this work we use Triton X-100 from Roche (Basel, Switzerland).
8. 6 μg/mL solution of muramidase enzyme in 20 mM sodium phosphate, pH 4.8. In this work we use cellosyl provided by Höchst AG (Frankfurt, Germany) which is not commercially available. However, cellosyl can be replaced by mutanolysin (Sigma, Dorset, UK) which has the same amino acid sequence as cellosyl (*see* **Note 2**).
9. A multichannel pipette capable of pipetting 18 μL (*see* **Note 3**).

3 Methods

3.1 Purification of the Peptidoglycan Synthase PBP1B

See Subheading 2.1, this method has been previously published [11] and is given below.

3.1.1 Overproduction of PBP1B and Its Solubilization from Cell Membranes

1. Prepare 2× 50 mL of LB medium (with 50 μg/mL Kanamycin) in 250 mL flasks.
2. Inoculate each flask with a 10 μL loop from a glycerol stock of *E. coli* BL21 pDML924 cells. Incubate the culture on an orbital shaker at 37 °C overnight for 14–16 h ensuring good aeration of the culture.
3. Prepare 10× 400 mL of LB medium (with 50 μg/mL kanamycin) in 2 L flasks.
4. Inoculate each flask with 5 mL of overnight-grown culture from **step 2**.
5. Incubate the culture on an orbital shaker at 37 °C to an OD_{578} of 0.6.
6. Induce the overproduction of His-tagged PBP1B by adding 0.8 mL of a 0.5 M IPTG stock solution per flask and continue growing the cells for 3 h.

7. Place the flask on ice for 10 min.
8. Harvest the cells by centrifugation at 8000 × *g* for 20 min at 4 °C.
9. Discard the supernatant and estimate the wet weight of the cell pellet by weighing the centrifuge tube on a balance tared with an equivalent, empty centrifuge tube.
10. All further steps are either done on ice or in a cold room at 4 °C.
11. Resuspend the cell pellet in PBP1B basic buffer (~5 mL for 1 g of wet cells weight).
12. Add a small amount of DNAse (~1–2 mg) to the resuspended cells.
13. Add the appropriate amounts of a 100 mM PMSF stock solution and of a 5 mg/mL protease inhibitor cocktail to bring the final concentrations to 0.1 mM PMSF and 5 μg/mL of protease inhibitor cocktail.
14. Disrupt the cells either by using a French press at 700 bar or by sonication with 6–8 pulses at 45 W (30 s "ON," 60 s "OFF"). Check for quantitative cell disruption by light microscopy.
15. Centrifuge the disrupted cell suspension at 130,000 × *g* for 1 h at 4 °C. Save 10 μL of the supernatant for SDS-PAGE analysis (**step 8** in Subheading 3.1.2) and discard the rest.
16. Resuspend the pellet in 30 mL of PBP1B extraction buffer and carefully homogenize the pellet with a Potter S homogenizer.
17. Stir the suspension at medium speed for at least 4 h at 4 °C.
18. Centrifuge the suspension at 130,000 × *g* for 1 h at 4 °C. Reserve the supernatant and discard the pellet.
19. The His-tagged PBP1B and other membrane proteins are present in the supernatant. Save a 10 μL aliquot of the supernatant for SDS-PAGE analysis (**step 8** in Subheading 3.1.2).

3.1.2 Ni-NTA Affinity Chromatography

1. Take a 5 mL suspension of Ni-NTA bead slurry (containing ~2.5 mL of beads) and place in a 50 mL tube. Wash them by centrifugation/suspension steps with 10 mL of water (3 times) and 10 mL of PBP1B extraction buffer with an added 15 mM imidazole (3 times).
2. Transfer the supernatant from **step 18** in Subheading 3.1.1 to the pellet of Ni-NTA beads in the 50 mL tube.
3. Incubate the bead suspension overnight at 4 °C on a rotary wheel.
4. Transfer the bead suspension into a disposable gravity column.
5. Collect the flow-through as the beads settle on the bottom of the column. Take a 10 μL aliquot for SDS-PAGE analysis (*see* **Note 4**).

6. Wash the beads with 5 mL of PBP1B extraction buffer with 15 mM imidazole added. Repeat this wash step ten times. Collect the flow through samples of each washing step and take a 10 μL aliquot of them for SDS-PAGE analysis (**step 8** below).
7. Elute the His-tagged PBP1B by adding 5 mL of PBP1B extraction buffer with 400 mM imidazole. Repeat this step twice. Collect the eluates and take a 10 μL aliquot of each fraction for SDS-PAGE analysis (**step 8** below).
8. Perform an 8 % or 10 % SDS-polyacrylamide gel electrophoresis to analyze samples from **steps 19** of Subheading 3.1.1, and **steps 5–7** above. Stain the gel with Coomassie Blue and identify the PBP1B-containing elution fractions. Verify the absence of large amounts of PBP1B in the wash steps.
9. Combine and transfer the PBP1B-containing elution fractions into a dialysis tube.
10. Dialyze the sample against 1 L of PBP1B dialysis buffer 1 for 1 h at 4 °C. Repeat this step four times. Save an aliquot of the dialyzed sample for SDS-PAGE (**step 14** below) (*see* **Note 5**).
11. Add 20 μL of 1.8 U/μL Thrombin into the dialysis tube and continue dialyzing against 1 L of PBP1B dialysis buffer 1 for 1 h at 4 °C.
12. Change the dialysis buffer and continue dialyzing overnight.
13. Change the dialysis buffer and continue dialyzing for 1 h.
14. Check for the completeness of the thrombin cleavage by analyzing ~1 μg PBP1B containing aliquots of the pre-thrombin cleaved sample (**step 10** above) and of the post thrombin cleaved sample (**step 13** above) by 8 % SDS-PAGE (*see* **Note 6**).

3.1.3 Ion Exchange Chromatography

1. Dialyze the PBP1B solution for 45 min against 1 L of PBP1B dialysis buffer 2. Repeat this step two times.
2. Dialyze the protein solution overnight against 1 L of PBP1B dialysis buffer 3.
3. Dilute the protein solution with the same volume of PBP1B dialysis buffer 4.
4. Dialyze the protein solution for 1 h against 1 L of PBP1B dialysis buffer 4. Repeat this step three times every half hour (*see* **Note 7**).
5. Prepare the HiTrap SP HP FPLC column by running at a flow rate of 5 mL/min 10 column volumes of water, 10 column volumes of PBP1B FPLC buffer B and 10 column volumes of PBP1B FPLC buffer A.
6. Reduce the flow rate to 1 mL/min and apply the protein solution from **step 4** above to the column.

7. Apply a linear, 70 mL gradient from 100 % PBP1B buffer A to 100 % PBP1B buffer B at a flow rate of 5 mL/min. Detect proteins at 280 nm. Collect fractions of 2 mL.
8. Analyze the fractions by 8 % or 10 % SDS-polyacrylamide gel electrophoresis.
9. Combine fractions containing PBP1B and transfer them into a dialysis tube.
10. Dialyze the PBP1B solution against 1 L of PBP1B dialysis buffer 5 for 1 h. Repeat this step three times.
11. Prepare 50–500 μL aliquots of the purified PBP1B solution and store them at −80 °C.
12. Determine the concentration of the purified PBP1B using a method compatible with Triton X-100, $MgCl_2$ and glycerol (*see* Subheading 2.1.3, **item 8**).

3.2 Glycosyltransferase Reaction

With the exception of the substrate, the glycosyltransferase reaction components are mixed directly in the well of the microplate to avoid loss during unnecessary transfers.

1. Pre-warm the plate reader to the desired temperature and ensure the measurement program is set up correctly (*see* **Note 8**).
2. Dry sufficient Dansyl-lipid II in a 200 μL microfuge tube to give a final concentration of 200 μM in 3 μL per reaction. This can be by simple air- or vacuum-drying. While the substrate is drying proceed with **steps 3–7**.
3. Prepare a master mix calculated to obtain final concentrations of 50 mM HEPES/NaOH, pH 7.5, 25 mM $MgCl_2$, and 0.5 μg/mL muramidase. For a total number of n reactions we mix n.2× 3 μL 1 M HEPES/NaOH, pH 7.5, n.2× 1.5 μL 1 M $MgCl_2$, and n.2× 5 μL 6 μg/mL muramidase (e.g., if 8 reactions are planned, prepare a master mix with 8.2× the volumes given). 9.5 μL of this master mix is pipetted directly into the appropriate well of the microplate.
4. Calculate the concentration of NaCl being introduced through the dilution of each of the protein components of the reaction. Add a sufficient volume of 2 M NaCl to bring the final concentration to 150 mM.
5. Calculate the concentration of Triton X-100 being introduced through the dilution of each protein component and add a sufficient volume of 0.2 % Triton X-100 solution to bring the final concentration within the range of 0.02–0.08 % (*see* **Note 9**).
6. Add sufficient ddH_2O to bring the final volume to 42 μL (not including the volume of protein stocks). Pipette 15 μL ddH_2O into the adjacent well, in the next row of the microplate (e.g., if reaction wells are A1–A8, the substrate solution is to be pipetted into B1–B8).

7. Add the protein(s). Mix by gentle orbital shaking by hand.
8. Dissolve the substrate in 3 μL ddH_2O per reaction and add 3 μL to the well with 15 μL of ddH_2O (*see* **Note 10**), making 18 μL of substrate solution.
9. Place the microplate into the plate reader and incubate for 5 min. During this time, load the measurement program and prepare it to start immediately.
10. Using the multichannel pipette, slowly pipette the 18 μL of substrate solution into the tips. Then, avoiding the introduction of air bubbles as much as possible, add the substrate to the reaction wells simultaneously and mix by pipetting up and down three to five times (*see* **Note 11**). Immediately after this addition, begin the measurement program (*see* **Note 12**).

3.3 Data Analysis

1. Export raw fluorescence measurements at each time point into a suitable data handling software, such as Microsoft Excel.
2. To allow direct comparison between samples data are normalized by taking the fluorescence reading at time 0 as 100 %. Each subsequent time point is given as a percentage of this value.
3. Plot the percentage relative fluorescence against time in seconds, giving a curve representing the reaction rate (Fig. 2).
4. For calculating the fold-effect of interaction partners, take the slope of the curve at its greatest rate for comparison (*see* **Note 13**).

4 Notes

1. We recommend storing Triton X-100 at 4 °C. At room temperature Triton X-100 accumulates peroxide radicals which can inactivate enzymes.
2. Mutanolysin should be effective in this assay given the sequence identity with cellosyl.
3. A multichannel pipette is used to simultaneously deliver the substrate solution to start the reactions, thus the number of pipette channels dictates the number of reactions that can be performed simultaneously.
4. Avoid drying out the beads by fitting a small tubing with a clamp, or other stopping device, at the outlet of the column which allows to stop the gravity flow.
5. This dialysis step is to remove the imidazole before Thrombin cleavage.
6. The difference in M.W. of the tagged and untagged protein is only ~2 kDa. Thus, if using SDS-polyacrylamide gels of ~5 cm

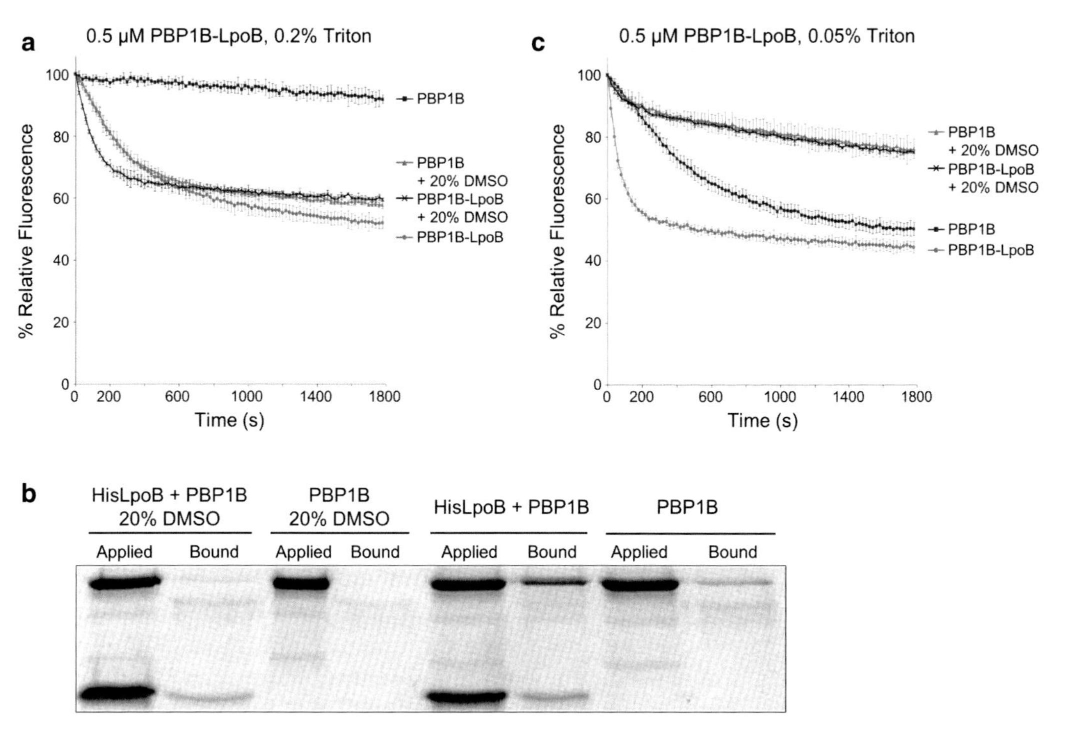

Fig. 2 Effects of DMSO and Triton X-100 on the activity and stimulation of PBP1B. (**a**) The GTase rate of 0.5 μM *E. coli* PBP1B with or without 0.5 μM LpoB at 0.2 % Triton X-100 (TX-100) in the presence or absence of 20 % DMSO. Data are plotted as fluorescence (expressed as a percentage of the initial value) against time (*s*), points are the mean ± SD of three independent experiments. *Dark grey line with squares*, PBP1B alone; *light grey line with circles*, PBP1B-LpoB; *light grey line with triangles*, PBP1B with 20 % DMSO; *black line with crosses*, PBP1B-LpoB with 20 % DMSO. (**b**) Coomassie stained SDS-PAGE gel of an in vitro cross-linking/pulldown experiment using a soluble version of LpoB with a hexahistidine tag [19]. HisLpoB (4 μM) and PBP1B (1 μM) were incubated in a buffer of 10 mM HEPES/NaOH, 4 mM $MgCl_2$, 150 mM NaCl, 0.05 % Triton X-100, pH 7.5 with and without 20 % DMSO. Interacting proteins were cross-linked by addition of 0.2 % w/v formaldehyde and incubation at 37 °C for 10 min. Samples were then applied to 50 μL equilibrated Ni-NTA beads and incubated o/n at 4 °C. Beads were extensively washed with 5× 1.5 mL buffer (as above without DMSO) containing 10 mM imidazole. Bound proteins were eluted by boiling the beads directly in SDS-PAGE loading buffer for 15 min, which simultaneously reverses cross-linking, and analyzed by SDS-PAGE. (**c**) The GTase rate of 0.5 μM *E. coli* PBP1B with or without 0.5 μM LpoB at 0.05 % Triton X-100 (TX-100) in the presence or absence of 20 % DMSO. Data are plotted as fluorescence (expressed as a percentage of the initial value) against time (*s*), points are the mean ± SD of three independent experiments. *Dark grey line with squares*, PBP1B alone; *light grey line with circles*, PBP1B-LpoB; *light grey line with triangles*, PBP1B with 20 % DMSO; *black line with crosses*, PBP1B-LpoB with 20 % DMSO

we recommend that the run persists until proteins of <50 kDa are off the bottom of the gel, increasing resolution of higher molecular weight proteins.

7. The protein may precipitate at low NaCl concentration, especially if the protein concentration is high, and therefore the dialysis steps are shorter than usual.

8. Microplate reader programming. Excitation and emission filters of 330 and 520 nm, respectively, are used. For collection of data, 90 cycles of 20 s are performed. Prior to measurement of fluorescence in each cycle the microplate is shaken by orbital shaking with a radius of 4 mm for 5 s. During initial assay setup the gain of the fluorimeter should be adjusted such that the signal from the substrate is in the mid-range. For example, our reader has a maximum of 50,000 U thus the gain is such that the initial fluorescence is ~25,000–30,000 U. This should not need to change unless the amount of substrate is varied.
9. During optimization of this assay we found that some glycosyltransferases are affected by concentrations of Triton X-100 greater than 0.02–0.08 %. During the previous optimization of this assay for microplate format the inhibitory effect of high detergent concentration was noted, but the addition of 25 % dimethylsulfoxide (DMSO) abrogated this detergent effect [18]. The organic solvent DMSO has traditionally been a component of in vitro peptidoglycan synthesis assays making use of its ability to solubilize the lipid II substrate without denaturing enzymes [20, 21]. Indeed, we confirmed that the addition of 20 % DMSO has a positive effect on the GTase rate of PBP1B, particularly at the higher Triton X-100 concentration of 0.2 % (Fig. 2a). However, DMSO at high concentration can potentially affect protein stability and protein-protein interactions. Indeed, in our hands the presence of 20 % DMSO severely impaired the interaction of PBP1B with LpoB (Fig. 2b). In contrast, at lower concentration of Triton X-100 (between 0.02 % and 0.08 %) the activity of PBP1B can be robustly assayed and its interactions with LpoB, CpoB, TolA and FtsN are maintained [14, 15, 19]. Consistent with these observations, at higher Triton X-100 concentration in the presence of 20 % DMSO the stimulation of PBP1B by LpoB is only 1.6-fold (Fig. 2a), comparable to the published 1.5-fold stimulation measured in the presence of 20 % DMSO [22]. This is a significantly smaller effect than the sevenfold stimulation measured at lower Triton X-100 concentration in the absence of DMSO (Fig. 2c). Consequently, we recommend to avoid using DMSO in in vitro peptidoglycan synthesis assays with protein interaction partners and to treat published data on the extent of the effect of Lpo proteins on PBP activity done in the presence of DMSO with caution.
10. To dissolve the Dansyl-lipid II, thoroughly pipette up and down ensuring the sides of the tube are contacted by the solvent. We avoid vortex mixing. Whether the substrate is successfully in solution can be visualized by observing fluorescence of the solution under UV light, which should be a homogeneous bright green/yellow color.

11. Without sufficient mixing, the first few minutes of the fluorescence readings are erratic and give an initial increase in fluorescence, up to 120% of the reading at t0.
12. Using the plate reader detailed (Subheading 2.2, **item 3**) there is a delay of ~20 s between inserting the microplate and beginning measurements. This delay can cause issues when using this assay to attempt to measure rapid reaction rates. For example; PBP1B at 1 μM when incubated with equal amounts of its regulator LpoB has such a rapid rate of reaction that between 30 and 50% of the available substrate is consumed before the reader begins its measurements. Thus, if this is the case, reaction conditions must be optimized to lower the reaction rates. To achieve this we have previously opted to reduce the temperature and/or enzyme concentration.
13. For this we use the analysis software to add a linear trendline to the data points of the curve at its greatest rate, which is not necessarily the initial time points. Comparison between the rate of a GTase with and without a regulator is done on the individual replicates of reactions within the same experiment series at the same conditions, and not on the averaged data. The fold-effect on activity is expressed as the mean ± the standard deviation.

Acknowledgements

This work was supported by the Wellcome Trust [101824/Z/13/Z]. We thank Andre Zapun (IBS, Grenoble, France) for assistance in the initial stages of our optimization of this assay.

References

1. Vollmer W, Blanot D, de Pedro MA (2008) Peptidoglycan structure and architecture. FEMS Microbiol Rev 32:149–167
2. Vollmer W, Höltje J-V (2004) The architecture of the murein (peptidoglycan) in Gram-negative bacteria: vertical scaffold or horizontal layer(s)? J Bacteriol 186:5978–5987
3. Vollmer W, Seligman SJ (2010) Architecture of peptidoglycan: more data and more models. Trends Microbiol 18:59–66. doi:10.1016/j.tim.2009.12.004, S0966-842X(09)00260-1 [pii]
4. Typas A, Banzhaf M, Gross CA, Vollmer W (2012) From the regulation of peptidoglycan synthesis to bacterial growth and morphology. Nat Rev Microbiol 10:123–136. doi:10.1038/nrmicro2677
5. Banzhaf M, van den Berg van Saparoea B, Terrak M, Fraipont C, Egan A, Philippe J, Zapun A, Breukink E, Nguyen-Disteche M, den Blaauwen T, Vollmer W (2012) Cooperativity of peptidoglycan synthases active in bacterial cell elongation. Mol Microbiol 85:179–194. doi:10.1111/j.1365-2958.2012.08103.x
6. Bertsche U, Kast T, Wolf B, Fraipont C, Aarsman ME, Kannenberg K, von Rechenberg M, Nguyen-Disteche M, den Blaauwen T, Höltje J-V, Vollmer W (2006) Interaction between two murein (peptidoglycan) synthases, PBP3 and PBP1B, in *Escherichia coli*. Mol Microbiol 61:675–690
7. Tipper DJ, Strominger JL (1965) Mechanism of action of penicillins: a proposal based on their structural similarity to acyl-D-alanyl-D-alanine. Proc Natl Acad Sci U S A 54:1133–1141
8. Wise EM Jr, Park JT (1965) Penicillin: its basic site of action as an inhibitor of a peptide cross-linking reaction in cell wall mucopeptide synthesis. Proc Natl Acad Sci U S A 54:75–81

9. Lovering AL, de Castro LH, Lim D, Strynadka NC (2007) Structural insight into the transglycosylation step of bacterial cell-wall biosynthesis. Science 315:1402–1405, doi: 315/5817/1402 [pii] 10.1126/science.1136611
10. Sung MT, Lai YT, Huang CY, Chou LY, Shih HW, Cheng WC, Wong CH, Ma C (2009) Crystal structure of the membrane-bound bifunctional transglycosylase PBP1b from *Escherichia coli*. Proc Natl Acad Sci U S A 106:8824–8829
11. Biboy J, Bui NK, Vollmer W (2013) In vitro peptidoglycan synthesis assay with lipid II substrate. Methods Mol Biol 966:273–288. doi:10.1007/978-1-62703-245-2_17
12. Bertsche U, Breukink E, Kast T, Vollmer W (2005) In vitro murein peptidoglycan synthesis by dimers of the bifunctional transglycosylase-transpeptidase PBP1B from *Escherichia coli*. J Biol Chem 280:38096–38101
13. Born P, Breukink E, Vollmer W (2006) In vitro synthesis of cross-linked murein and its attachment to sacculi by PBP1A from *Escherichia coli*. J Biol Chem 281:26985–26993
14. Müller P, Ewers C, Bertsche U, Anstett M, Kallis T, Breukink E, Fraipont C, Terrak M, Nguyen-Disteche M, Vollmer W (2007) The essential cell division protein FtsN interacts with the murein (peptidoglycan) synthase PBP1B in *Escherichia coli*. J Biol Chem 282:36394–36402
15. Egan AJF, Jean NL, Koumoutsi A, Bougault CM, Biboy J, Sassine J, Solovyova AS, Breukink E, Typas A, Vollmer W, Simorre JP (2014) Outer-membrane lipoprotein LpoB spans the periplasm to stimulate the peptidoglycan synthase PBP1B. Proc Natl Acad Sci U S A 111:8197–8202. doi:10.1073/pnas.1400376111
16. Typas A, Banzhaf M, van den Berg van Saparoea B, Verheul J, Biboy J, Nichols RJ, Zietek M, Beilharz K, Kannenberg K, von Rechenberg M, Breukink E, den Blaauwen T, Gross CA, Vollmer W (2010) Regulation of peptidoglycan synthesis by outer-membrane proteins. Cell 143:1097–1109. doi:10.1016/j.cell.2010.11.038
17. Schwartz B, Markwalder JA, Seitz SP, Wang Y, Stein RL (2002) A kinetic characterization of the glycosyltransferase activity of *Escherichia coli* PBP1b and development of a continuous fluorescence assay. Biochemistry 41:12552–12561
18. Offant J, Terrak M, Derouaux A, Breukink E, Nguyen-Disteche M, Zapun A, Vernet T (2010) Optimization of conditions for the glycosyltransferase activity of penicillin-binding protein 1a from *Thermotoga maritima*. FEBS J 277:4290–4298. doi:10.1111/j.1742-4658.2010.07817.x
19. Gray AN, Egan AJF, Van't Veer IL, Verheul J, Colavin A, Koumoutsi A, Biboy J, Altelaar MA, Damen MJ, Huang KC, Simorre JP, Breukink E, den Blaauwen T, Typas A, Gross CA, Vollmer W (2015) Coordination of peptidoglycan synthesis and outer membrane constriction during cell division. eLife 4. doi: 10.7554/eLife.07118
20. Terrak M, Ghosh TK, van Heijenoort J, Van Beeumen J, Lampilas M, Aszodi J, Ayala JA, Ghuysen JM, Nguyen-Disteche M (1999) The catalytic, glycosyl transferase and acyl transferase modules of the cell wall peptidoglycan-polymerizing penicillin-binding protein 1b of *Escherichia coli*. Mol Microbiol 34:350–364
21. Lupoli TJ, Lebar MD, Markovski M, Bernhardt T, Kahne D, Walker S (2014) Lipoprotein activators stimulate *Escherichia coli* penicillin-binding proteins by different mechanisms. J Am Chem Soc 136:52–55. doi:10.1021/ja410813j
22. Paradis-Bleau C, Markovski M, Uehara T, Lupoli TJ, Walker S, Kahne DE, Bernhardt TG (2010) Lipoprotein cofactors located in the outer membrane activate bacterial cell wall polymersases. Cell 143:1110–1120

Chapter 14

Analysis of Peptidoglycan Fragment Release

Ryan E. Schaub, Jonathan D. Lenz, and Joseph P. Dillard

Abstract

Most bacteria break down a significant portion of their cell wall peptidoglycan during each round of growth and cell division. This process generates peptidoglycan fragments of various sizes that can either be imported back into the cytoplasm for recycling or released from the cell. Released fragments have been shown to act as microbe-associated molecular patterns for the initiation of immune responses, as triggers for the initiation of mutualistic host–microbe relationships, and as signals for cell–cell communication in bacteria. Characterizing these released peptidoglycan fragments can, therefore, be considered an important step in understanding how microbes communicate with other organisms in their environments. In this chapter, we describe methods for labeling cell wall peptidoglycan, calculating the rate at which peptidoglycan is turned over, and collecting released peptidoglycan to determine the abundance and species of released fragments. Methods are described for both the separation of peptidoglycan fragments by size-exclusion chromatography and further detailed analysis by HPLC.

Key words Peptidoglycan, PG, Murein, Peptidoglycan fragments, Peptidoglycan turnover, Pulse-chase, Size-exclusion chromatography, HPLC

1 Introduction

Peptidoglycan is a critical structural macromolecule that protects bacterial cells from osmotic rupture and determines cell shape [1]. In most bacteria, peptidoglycan is composed of a lattice made up of strands of repeating units of *N*-acetylglucosamine (GlcNAc) and *N*-acetylmuramic acid (MurNAc) sugars cross-linked by three-to-five amino acid peptide stems covalently attached to the MurNAc moiety. This dynamic structure is constantly expanded during cell growth, broken during cell division, and remodeled to accommodate the assembly of large membrane-spanning structures [2]. Both gram-positive and gram-negative bacteria break down a significant portion of their peptidoglycan in the course of each cell cycle. In *E. coli*, almost 50 % of the cell's peptidoglycan is turned over each generation [3], with the vast majority being reimported into the bacterial cytoplasm via the muropeptide permease AmpG and processed for recycling. Gram-positive bacteria lack an AmpG

Hee-Jeon Hong (ed.), *Bacterial Cell Wall Homeostasis: Methods and Protocols*, Methods in Molecular Biology, vol. 1440,
DOI 10.1007/978-1-4939-3676-2_14,

homolog, and most do not recycle peptidoglycan [4], though there are exceptions [5]. Despite the presence of recycling pathways for muropeptides, the assembly and disassembly of cell wall does not represent a closed system, as some fragments of peptidoglycan escape into the environment. These fragments can induce inflammatory responses in hosts, provide protective immune modulatory signals, aid in initiating mutualism, and coordinate bacteria–bacteria interactions (reviewed in Refs. [6, 7]).

There is a growing appreciation for the ability of bacteria to release soluble peptidoglycan fragments. Among gram-negative bacteria, the release of peptidoglycan fragments was thought to only be a significant feature of *Bordetella pertussis* and *Neisseria gonorrhoeae* (Fig. 1), which release large amounts of inflammatory peptidoglycan monomers [8, 9]. Soluble, released peptidoglycan fragments typically include those generated by the activity of specific peptidoglycanases, including lytic transglycosylases, endopeptidases, and *N*-acetylmuramyl-L-alanine amidases [10]. Different bacterial species are equipped with varying numbers of peptidoglycan-degrading enzymes with different specificities. Bacteria also vary in the efficiency of peptidoglycan recycling, making it difficult to predict exactly which peptidoglycan fragments are released by a given bacterial species or strain. For this reason, sensitive methods

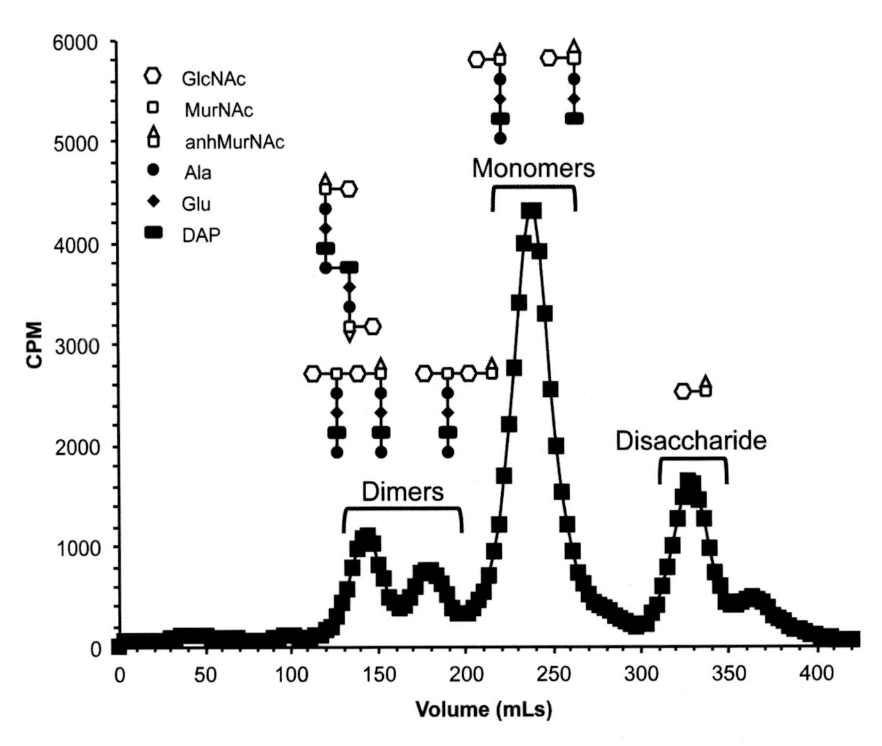

Fig. 1 Fragment release from wild-type *Neisseria gonorrhoeae* following size-exclusion chromatography using [^{3}H]-glucosamine to label peptidoglycan. The elution of peptidoglycan fragments containing sugars make up clearly defined peaks

capable of distinguishing released sugar and peptide moieties of peptidoglycan must be utilized in order to determine the composition of peptidoglycan fragments released by bacteria and what role these fragments play in the interaction of bacteria with their environment.

The following protocols describe how to utilize radioactive precursor molecules for labeling peptidoglycan sacculi in order to achieve highly sensitive detection of released fragments and monitor peptidoglycan turnover [11]. Metabolic pulse-labeling using tritiated peptidoglycan precursors, such as D-glucosamine or *meso*-2,6-diaminopimelic acid allow incorporation of traceable radioactivity into the sugar backbone or peptide chain of peptidoglycan, respectively. Radiolabeling provides a quantifiable way to measure the abundance of particular released fragments within and between bacterial strains. Bacterial species differ in the fragments of peptidoglycan they release (Fig. 2) and can include: free peptides, free GlcNAc-MurNAc disaccharides, monosaccharides, monomers (a single disaccharide and single peptide stem), and/or dimers (two disaccharide subunits and one or two peptide stems). Since all of these molecules exist within a molecular weight range of 200–2000 Da, we present a size-exclusion chromatography methodology to differentiate between released fragments within this size range.

More detail is often desired in characterizing released fragments, with HPLC analysis as the standard in the field for the discrimination of chemical differences between similarly sized muropeptides [12]. Analysis by HPLC can be used to determine such characteristics as the length of peptide chains, the presence of modifications such as acetylation, or the type of bond present on the MurNAc sugar (reducing vs. anhydro), all of which can have implications for the biological function of the released molecules [9, 13, 14]. For this purpose, we present a typical application of

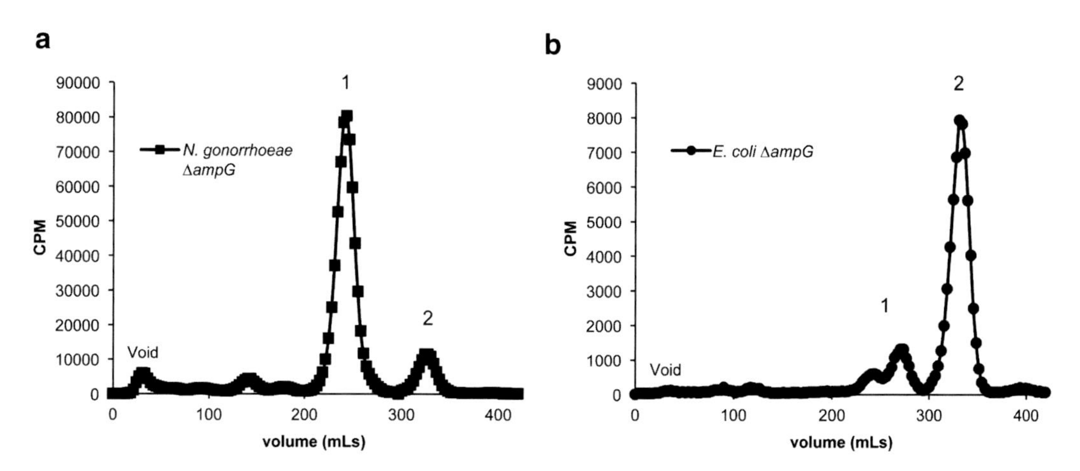

Fig. 2 Peptidoglycan fragments released from *N. gonorrhoeae* Δ*ampG* (**a**) and *E. coli* Δ*ampG* (**b**). *N. gonorrhoeae* lacking *ampG* release mostly peptidoglycan monomer (1), whereas *E. coli* lacking *ampG* release mostly disaccharide (2)

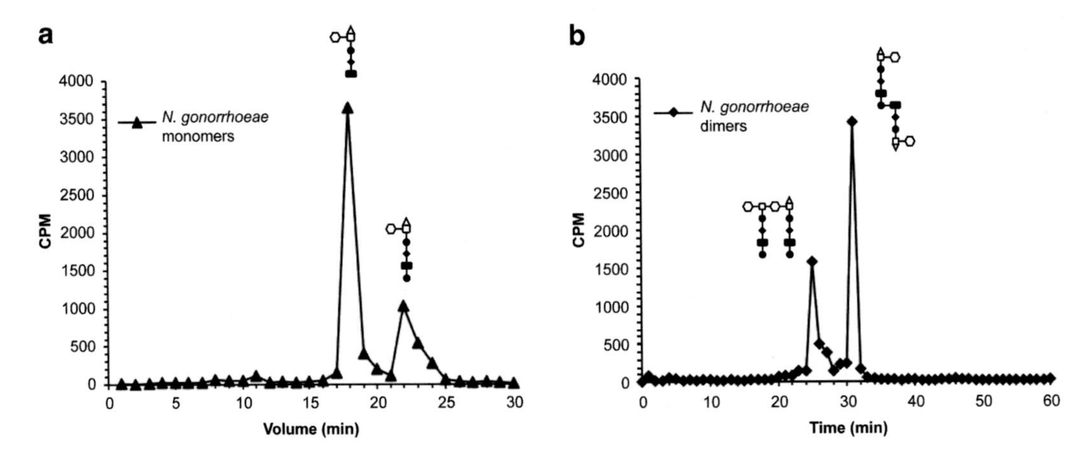

Fig. 3 Reversed-phase HPLC of radiolabeled fragments. Pooled fractions from size-exclusion chromatography of wild-type *N. gonorrhoeae* supernatant were analyzed by HPLC with a C18 column. (**a**) Monomer fractions were run using a 4–13 % gradient of acetonitrile with 0.05 % TFA at 1 mL/min for 30 min. Peaks with retention times of 18 min and 22 min correspond to 1,6-anhydrodissacharide-tripeptide and -tetrapeptide, respectively. Quantification of peaks reveals a 3:1 ratio of tripeptide to tetrapeptide peptidoglycan monomers released. (**b**) Dimer fractions were run over a 5–25 % gradient of acetonitrile with 0.05 % TFA at 1 mL/min for 60 min. Peaks with retention times of 25 and 31 min correspond to glycosidically linked and peptide-linked dimers, respectively

HPLC utilizing a C18 column for separating two related species of peptidoglycan monomer (Fig. 3a) and two configurations of peptidoglycan dimer (Fig. 3b), both of which elute as single peaks from size-exclusion columns. Our goal is to provide these protocols as a framework for the exploration of peptidoglycan fragment release in many species. Determining the peptidoglycan fragments released by various bacteria has the potential to reveal not only novel information about host–bacteria and bacteria–bacteria interactions but also help to better define the basic biology of peptidoglycan metabolism.

2 Materials

Prepare all media and solutions with deionized water and autoclave or filter-sterilize. The media listed below was originally adapted for *N. gonorrhoeae* (GC), but has also been successfully used for radiolabeling *E. coli* peptidoglycan. For species with specific nutritional requirements, other growth media can be employed provided a version not containing glucose is used during pulse-labeling.

2.1 [^{3}H]-Glucosamine Labeling of Peptidoglycan Sacculi

1. GC medium base (GCB): 1.5 % proteose peptone no. 3, 0.4 % K_2HPO_4, 0.1 % KH_2PO_4, 0.1 % NaCl, 0.1 % corn starch, and 1.0 % agar.
2. GC medium base broth (GCBL): 1.5 % proteose peptone no. 3, 0.4 % K_2HPO_4, 0.1 % KH_2PO_4, 0.1 % NaCl, and 0.042 % sodium bicarbonate.

3. GC medium base broth with glucose supplements (cGCBL): GCBL with Kellogg supplements I (22.2 mM glucose, 0.68 mM glutamine, 0.45 mM cocarboxylase) and II (1.23 mM $Fe(NO_3)_3$).
4. GCBL with pyruvate supplements. GCBL with Kellogg supplements I (36.35 mM pyruvate, 0.68 mM glutamine, 0.45 mM cocarboxylase) and II (1.23 mM $Fe(NO_3)_3$).
5. GCB plates: GC medium base with Kellogg supplements I and II added.
6. CO_2 incubator.
7. Petri dishes.
8. Tube rotator.
9. Polyester-tipped applicator swabs.
10. Pipette with disposable tips.
11. Vortex mixer.
12. [6-^{3}H]-glucosamine.
13. Sterile 15 mL conical vials.
14. Clinical centrifuge.
15. Microcentrifuge.
16. Microcentrifuge tubes.
17. 10 mL syringes.
18. 0.2 μm syringe filters.
19. Scintillation counter.
20. Scintillation vials.
21. Scintillation fluid.
22. Boiling water bath.
23. 50 mM sodium acetate pH = 5.
24. 8 % (w/v) sodium dodecyl sulfate solution.
25. Microcentrifuge cap locks (for boiling steps).

2.2 Size-Exclusion Chromatography

1. 1 L Erlenmeyer flasks for reconstituting and autoclaving beads.
2. Two 2.5 cm × 75 cm glass columns (e.g., Bio-Rad Econo-Columns).
3. Column buffer reservoir (500 mL–1 L capacity).
4. 3.2 mm (internal dimension) silicone tubing.
5. Stopcocks.
6. Polyacrylamide beads with low molecular weight exclusion limit (e.g., Bio-Rad Bio-Gel P6, medium size, with 6000 Da nominal exclusion limit).
7. Polyacrylamide beads with high molecular weight exclusion limit (e.g., Bio-Rad Bio-Gel P30, medium size, with 40,000 Da nominal exclusion limit).

8. 0.1 M lithium chloride (LiCl), autoclaved (1–2 L/column run plus 5 L if pouring new columns).
9. Disposable pipettes and pipetting device.
10. Automated fraction collector.
11. Disposable borosilicate glass tubes to fit fraction collector.
12. Scintillation vials with lids.
13. Scintillation fluid cocktail (i.e., PerkinElmer Ultima-Flo AP).
14. Repeating scintillation fluid dispenser (recommended).
15. Vortex mixer.
16. Scintillation counter capable of measuring tritiated [^{3}H] samples.
17. 2.5 cm × 20 cm (98 mL volume) glass column for desalting.
18. Speed-Vac, lyophilizer, rotary evaporator, or other concentrating device.

2.3 Reversed-Phase HPLC

1. HPLC, UPLC or other liquid chromatography system with loading loop and trigger-style injector.
2. C18 analytical column with in-line guard column (protocols described are optimized for a 250 mm × 4.6 mm column with 5 μm particle size).
3. Water, HPLC-grade, submicron filtered.
4. Trifluoroacetic acid (TFA).
5. Acetonitrile (ACN), HPLC grade.
6. 500–1000 mL glass bottles thoroughly cleaned with ultrapure or HPLC-grade water.
7. Automated fraction collector for HPLC.
8. Disposable borosilicate glass tubes to fit fraction collector.

3 Methods

3.1 Quantitative Metabolic Radiolabeling of Bacteria and Collection of Released Peptidoglycan Fragments

Metabolic radiolabeling of bacteria is accomplished under conditions that promote the rapid incorporation of labeled precursors into peptidoglycan, typically conditions of exponential growth. The conditions listed here have been refined for labeling of *Neisseria* species, but these conditions also support labeling of *E. coli* and should be considered adaptable to suit the particular growth requirements of most bacteria (Fig. 2). In general, bacteria should be cultured at a temperature promoting rapid growth (37 °C with aeration for *E. coli* and *Neisseria*). All media should be prepared and warmed in advance to avoid cold-shock and autolysis. Manipulations of cultures can be done at room temperature, but work should be done quickly.

1. Streak out frozen stocks of strains to be labeled to single colonies on GCB plates. Incubate plates overnight in a 37 °C incubator with 5 % CO_2.
2. The following day, streak colonies onto GCB plates. Grow plates overnight at 37 °C with 5 % CO_2. Make liquid media: GCBL, GCBL with glucosamine supplements (cGCBL), and GCBL with pyruvate supplements.
3. Aliquot media needed for the next day's experiments into 15 mL conical tubes and place at 37 °C to warm.
4. On the following day, begin cultures for radiolabeling. Swab each overnight plate into 3 mL of warm cGCBL, vortex, and measure OD_{540}.
5. Use swabbed cultures to create 2× 3 mL cultures at $OD_{540} = 0.25$ in cGCBL in 15 mL conical tubes. Work quickly to avoid cold-shock and autolysis.
6. Grow cultures at 37 °C in a roller drum for 3 h or until late log phase.
7. After 3 h, measure the OD_{540} of cultures to determine the appropriate volume of culture to obtain 2× 3 mL cultures at $OD_{540} = 0.2$ (approximately 1×10^8 cfu/mL) and transfer the culture volume necessary for seeding each culture into a microcentrifuge tube(s).
8. Centrifuge cells at 15,000 × *g* for 1 min, discard supernatant, and wash cells with 1 mL of warm GCBL.
9. Transfer cells to 15 mL conical tubes and bring to a total volume of 3 mL in warm GCBL with pyruvate supplements.
10. Add 10 μCi/mL of [^{3}H]-glucosamine to each 3 mL culture (2 cultures per strain) and grow as above for 45 min to pulse-label. The pulsing time can be extended to label more peptidoglycan, but longer times will eventually lead to the labeling of other cellular components. If labeling efficiency is determined to be too low, the concentration of [^{3}H]-glucosamine per culture or the total number bacteria labeled can be increased.
11. At the conclusion of pulse-labeling, centrifuge cultures in a clinical centrifuge for 5 min at 3500 × *g*. Discard media and use 1 mL warm GCBL to transfer bacteria to a microcentrifuge tube. Centrifuge bacteria for 1 min at 15,000 × *g*. Remove supernatant and wash pellet with 1 mL warm GCBL, centrifuging again for 1 min at 15,000 × *g* and discarding media to remove unincorporated label. Suspend each strain in 6 mL warm cGCBL.
12. For quantitative comparisons between strains, immediately take 60 μL of each 6 mL culture for scintillation counting. Place strains at 37 °C while measurement is taking place. Using scintillation measurements, calculate the CPM/mL of each culture and normalize the CPM to match the culture

with the lowest reading by removing volume from cultures with higher readings. If only one strain is being labeled or if quantitative comparison is not desired, this step can be skipped (*see* **Note 1**).

13. Split 6 mL volume of each strain into 2× 3 mL cultures (to provide sufficient aeration) and return cultures to the roller drum at 37 °C to continue growth for 2.5 h.
14. After the chase period is completed, centrifuge cultures in a clinical centrifuge for 5 min at 3500 × *g*. Remove supernatant from each strain and filter through a 0.2 μm syringe filter. If desired, remove 60 μL of filtered supernatants and measure CPM by scintillation counting to determine the percentage of peptidoglycan released relative to the beginning of the chase period.
15. Store supernatants at −20 °C for further analysis by size-exclusion chromatography and/or HPLC.

3.2 Analysis of Peptidoglycan Turnover

The rate of peptidoglycan turnover is determined from the amount of radiolabeled peptidoglycan remaining in the sacculi during a chase period following pulse-labeling. This method measures the amount of radiolabel in the sacculi as peptidoglycan is removed from the cell wall and either metabolized into other cellular components or released into the environment.

1. To assess the rate of peptidoglycan turnover, begin by metabolically labeling cultures and normalizing culture volumes to total CPM by completing Subheading 3.1, **steps 1–12**.
2. Return 6 mL cultures to the roller drum at 37 °C to continue growth. At desired time points (e.g., 0, 0.5, 1, 2, 4 h), remove 1 mL from each culture and transfer to a microcentrifuge tube. Centrifuge for 1 min at 15,000 × *g* and remove supernatant.
3. To isolate macromolecular peptidoglycan, suspend bacterial pellets in 165 μL of 50 mM sodium acetate (pH 5) and 165 μL 8% sodium dodecyl sulfate. Boil samples for 30 min using microcentrifuge cap locks.
4. Add 800 μL unlabeled carrier (*see* **Note 2**) and collect insoluble macromolecular peptidoglycan by centrifugation for 30 min at 17,000 × *g*, 15 °C or room temperature (SDS will precipitate at lower temperatures).
5. Carefully remove the supernatant and suspend the insoluble peptidoglycan pellet in 200 μL of sterile water. Measure CPM from each time point to calculate the amount of labeled peptidoglycan remaining at each time point compared to $T = 0$. The rate of radioactivity loss from the sacculi provides the peptidoglycan turnover rate. The entire peptidoglycan turnover procedure should be repeated for a total of three independent experiments to calculate the significance of observed differences in peptidoglycan turnover.

3.3 Size-Exclusion Chromatography

Released peptidoglycan fragments in conditioned medium from bacterial culture must be separated from any extraneous labeled sugars, large fragments of lysed sacculi, and components of the culture medium that could impede downstream analysis of fragments. The following method describes the use of size-exclusion chromatography to separate and collect fractions containing commonly observed peptidoglycan fragments. The amount and relative proportions of fragments can be determined by quantifying total CPM of each peak that corresponds to certain released fragments.

3.3.1 Pouring New Size-Exclusion Columns

1. If pouring a new tandem size-exclusion chromatography column, hydrate and sterilize both the low and high molecular weight exclusion limit beads prior to use. A general rule for packing a column is to use twice the buffer volume as the total column volume for bead hydration. For the Bio-Rad Bio-Gel beads, hydrate 52 g of P30 and 70 g of P6 beads (separately) in 500 mL of 0.1 M LiCl overnight. The next day, autoclave beads on a 30 min liquid cycle and let cool to room temperature.
2. Once beads have settled, decant as much liquid as possible and add 500 mL of fresh, autoclaved 0.1 M LiCl, swirling to mix. Let settle and repeat the decanting and filling process four times.
3. Prior to filling, sterilize columns by autoclaving at 121 °C for 15 min or by rinsing columns with either 2 N NaOH or 100% ethanol followed by two rinses with 0.1 M LiCl. Securely attach a stopcock or other on/off valve directly to the bottom of each column.
4. Once the beads have settled, decant ~150 mL of the excess buffer, swirl beads to resuspend, and pour the P6 beads into one column and the P30 beads into the other, making sure to keep the stopcocks open during filling and adding buffer as needed to keep columns hydrated. Save any beads (especially P6) that do not fit initially. Each column will hold ~350 mL of hydrated beads. Stop column flow by closing the stopcocks when columns are filled. Allow beads to settle 12–18 h.
5. Following settling, pipette off any excess LiCl from the top of the P6 column and fill to the top with any remaining hydrated P6 beads to avoid leaving any air/buffer pocket.
6. To connect the columns, mount the P30 column above the P6 column in a standing clamp apparatus (a minimum 8 feet of clearance floor-to-ceiling is recommended for proper installation). Remove the stopcock from the column filled with P30 beads and use silicone tubing to attach the bottom of the P30 column to the top of the P6 column. When complete, attach a reservoir to the top of the P30 column. Additional tubing will be needed at the bottom of the P6 column to connect the outflow to an automated fraction collector.

7. The reservoir at the top of the tandem column apparatus should always be kept with autoclaved 0.1 M LiCl buffer above the level of the beads when the column is not in use. Beads should never be allowed to dry after being hydrated. If a column does dry (typically the upper column), repeat Subheading 3.3.1, **steps 1–7** for the dried column. Used beads can be repoured following rehydration and autoclaving.

3.3.2 Running Samples on Size-Exclusion Columns

1. To introduce sample to prepared columns, carefully remove buffer from the reservoir with a pipette, avoiding disturbing the beads at the top of the column. To completely remove the liquid, excess buffer can be drained off by opening the stopcock.
2. Once buffer has drained to expose the top of the beads, apply filtered culture supernatant (Subheading 3.1, **step 15**) to the top of the column by decanting or pipetting a single labeled sample onto the column. Immediately begin a timer to start tracking the void time (this must be determined empirically but is typically 2–4 h) (*see* **Note 3**). Allow the sample to soak into the column.
3. Once the sample has entered the column, apply a small amount of 0.1 M LiCl to the top of the column (5–10 mL), and allow this buffer to soak in as above. Adding a small amount of buffer here keeps the column flowing and hydrated, without risking dilution of your sample.
4. Once a small amount of buffer has flowed into the column, the reservoir can be filled to capacity and covered with a vented lid to avoid contamination. During the column run (12–18 h) this reservoir should always contain 0.1 M LiCl, which can be achieved through the use of a large reservoir (>1 L) or by establishing a gravity-fed continuous flow or siphon to the column-mounted reservoir from a larger source bottle or tank (*see* **Note 4**).
5. At the conclusion of the void time, initiate collection of fractions, which can be collected based on time or volume. Automated fraction collectors that hold ≥175 collection tubes are recommended to avoid the need to switch collection drums or racks during collection, or collect manually. Automated units that collect volumes typically count drops via an electronic eye and may require some testing to refine the desired volume. For the analysis of peptidoglycan fragments from gonococci, 150–170× 3 mL fractions (at ~75 drops/fraction) are needed following the void volume, with the exact number based upon the labeling technique (*see* **Notes 5** and **6**). At the conclusion of fraction collecting, it may be desirable to run the column with buffer for an additional 2–3 h to assure the voiding of all labeled material prior to the next run. When complete, be sure to stop column flow and cap the column (*see* **Note 7**).

6. To measure radiolabeled peptidoglycan fragments, a portion of each collected fraction should be mixed with scintillation fluid. For [^{3}H]-glucosamine-labeled fragments, a mixture consisting of 0.5 mL of fraction volume and 3 mL scintillation fluid (i.e., Ultima-Flo AP—PerkinElmer) is sufficient for detection. To increase detection, a mixture of 1 mL of fraction volume and 3 mL of scintillation fluid can be used.
7. Detection of radiation is performed by scintillation counting and CPM/fraction can be graphed to determine when fragments elute and in what proportions (Fig. 1). Samples labeled together (quantitatively) in a single experiment should be analyzed in successive runs on the same sizing column and can be initiated at the conclusion of the previous run. Graph CPM per fraction, or CPM per column volume, to determine which fractions make up peaks containing peptidoglycan fragments of interest (Fig. 1)
8. Save the remaining unmeasured portion of each fraction for additional analysis. As peaks of interest are identified from scintillation counting, fractions that make up those peaks can be pooled and stored at −20 °C.

3.3.3 Concentration and Desalting of Collected Fractions for HPLC

Analysis of radiolabeled fractions requires only a portion of the total fraction volume, leaving sufficient material for further analysis. Since downstream HPLC analysis is often desired and the injection volume for most HPLC instrumentation is small (10 μL–2 mL, depending on the column size and injection loop), it is often necessary to concentrate all fractions comprising a single peak into a smaller volume. Concentration of fractions, however, will result in concentration of the LiCl from the column running buffer. It is recommended to remove this salt prior to separation by HPLC, though a shortcut to increase the throughput of analysis is described below (*see* **Note 8**).

1. A 2.5 cm × 20 cm glass column (98 mL volume) should be prepared with hydrated, sterilized P6 beads as above, except that Ultrapure or HPLC-grade water should be substituted for LiCl, since this column will be used for desalting. Column should be fitted with a stopcock on the bottom and reservoir on the top.
2. Pooled fractions of interest from Subheading 3.3.2, **step 8** should be reduced in volume by dehydration (typically in a Speed-Vac or lyophilizer apparatus), in equipment approved for radioactive materials.
3. Once dehydrated and suspended in a smaller volume (<3 mL), sample can be applied to the top of the column (as in Subheading 3.3.2, **steps 2–4**), using water to fill the reservoir rather than LiCl. Once sample has been applied, collect the void volume into a graduated cylinder or other container with

volume markings. Depending on the size and chemical composition of the fragment of interest (peptidoglycan monomer, peptidoglycan dimer, peptides, etc.), different combinations of void volume and fractions collected may be required and should be determined empirically. Generally for a 2.5× 20 cm column, collection of a ~20 mL void is followed by manual collection of 15–20× 1 mL fractions to retrieve peptidoglycan monomers.

4. Each desalted fraction should be sampled (~100 μL out of 1 mL), mixed with 3 mL scintillation fluid, and measured by scintillation counting to confirm that the radioactive fraction applied to the column was recovered. Peptidoglycan fragments should elute as distinct peaks spread over several fractions.
5. Aqueous products containing detectable radiation should then be reduced in volume as above and suspended in a small volume (100–500 μL). This final suspension should be measured for levels of radioactivity prior to storage (at −20 °C) and is suitable for analysis by HPLC.

3.4 Analysis of Fragments by Reversed-Phase HPLC

Size-exclusion chromatography provides information on the size and quantity of released peptidoglycan fragments, allowing quantitative comparisons of released fragments within a single sample or between various strains, species, or conditions. Details such as the exact peptide stem length among monomers, the linkages present within dimers, and whether fragments have reducing or 1,6-anhydro ends are better explored through additional chromatography techniques. Reversed-phase HPLC using a C18 column is a common approach for analyzing peptidoglycan fragments (and is described here). Other columns including those with different carbon-chain length bonded phases, size-exclusion columns, cation/anion exchange columns, and a variety of length and particle size options could be considered depending on the experimental question (*see* **Note 9**).

1. Prior to separation of peptidoglycan fragments by HPLC, necessary buffers should be made and degassed (under vacuum) in advance. Many effective separations of peptidoglycan monomers, dimers, and peptides on C18 columns can be accomplished with two buffers: (a) HPLC-grade water + 0.05 % TFA, and (b) 25 % acetonitrile (in HPLC-grade water) + 0.05 % TFA. These buffers should ideally be made fresh or used within 1 week. Prior to the first use of the day, prime the HPLC lines and clean the column first by flushing with Buffer B then with Buffer A. A blank sample (i.e., buffer only) should be run before each day of use.
2. While the analysis of radiolabeled fragments will typically involve detection by scintillation counting of fractions, peptidoglycan

fragments can also be detected at 206 nm. UV detection should be performed any time fragments are separated by HPLC to assure that the HPLC is operating properly. Turn on the UV lamp and allow to warm for at least 1 h to stabilize readings.

3. Create an HPLC program for the samples to be analyzed. For separation of monomers, a gradient of 4–13% acetonitrile over 30 min at 1 mL/min allows separation of 1,6-anhydrodisaccharide-tripeptide monomer from 1,6-anhydrodisaccharide-tetrapeptide monomer (Fig. 3a). Reducing-end fragments have a shorter retention time compared to fragments with anhydro linkages.
4. Prior to starting the run, prepare the sample from Subheading 3.3.3, **step 5** so that at least 1000 CPM will be injected into the column. Samples can be prepared for loading either via the "partial fill" method (<1/2 loop volume) if sample is limiting, or the "complete fill" method (2–5 loop volumes) for greatest precision.
5. Immediately following injection, begin fraction collection using an automated fraction collector set to collect fractions using ≤1 min increments over the entire run time.
6. At the conclusion of the run, combine the entire volume of each fraction with 3 mL scintillation cocktail, vortex thoroughly to mix, and measure by scintillation counting (Fig. 3). Clean column after each run by running 25% acetonitrile (100% of Buffer B) to elute any remaining material from the column, then flushing with Buffer A.

4 Notes

1. If quantitative labeling is desired (in Subheading 3.1, **step 12**) but there is not immediate access to a scintillation counter, collect 60 μL of culture as above and take CPM readings when possible. At the conclusion of the chase period, take another 60 μL sample from the filtered supernatant and measure CPM. It is then possible to normalize the counts from the released fragments to the total counts at the beginning of the chase period. Although this method is not preferred, fractions can be characterized as % of total CPM.
2. To increase the efficiency of sacculi recovery and make centrifuged material more easily visible, it is common to use unlabeled (cold) carrier when working with small quantities of radiolabeled peptidoglycan. Carriers can be in the form of unlabeled macromolecular peptidoglycan or even whole bacterial cells. One commercially available option is lyophilized preparations of *Micrococcus luteus* peptidoglycan.

3. When establishing a new size-exclusion chromatography system, or repouring columns, it is recommended to determine the time required for samples to pass entirely through the system (void time), since this will impact the time range for collecting relevant fractions. To determine the void time, prepare a 3–6 mL solution of Blue Dextran (Sigma-Aldrich) in water and apply to the top of the sizing column as in Subheading 3.3.2. Proceed as if running a labeled sample, timing the movement of the blue dextran until it exits the bottom of the columns. The time from addition of blue dextran solution until it exits the column is the void time.
4. Always monitor a size-exclusion chromatography column for which no large-volume reservoir is providing continuous flow. Never allow a column to run dry. Cracked or caked beads that result from significant drying are signs that the column should be emptied of beads and repoured. Over time, prolonged use will cause the beads in the column to compact, slowing the flow of samples and changing the void time. If unacceptable slowing of runs or broadening of known peaks is observed, the column(s) should be emptied of beads and repoured.
5. As an alternative to labeling the peptidoglycan sugar backbone using [^{3}H]-glucosamine, [^{3}H]-*rac*-2,6-diaminopimelic acid ([^{3}H]-DAP) can be used to label peptide stems. Labeling with [^{3}H]-DAP can be done by substituting GCBL with pyruvate supplements for Dulbecco's Modified Eagle's Medium (DMEM) without cysteine. At Subheading 3.1, **step 9**, wash and dilute cells into DMEM without cysteine supplemented with 100 μg/mL methionine and 100 μg/mL threonine. Due to less efficient labeling by [^{3}H]-DAP, add 20 μCi/mL [^{3}H]-DAP to each 3 mL culture at Subheading 3.1, **step 10** and proceed as noted above.
6. Different peptidoglycan precursors and different isotopes are available that can be used to radiolabel peptidoglycan. The best peptidoglycan precursor to use will depend of the species and strain of bacteria to be labeled. For example, *meso*-DAP is an amino acid specific to peptidoglycan but is only found in gram-negatives and some gram-positive rods (gram-positives typically use L-Lysine in place of *meso*-DAP). Combinations of precursors and isotopes, such as [^{14}C]-glucosamine in combination with [^{3}H]-DAP, can be used to measure the release of both sugars and peptides simultaneously.
7. Care should always be taken to avoid spills and overflows of radioactive material. Column void and flow-through material should be collected in a sufficiently sized container placed within secondary containment and surrounded by absorbent material. All fractions and flow-through should be collected, tested, and disposed of in accordance with local and institutional disposal regulations.

8. HPLC is routinely used for the desalting of nucleotides and proteins and some columns can handle the introduction of soluble salt without changing retention times. To increase the throughput of fraction analysis from size-exclusion chromatography, it is possible to run a portion of pooled fractions without the lengthy concentration-desalting-concentration protocol. To skip directly from obtaining column fractions to analysis by HPLC, the HPLC injector should be fitted with a larger volume loop (about 2 mL). Pooled fractions from size-exclusion (typically 25 mL from the center of the peak of interest) must also be sufficiently radioactive that >1000 CPM are available in your loaded volume to achieve reliable detection of products. Pooled fractions taken directly from a sizing column (Subheading 3.3.2, **step 8**) can then be run as in Subheading 3.4, **step 2**, though caution is advised when interpreting comparisons between these runs and any analyses done with primarily aqueous samples loaded from smaller-volume loops.
9. In complex samples analyzed by HPLC, it is possible that multiple products can elute simultaneously. These problems can occasionally be resolved by executing a longer run with a slower ramp or modifying the beginning or ending concentrations of acetonitrile. In some cases it may be necessary to use a different HPLC column for full separation of certain products.

Acknowledgements

This work was supported by the National Institutes of Health through grants AI097157 and AI099539.

References

1. Höltje JV (1998) Growth of the stress-bearing and shape-maintaining murein sacculus of *Escherichia coli*. Microbiol Mol Biol Rev 62(1):181–203
2. Typas A, Banzhaf M, Gross CA, Vollmer W (2012) From the regulation of peptidoglycan synthesis to bacterial growth and morphology. Nat Rev Micro 10(2):123–136. doi:10.1038/nrmicro2677
3. Goodell EW (1985) Recycling of murein by *Escherichia coli*. J Bacteriol 163(1): 305–310
4. Reith J, Mayer C (2011) Peptidoglycan turnover and recycling in Gram-positive bacteria. Appl Microbiol Biotechnol 92(1):1–11. doi:10.1007/s00253-011-3486-x
5. Litzinger S, Duckworth A, Nitzsche K, Risinger C, Wittmann V, Mayer C (2010) Muropeptide rescue in *Bacillus subtilis* involves sequential hydrolysis by beta-*N*-acetylglucosaminidase and *N*-acetylmuramyl-L-alanine amidase. J Bacteriol 192(12):3132–3143. doi:10.1128/JB.01256-09
6. Cloud-Hansen KA, Peterson SB, Stabb EV, Goldman WE, McFall-Ngai MJ, Handelsman J (2006) Breaching the great wall: peptidoglycan and microbial interactions. Nat Rev Microbiol 4(9):710–716. doi:10.1038/nrmicro1486
7. Dworkin J (2014) The medium is the message: interspecies and interkingdom signaling by peptidoglycan and related bacterial glycans. Annu Rev Microbiol 68:137–154. doi:10.1146/annurev-micro-091213-112844
8. Cookson BT, Cho HL, Herwaldt LA, Goldman WE (1989) Biological activities and chemical composition of purified tracheal cytotoxin of *Bordetella pertussis*. Infect Immun 57(7):2223–2229

9. Sinha RK, Rosenthal RS (1980) Release of soluble peptidoglycan from growing conococci: demonstration of anhydro-muramyl-containing fragments. Infect Immun 29(3):914–925
10. Vollmer W, Joris B, Charlier P, Foster S (2008) Bacterial peptidoglycan (murein) hydrolases. FEMS Microbiol Rev 32(2):259–286. doi:10.1111/j.1574-6976.2007.00099.x
11. Jacobs C, Huang LJ, Bartowsky E, Normark S, Park JT (1994) Bacterial cell wall recycling provides cytosolic muropeptides as effectors for beta-lactamase induction. EMBO J 13(19): 4684–4694
12. Desmarais SM, de Pedro MA, Cava F, Huang KC (2013) Peptidoglycan at its peaks: how chromatographic analyses can reveal bacterial cell wall structure and assembly. Mol Microbiol 89(1):1–13. doi:10.1111/mmi.12266
13. Magalhaes JG, Philpott DJ, Nahori MA, Jehanno M, Fritz J, Le Bourhis L, Viala J, Hugot JP, Giovannini M, Bertin J, Lepoivre M, Mengin-Lecreulx D, Sansonetti PJ, Girardin SE (2005) Murine Nod1 but not its human orthologue mediates innate immune detection of tracheal cytotoxin. EMBO Rep 6(12):1201–1207. doi:10.1038/sj.embor.7400552
14. Veyrier FJ, Williams AH, Mesnage S, Schmitt C, Taha MK, Boneca IG (2013) De-O-acetylation of peptidoglycan regulates glycan chain extension and affects in vivo survival of *Neisseria meningitidis*. Mol Microbiol 87(5): 1100–1112. doi:10.1111/mmi.12153

Chapter 15

Analysis of Cell Wall Teichoic Acids in *Staphylococcus aureus*

Gonçalo Covas, Filipa Vaz, Gabriela Henriques, Mariana G. Pinho, and Sérgio R. Filipe

Abstract

Most bacterial cells are surrounded by a surface composed mainly of peptidoglycan (PGN), a glycopolymer responsible for ensuring the bacterial shape and a telltale molecule that betrays the presence of bacteria to the host immune system. In *Staphylococcus aureus*, as in most gram-positive bacteria, peptidoglycan is concealed by covalently linked molecules of wall teichoic acids (WTA)—phosphate rich molecules made of glycerol and ribitol phosphates which may be tailored by different amino acids and sugars.

In order to analyze and compare the composition of WTA produced by different *S. aureus* strains, we describe methods to: (1) quantify the total amount of WTA present at the bacterial cell surface, through the determination of the inorganic phosphate present in phosphodiester linkages of WTA; (2) identify which sugar constituents are present in the assembled WTA molecules, by detecting the monosaccharides, released by acid hydrolysis, through an high-performance anion exchange chromatography analysis coupled with pulsed amperometric detection (HPAEC-PAD) and (3) compare the polymerization degree of WTA found at the cell surface of different *S. aureus* strains, through their different migration in a polyacrylamide gel electrophoresis (PAGE).

Key words *Staphylococcus aureus*, Bacterial cell wall, Bacterial cell surface, Wall teichoic acids (WTA), Bacterial glycopolymers, Monosaccharide analysis, PAGE, HPLC, HPAEC-PAD

1 Introduction

The cell envelope of gram-positive bacteria contains several cell-wall glycopolymers (CWG) that surround a thick multilayer peptidoglycan (PGN) matrix. PGN is a polymer of glycan chains composed of alternated residues of *N*-acetylglucosamine and *N*-acetylmuramic acid, with short peptide chains cross-linking different glycan chains. Despite the fact that bacterial PGNs have a relatively conserved composition and architecture [1], structures of the CWG are highly diverse and often species specific [2]. Among these CWG, two classes of anionic phosphate rich polysaccharides are frequently found: wall teichoic acids (WTA), which

Hee-Jeon Hong (ed.), *Bacterial Cell Wall Homeostasis: Methods and Protocols*, Methods in Molecular Biology, vol. 1440, DOI 10.1007/978-1-4939-3676-2_15,

can be attached to the peptidoglycan, and lipoteichoic acids (LTA), that are linked to a membrane lipid.

Teichoic acids (TA) contribute to a variety of processes in the bacterial metabolism, including resistance to environmental stresses, such as heat [3] or low osmolarity [4], to antimicrobial peptides [5], and to lytic enzymes produced by the host, such as lysozyme [6, 7]. TA can also act as receptors for phage particles [8], and provide a reservoir of cationic ions close to the bacterial surface, particularly of magnesium ions that may be important for the activity of different bacterial enzymes [9]. Using *Staphylococcus aureus* as a bacterial model, we have further shown that WTA ensure the localization of proteins involved in the synthesis of a highly polymerized PGN [10] and are capable of concealing the PGN at the bacterial cell surface from detection by host immune PGN receptors [11].

S. aureus, as most gram-positive bacteria, usually contains only one type of WTA and one type of LTA [2]. In these microorganisms, WTA is anchored to the *N*-acetylmuramic acid residue in the PGN, through a phosphodiester bond than connects to the WTA linker unit composed of *N*-acetylglucosamine and *N*-acetylmannosamine. This linker is coupled to a polymer of phosphodiester linked ribitol units by two glycerol-phosphate (GroP) units (Fig. 1a). The ribitol-phosphate monomers can be repeated up to 40 times and may be linked to D-alanine and/or *N*-acetylglucosamine residues [12]. WTA molecules can be extracted for analysis through the disruption of the phosphodiester bonds that connect WTA to the surface of bacteria, by alkaline hydrolysis [13].

In this chapter, we describe methods to analyze the WTA present at the surface of different *S. aureus* strains. We describe how to determine the total amount of inorganic phosphate present in WTA extracts [14], which may be used to assess the total amount of WTA produced by different staphylococcal strains. We also describe how to determine the monosaccharide composition of WTA extracts. This is done using high performance anion exchange chromatography coupled with pulsed amperometric detection (HPAEC-PAD) analysis [15] to analyze sugar residues released by acidic hydrolysis of bacterial WTA [16]. Finally, we describe how to analyze the degree of polymerization of the extracted WTA molecules [10, 11, 17], by Native-PAGE followed by alcian blue–silver staining [18, 19].

2 Materials

Prepare all solutions using ultrapure water with a resistivity ≥18 MΩ cm at 25 °C and analytical grade reagents. Prepare and store all reagents at room temperature (unless indicated otherwise). Diligently follow all recommended safety guidelines and waste disposal regulations recommended by your host institution.

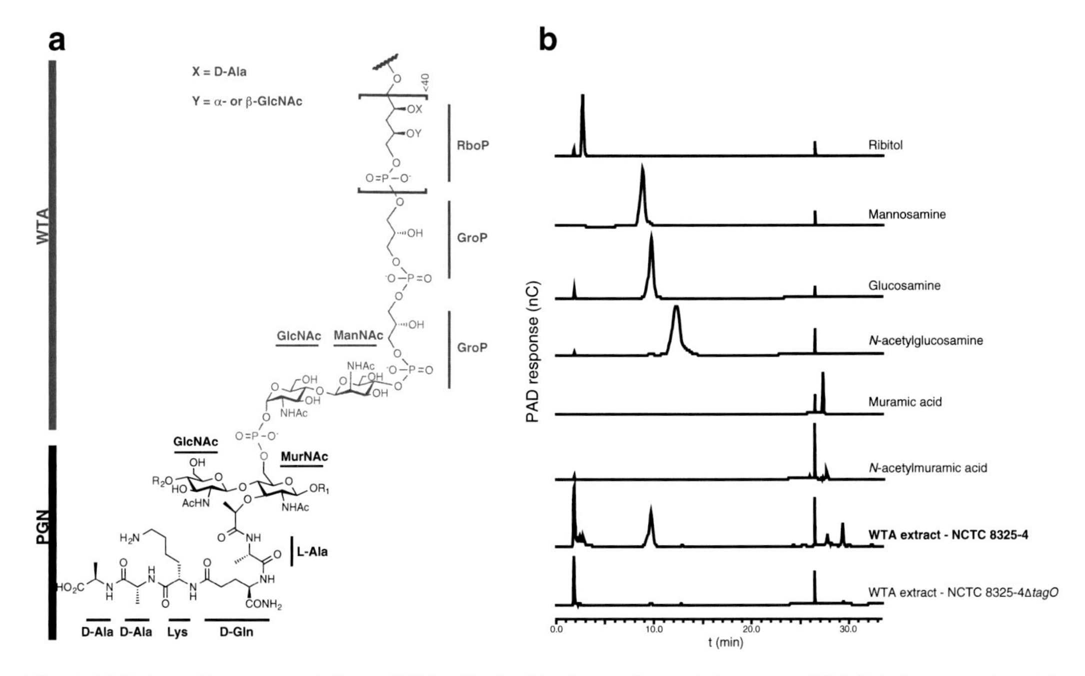

Fig. 1 (**a**) Schematic representation of WTA attached to the surface of *S. aureus*. WTA (*blue*) are anchored to the PGN macromolecule (*black*) through phosphodiester bond that links PGN *N*-acetylmuramic acid residue (Mur*N*Ac) to the WTA linker unit that is composed of *N*-acetylglucosamine (GlcNAc) and *N*-acetylmannosamine (ManNAc) residues. Two glycerol-phosphate (GroP) units connect the WTA linker to a polymer of phosphodiester linked ribitol units (RboP). These ribitol-phosphate monomers can be repeated up to 40 times and may be linked to D-alanine (D-Ala) and/or *N*-acetylglucosamine residues. The PGN stem peptide, which contains alanine, glutamine (D-Gln), lysine (Lys), is also represented. (**b**) Analysis of the WTA monosaccharide composition by HPAEC-PAD. Chromatograms showing the monosaccharide composition of WTA extracted from the NCTC 8325-4 Δ*tagO* mutant and parental strain are shown. As expected, WTA extracts from NCTC 8325-4 parental strain contained glucosamine and ribitol but no muramic acid residues. Glucosamine and ribitol residues were absent in WTA extracted from the Δ*tagO* mutant strain that is unable to produce WTA [11]

2.1 Extraction of WTA

1. Tryptic Soy Broth: 30 g in 1 L of double-distilled water. Autoclave at 121 °C for 15 min.
2. Buffer 1: 50 mM 2-(*N*-morpholino)ethanesulfonic acid (MES), pH 6.5. Dissolve 10.86 g of MES sodium salt in 900 mL of water. Adjust the pH with HCl and add water to 1 L.
3. Buffer 2: 50 mM MES, pH 6.5, 4 % (w/v) sodium dodecyl sulfate (SDS). Same as buffer 1 plus 40 g of SDS for 1 L of final volume (*see* **Note 1**).
4. Buffer 3: 50 mM MES, pH 6.5, 2 % (w/v) NaCl. Same as buffer 1 plus 20 g of NaCl for 1 L of final volume.
5. Buffer 4: 20 mM Tris–HCl, pH 8.0, 0.5 % (w/v) SDS. Dissolve 2.42 g of Tris and 5 g of SDS in 900 mL of water. Adjust the pH with HCl and add water to 1 L.

6. Proteinase K stock solution: Dissolve 20 mg of Proteinase K from *Tritirachium album* (Sigma-Aldrich) in 1 mL of water. Store at –20 °C.
7. Sodium hydroxide (NaOH) hydrolysis solution: 0.1 mM NaOH. Add 104.6 μL of 50% NaOH (ion chromatography grade, Fluka) to a final volume of 20 mL of water. Prepare fresh before use.
8. 250 mL conical flasks.
9. 50 mL Falcon tubes.
10. 2 mL Eppendorf tubes.
11. Centrifuge with rotors for 2 mL Eppendorf and 50 mL Falcon tubes.
12. Thermomixer equipped with heating block for 2 mL tubes.

2.2 WTA Phosphate Quantification

1. Phosphate (PO_4^{3-}) standard solution: 100.7 μg/mL. Dissolve 10.07 mg of KH_2PO_4 in 100 mL of water.
2. Vacuum concentrator (e.g., SpeedVac®).
3. Perchloric acid ($HClO_4$) solution: 70% (v/v) $HClO_4$.
4. Sulfuric acid (H_2SO_4) solution: 3 M. Add 31.9 mL of concentrated sulfuric acid (18.8 M) to a final volume of 200 mL of water.
5. Ascorbic acid solution: 3.3% (w/v). Dissolve 330 mg of ascorbic acid in 10 mL of water. Prepare fresh before use.
6. Ammonium molybdate solution: 2.5% (w/v). Dissolve 250 mg of ammonium molybdate in 10 mL of water. Prepare fresh before use.
7. P-reagent: Mix 6 mL of Ascorbic Acid solution, 1 mL of sulfuric acid solution and 2 mL of ammonium molybdate solution. Prepare fresh before use.
8. UV–Vis spectrophotometer.
9. Fume hood with a Bunsen burner.

2.3 Analysis of WTA Monosaccharides

1. Hydrochloric acid (HCl) hydrolysis solution: 3 M HCl. Add 5 mL of 37% (v/v) HCl to 15 mL of water.
2. 1 M NaOH solution: Add 104.6 mL of 50% NaOH (ion chromatography grade) to a final volume of 2 L of water. Filter with a 0.2 μm Stericup and Steritop. Prepare fresh before use and degas by incubating in an ultrasound bath for 15 min (*see* **Note 2**).
3. 1 M $NaCH_3COO$ solution: Add 164.06 g of $NaCH_3COO$ (HPLC grade) to 2 L of water. Filter with a 0.2 μm Stericup and Steritop. Degas by incubating in an ultrasound bath for 15 min.

4. 2 L of water with a resistivity ≥18 MΩ filtered with a 0.2 μm Stericup and Steritop and degased by incubating in an ultrasound bath for 15 min.
5. Thermomixer (e.g., Eppendorf®).
6. Vacuum concentrator with trap resistant to acidic and organic solvents (e.g., SpeedVac®).
7. Vortex.
8. Centrifuge with rotors for 2 mL Eppendorf and 50 mL Falcon tubes.
9. Ion chromatography system with Helium blanketing system, pulsed amperometric detector and disposable Electrodes for Carbohydrates (e.g., Dionex® ICS-5000).
10. CarboPac PA10 Analytical Column 4×250 mm (Dionex®).
11. Amino Trap Column 4×50 mm (Dionex®).
12. Borate Trap Column (Dionex®).
13. Stericup and Steritop filters (e.g., Millipore®).

2.4 Analysis of WTA Extracts

1. Native gel buffer: 3.0 M Tris–HCl, pH 8.5. Dissolve 363.42 g of Tris in 900 mL of water. Adjust the pH with HCl and add water to 1 L.
2. Acrylamide stock solution: 30% [T] acrylamide cross-linked with 2.7% bisacrylamide [C].
3. Ammonium persulfate (APS): 10% (w/v) in water. Prepare fresh before use.
4. *N*,*N*,*N*,*N*-tetramethylethylenediamine (TEMED).
5. Isopropanol.
6. Native PAGE running buffer: 0.1 M Tris–HCl, pH 8.2, 0.1 M Tricine. Dissolve 12.11 g of Tris and 17.91 g of Tricine in 900 mL of water. Adjust the pH with HCl and add water to 1 L.
7. 3× loading buffer: 0.3 M Tris–HCl, pH 8.2, 0.3 M Tricine, 30% (v/v) glycerol, 0.15% (w/v) bromophenol blue. Dissolve 3.6 g of Tris, 5.4 g of Tricine in 60 mL of water. Adjust the pH with HCl (*see* **Note 3**). Add 0.15 g of bromophenol blue and make up to 70 mL of water. Add 30 mL of glycerol. Mix well.
8. One time loading buffer: Mix 1 mL of 3× loading buffer with 2 mL of water.
9. Alcian blue staining solution: 5% (v/v) acetic acid, 40% (v/v) ethanol, 0.1% (w/v) alcian blue. Dissolve 1 g of alcian blue in a solution containing 50 mL of acetic acid, 400 mL of ethanol and 550 mL of water.
10. Alcian blue destaining solution: 10% (v/v) acetic acid, 20% (v/v) ethanol. Add 100 mL of acetic acid and 200 mL of ethanol to 700 mL of water.

11. Silver stain plus kit (e.g., Bio-Rad®).
12. Ultrasound bath.
13. Electrophoresis system compatible with 20 cm × 16 cm × 1.0 cm cassettes.
14. Orbital shaker.

3 Methods

Carry out all procedures at room temperature unless otherwise specified.

3.1 Extraction of WTA

1. Inoculate a single colony of *Staphylococcus aureus* into 40 mL of TSB in a 250 mL conical flask and incubate at 30 °C with aeration (200 rpm) overnight or until stationary growth is reached.
2. Collect 20 mL (*see* **Note 4**) of culture into a 50 mL Falcon tube by centrifugation (7000 × *g*, 15 min).
3. Resuspend the pellet in 20 mL of buffer 1 and centrifuge (7000 × *g*, 15 min).
4. Resuspend the pellet in 20 mL of buffer 2 and incubate in boiling water for 1 h (*see* **Note 5**).
5. Centrifuge the sample (7000 × *g*, 15 min) and resuspend the pellet in 2 mL of buffer 2.
6. Transfer the sample to a 2 mL Eppendorf tube and centrifuge (16,000 × *g*, 5 min).
7. Resuspend the pellet in 2 mL of buffer 2 and centrifuge (16,000 × *g*, 5 min).
8. Repeat the previous step first with 2 mL of buffer 3 and then with 2 mL of buffer 1.
9. Resuspend the pellet in 2 mL of buffer 4, add 2 μL of Proteinase K stock solution (*see* **Note 6**) and incubate at 50 °C for 4 h with shaking (1000 rpm).
10. Centrifuge the sample (16,000 × *g*, 5 min) and resuspend the pellet in 2 mL of buffer 3.
11. Repeat the previous step three times resuspending the pellet in 2 mL of water.
12. Resuspend the pellet in 1 mL of sodium hydroxide hydrolysis solution (*see* **Note 7**) and incubate at 25 °C for 16 h with shaking (1000 rpm).
13. After centrifugation (16,000 × *g*, 15 min), collect the supernatant containing the extractable WTA.
14. Store the samples at 4 °C until further analysis (*see* **Note 8**).

3.2 WTA Phosphate Quantification

Rinse all glassware with water before use (*see* **Note 9**).

1. Prepare triplicates of a phosphate calibration curve with 0, 10, 20, 40, 60, 80, and 100 μL of phosphate standard solution and lyophilize until dry.
2. Lyophilize 1, 2, and 5 μL of WTA extract to dryness (*see* **Note 10**).
3. Add 50 μL of perchloric acid solution to each sample and calibration points and transfer them to glass test tubes.
4. Gently heat each test tube near the flame in a fume hood. When fumes are released, remove tube from the flame and allow released fumes to condense inside the tube. Repeat this step three more times.
5. Add 950 μL of water and 500 μL of P-reagent.
6. Cap the test tubes and mix by vortexing.
7. Incubate at 37 °C for 90 min.
8. Read absorbance at 820 nm. Typical values of phosphate per μL of WTA extract are shown on Fig. 2c.

3.3 Analysis of WTA Monosaccharides

1. Start the acid hydrolysis by adding 20 μL of WTA extract to 1 mL of hydrochloric hydrolysis solution in a 1 mL Eppendorf tube. Incubate at 90 °C for 2 h with shaking (1000 rpm).
2. After hydrolysis, puncture the Eppendorf tube cap and lyophilize the sample until completely dry in a vacuum concentrator.
3. Add 1 mL of water to the Eppendorf tube and gently vortex to resuspend the sample.
4. Lyophilize until complete dryness in a vacuum concentrator to eliminate remains of acid solution.
5. Add 150 μL of water to the Eppendorf tube and gently vortex to resuspend the sample.
6. Centrifuge (16,000 × *g*, 5 min) and transfer the supernatant containing the WTA hydrolyzed extract to a clean sample vial.
7. Start the Dionex ICS chromatography system and assemble the CarboPac PA10 column with an Amino Trap pre-column and a Borate Trap Column before the injection loop (*see* **Note 11**).
8. Wash the system with 18 mM NaOH (1.8% (v/v) of a 1 M NaOH solution in water) for 30 min at a flow of 1.0 mL/min.
9. Inject 10 μL of water as a blank.
10. Analyze 10 μL of each WTA hydrolyzed extract by HPAEC-PAD using the elution method depicted in Table 1 (*see* **Note 12**) and the detection method presented in Table 2. For a typical chromatogram of a WTA extract, please refer to Fig. 1b.
11. Run a standard of known concentration for each monosaccharide of interest (*see* **Note 13**).

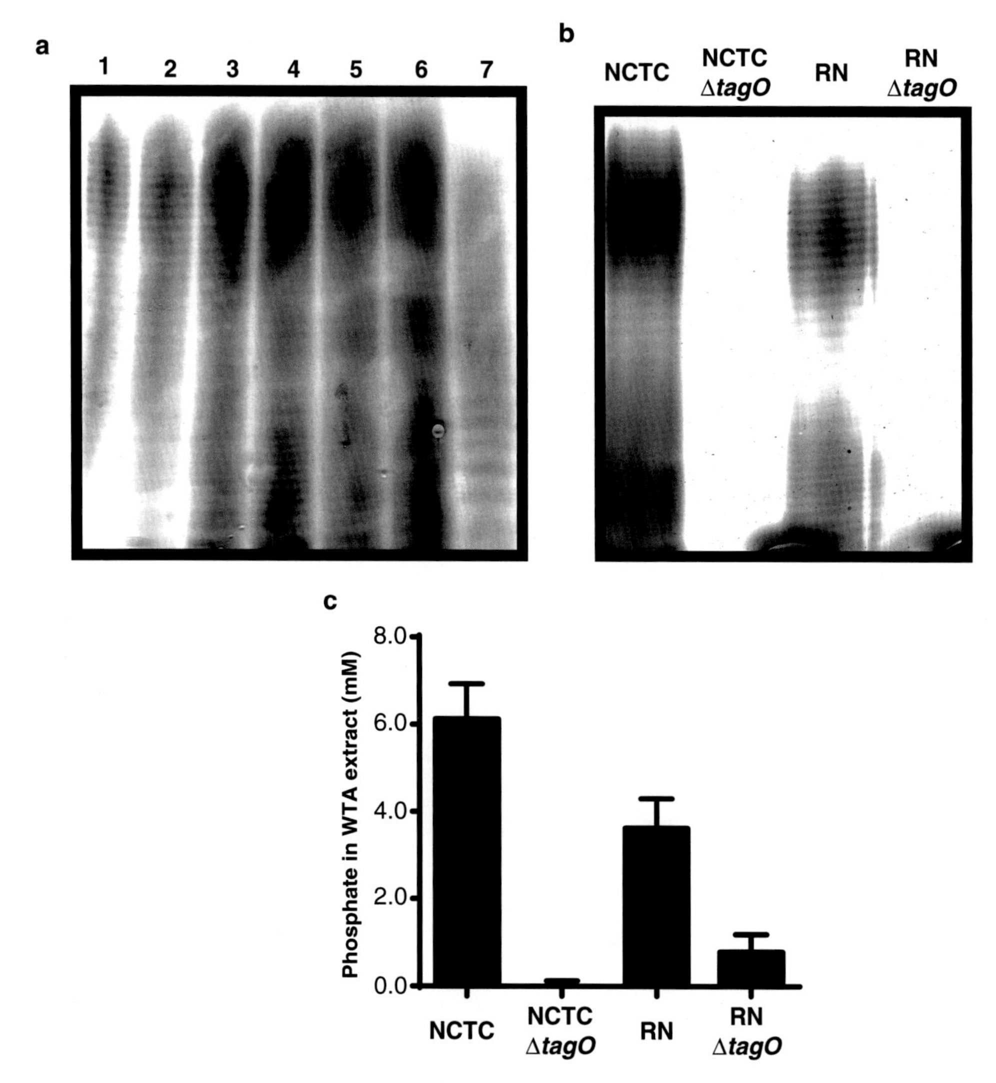

Fig. 2 (**a**) Optimization of the extraction of WTA from *S. aureus*. WTA, which were extracted from *S. aureus* RN4220 strain using different hydrolysis conditions, were analyzed by Native PAGE and detected by alcian blue–silver staining. WTA were extracted using a recently prepared 0.1 M NaOH solution (ion chromatography grade) at 4 °C (*lane 1*), 18 °C (*lane 2*), 25 °C (*lane 3*), and 37 °C (*lane 4*). A lower yield of extracted WTA may be observed when the NaOH solution has not been prepared recently (*lane 5*) or when NaOH pellets of analytical grade were used to prepared the NaOH solution (*lane 6*). Moreover, WTA were degraded when stored in NaOH solution, at 4 °C, for periods longer than 48 h (*lane 7*). (**b**) WTA are not observed in *S. aureus* strains that lack *tagO* when analyzed by Native PAGE. WTA were isolated from *S. aureus* NCTC 8325-4 (*lane 1*) and *S. aureus* RN4220 (*lane 3*). However, no WTA were detected in *S. aureus* NCTC 8325-4 Δ*tagO* (*lane 2*), *S. aureus* RN4220 Δ*tagO* (*lane 4*) mutant strains. (**c**) Absence of WTA in *S. aureus tagO* null mutants also results in lack of phosphate in WTA extracts. WTA extracts from *S. aureus* NCTC 8325-4 strain had higher phosphate content than the WTA extract from the *S. aureus* RN4220 strain. This is in accordance with results observed by Native-PAGE analysis (panel **b**). The levels of phosphate found in WTA extracts from *S. aureus* NCTC Δ*tagO* and RN4220 Δ*tagO* mutant strains were negligible or significant reduced, respectively. Results are shown as the mean and standard deviation of three technical replicates

Table 1
Monosaccharide separation method eluent concentration profile

Time/min	NaOH (mM) [% (v/v) 1 M NaOH]	$NaCH_3COO$ (mM) [% (v/v) 1 M $NaCH_3COO$]
00	18 [1.8]	0 [0]
20	18 [1.8]	0 [0]
25	18 [1.8]	200 [20]
30	18 [1.8]	200 [20]
35	18 [1.8]	800 [80]
40	18 [1.8]	800 [80]
45	18 [1.8]	0 [0]
65	18 [1.8]	0 [0]

All eluent gradients are linear and the flow is 1.0 mL/min at 30 °C

Table 2
Waveform used to specifically detect the monosaccharides while suppressing the signal from amino acids that may contaminate the solution

Time (s)	Potential (V)	Integration
0.00	+0.1	Off
0.20	+0.1	On
0.40	+0.1	Off
0.41	-2.0	Off
0.42	-2.0	Off
0.43	+0.6	Off
0.44	-0.1	Off
0.50	-0.1	Off

3.4 Analysis of WTA Extracts

1. Prepare a 20% acrylamide separating gel by mixing 10 mL of native gel buffer with 20 mL of acrylamide stock solution per gel.
2. Prepare a 3% stacking gel by mixing 1.5 mL of native gel buffer with 3 mL of water and 0.5 mL of acrylamide stock solution per gel.
3. To degas the separating gel and stacking gel solutions incubate for 5 min in an ultrasound bath (*see* **Note 14**).

4. Add 300 μL of 10% APS and 30 μL of TEMED to the separating gel solution and cast the separating gel into a 20 cm× 16 cm×1 cm cassette (*see* **Note 15**). Save approximately 5 cm for the stacking gel on the cassette.
5. Gently add isopropanol to the cassettes until it covers the gel solution undergoing polymerization (*see* **Note 16**).
6. After polymerization of the separating gel (usually around 10 min) remove the isopropanol layer, add 50 μL of 10% APS and 5 μL of TEMED to the stacking gel solution and then cast the staking gel on top of the separating gel.
7. Immediately after, insert a 1.0 cm 10-well comb into the cassette without introducing air bubbles into the gel.
8. After polymerization, remove the comb from the cassette and mount the electrophoresis system. Fill the cathode and anode chambers with native PAGE running buffer.
9. Prepare the WTA samples by adding 10 μL of WTA extract to 20 μL of 3× loading buffer. Mix this sample with 50 μL of 1× loading buffer (*see* **Note 17**).
10. After loading the samples into the gel, separate the WTA extracts by electrophoresis at 4 °C, under a constant 20 mA current per gel, until the loading buffer dye reaches the end of the gel (17–18 h).
11. After electrophoresis, carefully remove the gel from the cassette (*see* **Note 18**) and incubate the gel for 5 min in 500 mL of water under mild orbital agitation (160 rpm).
12. Stain the gel with 500 mL alcian blue staining solution for 1 h at room temperature and mild agitation (160 rpm).
13. Recover the alcian blue staining solution to reuse (*see* **Note 19**). Remove the dye in excess by incubating the gel in 500 mL of alcian blue destaining solution. Renew the destaining solution as needed until a clear background is obtained.
14. Incubate the gel in 1 L of water for 30 min.
15. Perform silver staining of the gel with the Silver stain plus kit according to the manufacturer instructions (*see* **Note 20**) until a pattern similar to that shown in Fig. 2a or b is visible.

4 Notes

1. SDS in solution precipitates at low temperatures (≤10 °C). Keep solutions with SDS at room temperature.
2. Ionic chromatography is very sensitive to water purity. It is strongly recommended to use ultrapure water with a resistivity ≥18 MΩ cm at 25 °C and Ionic chromatography grade

reagents. Avoid using analytical grade NaOH or $NaCH_3COO$, since they may contain carbonate anions, which interfere with the sensitivity of the methods and its ability to resolve two different compounds. Also degas thoroughly all solvents for Ionic chromatography and keep them under helium blanketing. Dissolved atmospheric CO_2 will be converted to carbonate at high pH.

3. An equimolar combination of Tris and Tricine buffers is expected to result in a pH value very close to 8.2. Usually there is no need to adjust the pH.
4. If you wish to compare different strains of *S. aureus* adjust the volume of culture so that you collect the same number of cells per strain.
5. In order to prevent hydrolysis of the sample by autolysins and other lytic enzymes, it is important that the SDS boiling step is carried out correctly. Also this step promotes cell membrane disruption and release of the cellular content.
6. Incubation with proteinase K degrades cell wall associated proteins.
7. It is important to use a freshly prepared sodium hydroxide solution and of the highest purity available. An aged sodium hydroxide solution or a sodium hydroxide solution of lower purity can impair the hydrolysis yield as seen in Fig. 2a.
8. The WTA samples are perishable even if stored at 4 °C or −20 °C for more than 48 h. Lane 7 in Fig. 2a shows that a WTA extract stored at 4 °C for 2 weeks displays signs of degradation.
9. Left-over phosphate and other ions from inefficient desalting of the glassware after washing can interfere with phosphate quantification.
10. Do triplicates of 1, 2, and 5 μL of each WTA extract. Since the final concentration of WTA is hard to predict it is advisable to analyze several volumes of extract.
11. Use of an Amino Trap pre-column delays the elution of amino acids and thus prevents their interference with the detection of monosaccharides. A Borate Trap Column before the injection loop lowers the concentration of borate anions, a common contaminant in ionic chromatography.
12. The acidic hydrolysis conditions used promote de-*N*-acetylation of the samples. Therefore, we expect to detect glucosamine and not *N*-acetylglucosamine. The same applies to other studied monosaccharides.
13. We typically run 10 μL of a ≤ 1 mM solution of the monosaccharide. However, the monosaccharide detection sensitivity is highly affected by its REDOX potential and the detector's waveform.

14. Degassing the acrylamide solution removes dissolved oxygen that impairs polymerization. Also, it prevents the formation of air bubbles within the gel matrix during casting that will impair the sample resolution.
15. A high acrylamide percentage gel polymerizes quite fast after addition of the initiators. Be quick and gentle on transferring the acrylamide solution to the cassette in order to prevent an uneven or unleveled gel matrix.
16. Use isopropanol to seal the resolving gel matrix and thus prevent oxygenation of the acrylamide solution. Also the isopropanol–acrylamide solution interface promotes a leveled resolving gel surface.
17. Loading smaller volumes may result in an uneven distribution of the sample within the well.
18. To avoid tearing the gels, always wet your gloves before handling the gel. Also make sure that you are using clean gloves or tweezers when handling the gel. It is also advisable to stain each gel in a separate container. Staining multiple gels in the same container can promote uneven staining.
19. The alcian blue staining solution is reusable. However, it is photosensitive. At first sign of precipitation discard and prepare fresh. To extend the alcian blue solution shelf-time protect it from light exposure.
20. It may take up to 15–20 min for first bands to appear.

Acknowledgements

This work was supported by PhD fellowships from the Portuguese Science Agency (FCT): SFRH/BD/52207/2013 (GC) and SFRH/BD/78748/2011 (FV). Research work described was also supported by grants PTDC/BIA-PLA/3432/2012 and IF/01464/2013 (SF).

References

1. Vollmer W, Blanot D, de Pedro MA (2008) Peptidoglycan structure and architecture. FEMS Microbiol Rev 32(2):149–167. doi:10.1111/j.1574-6976.2007.00094.x
2. Weidenmaier C, Peschel A (2008) Teichoic acids and related cell-wall glycopolymers in Gram-positive physiology and host interactions. Nat Rev Microbiol 6(4):276–287. doi:10.1038/nrmicro1861
3. Vergara-Irigaray M, Maira-Litran T, Merino N, Pier GB, Penades JR, Lasa I (2008) Wall teichoic acids are dispensable for anchoring the PNAG exopolysaccharide to the *Staphylococcus aureus* cell surface. Microbiology 154(Pt 3): 865–877. doi:10.1099/mic.0.2007/013292-0
4. Oku Y, Kurokawa K, Matsuo M, Yamada S, Lee BL, Sekimizu K (2009) Pleiotropic roles of polyglycerolphosphate synthase of lipoteichoic acid in growth of *Staphylococcus aureus* cells. J Bacteriol 191(1):141–151. doi:10.1128/JB.01221-08
5. Peschel A, Otto M, Jack RW, Kalbacher H, Jung G, Gotz F (1999) Inactivation of the dlt operon in *Staphylococcus aureus* confers

sensitivity to defensins, protegrins, and other antimicrobial peptides. J Biol Chem 274(13): 8405–8410

6. Bera A, Biswas R, Herbert S, Kulauzovic E, Weidenmaier C, Peschel A, Gotz F (2007) Influence of wall teichoic acid on lysozyme resistance in *Staphylococcus aureus*. J Bacteriol 189(1):280–283. doi:10.1128/JB.01221-06
7. Collins LV, Kristian SA, Weidenmaier C, Faigle M, Van Kessel KP, Van Strijp JA, Gotz F, Neumeister B, Peschel A (2002) *Staphylococcus aureus* strains lacking D-alanine modifications of teichoic acids are highly susceptible to human neutrophil killing and are virulence attenuated in mice. J Infect Dis 186(2):214–219. doi:10.1086/341454
8. Chatterjee AN (1969) Use of bacteriophage-resistant mutants to study the nature of the bacteriophage receptor site of *Staphylococcus aureus*. J Bacteriol 98(2):519–527
9. Heptinstall S, Archibald AR, Baddiley J (1970) Teichoic acids and membrane function in bacteria. Nature 225(5232):519–521
10. Atilano ML, Pereira PM, Yates J, Reed P, Veiga H, Pinho MG, Filipe SR (2010) Teichoic acids are temporal and spatial regulators of peptidoglycan cross-linking in *Staphylococcus aureus*. Proc Natl Acad Sci U S A 107(44):18991–18996. doi:10.1073/pnas.1004304107
11. Atilano ML, Yates J, Glittenberg M, Filipe SR, Ligoxygakis P (2011) Wall teichoic acids of *Staphylococcus aureus* limit recognition by the drosophila peptidoglycan recognition protein-SA to promote pathogenicity. PLoS Pathog 7(12):e1002421. doi:10.1371/journal.ppat.1002421
12. Brown S, Zhang YH, Walker S (2008) A revised pathway proposed for *Staphylococcus aureus* wall teichoic acid biosynthesis based on in vitro reconstitution of the intracellular steps. Chem Biol 15(1):12–21. doi:10.1016/j.chembiol.2007.11.011
13. Endl J, Seidl HP, Fiedler F, Schleifer KH (1983) Chemical composition and structure of cell wall teichoic acids of staphylococci. Arch Microbiol 135(3):215–223
14. Chen PS, Toribara TY, Warner H (1956) Microdetermination of phosphorus. Anal Chem 28(11):1756–1758. doi:10.1021/Ac60119a033
15. Engel A, Händel N (2011) A novel protocol for determining the concentration and composition of sugars in particulate and in high molecular weight dissolved organic matter (HMW-DOM) in seawater. Mar Chem 127: 180–191.doi:10.1016/j.marchem.2011.09.004
16. Carvalho F, Atilano ML, Pombinho R, Covas G, Gallo RL, Filipe SR, Sousa S, Cabanes D (2015) L-Rhamnosylation of *Listeria monocytogenes* wall teichoic acids promotes resistance to antimicrobial peptides by delaying interaction with the membrane. PLoS Pathog 11(5): e1004919. doi:10.1371/journal.ppat.1004919
17. Atilano ML, Pereira PM, Vaz F, Catalao MJ, Reed P, Grilo IR, Sobral RG, Ligoxygakis P, Pinho MG, Filipe SR (2014) Bacterial autolysins trim cell surface peptidoglycan to prevent detection by the Drosophila innate immune system. eLife 3:e02277. doi:10.7554/eLife.02277
18. Meredith TC, Swoboda JG, Walker S (2008) Late-stage polyribitol phosphate wall teichoic acid biosynthesis in *Staphylococcus aureus*. J Bacteriol 190(8):3046–3056. doi:10.1128/JB.01880-07
19. Min H, Cowman MK (1986) Combined alcian blue and silver staining of glycosaminoglycans in polyacrylamide gels: application to electrophoretic analysis of molecular weight distribution. Anal Biochem 155(2):275–285

Chapter 16

Analysis of Bacterial Cell Surface Chemical Composition Using Cryogenic X-Ray Photoelectron Spectroscopy

Madeleine Ramstedt and Andrey Shchukarev

Abstract

This chapter describes a method for measuring the average surface chemical composition with respect to lipids, polysaccharides, and peptides (protein + peptidoglycan) for the outer part of the bacterial cell wall. Bacterial cultures grown over night are washed with a buffer or saline at controlled pH. The analysis is done on fast-frozen bacterial cell pellets obtained after centrifugation, and the analysis requires access to X-ray photoelectron spectroscopy instrumentation that can perform analyses at cryogenic temperatures (for example using liquid nitrogen). The method can be used to monitor changes in the cell wall composition following environmental stimuli or genetic mutations. The data obtained originate from the outermost part of the cell wall. Thus, it is expected that for gram-negative bacteria only the outer membrane and part of the periplasmic peptidoglycan layer is probed during analysis, and for gram-positive bacteria only the top nanometers of the peptidoglycan layer of the cell wall is monitored.

Key words X-ray photoelectron spectroscopy, Cell wall composition, C 1s spectra, Bacterial cells

1 Introduction

X-ray photoelectron spectroscopy (XPS) is a surface analysis technique widely used in material science to determine the chemical composition of the near-surface of a sample (less than 10 nm analysis depth). During analysis, the sample is exposed to X-ray photons of a specific energy. The photons will cause electrons in the atoms at the sample surface to be emitted, and the kinetic energies of these electrons are measured after leaving the surface. Taking into account the work function (ϕ) of the spectrometer, which is a constant, the difference between the energy of the incoming photon ($h\nu$) and the kinetic energy (KE) of the outgoing electron can be calculated. It is the energy with which the electron was bound in the atom at the surface, i.e., the binding energy (BE), and it is measured in electron volts (eV).

$$\mathrm{BE} = h\nu - \mathrm{KE} - \phi$$

Hee-Jeon Hong (ed.), *Bacterial Cell Wall Homeostasis: Methods and Protocols*, Methods in Molecular Biology, vol. 1440, DOI 10.1007/978-1-4939-3676-2_16,

Each element has a set of orbital binding energies specific for that element. This allows for characterization of surface elemental composition using XPS (Fig. 1). Furthermore, different oxidation states or neighboring atoms that, for example, have a high electronegativity give rise to changes in the binding energy of core electrons. For example: sulfur atoms in thiol-groups and sulfur atoms in sulfate will have different binding energies (162.8–164.3 eV vs. 168.6–171.0 eV) [1], and a carbon atom in an aliphatic carbon chain has a lower binding energy compared to a carbon atom with a nitrogen or oxygen nearest neighbor (285.0 eV vs. 286.0 eV and 286.5 eV respectively) [2, 3] (Fig. 2). These so-called chemical shifts enable determination of a more detailed chemical composition of the sample surface. If an emitted electron collides with gas

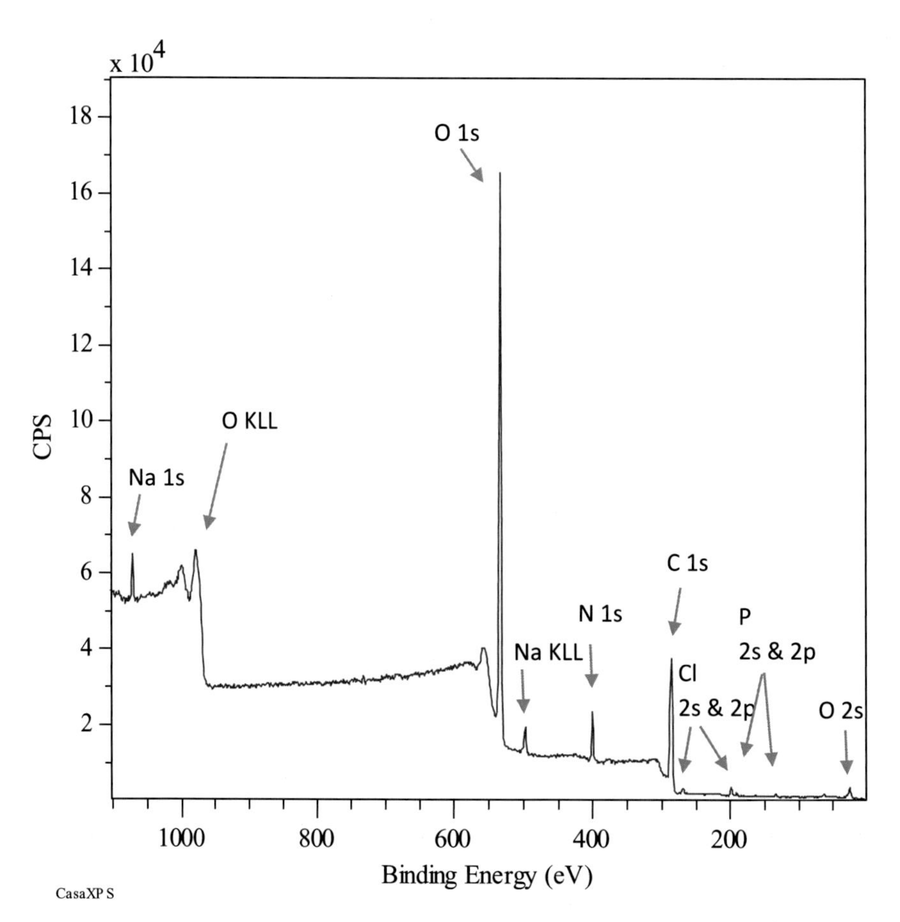

Fig. 1 XPS survey spectrum of *E. coli* with O-antigen showing peaks from Na, O, N, C, Cl, and P. Na and Cl, as well as part of the P signal, originate from PBS that was used for washing the bacterial cell pellet. Spectrum acquired using Kratos Axis Ultra DLD spectrometer with monochromatic Al Kα radiation ($h\nu = 1486.6$ eV, 150 W) and liquid nitrogen cooling. An analysis area of 0.3×0.7 mm^2, pass energy of 160 eV, and three sweeps with a total acquiring time of 180 s were used to obtain this survey spectrum. Peaks marked with KLL are so-called Auger peaks and originate from electrons emitted during an atomic relaxation process following the photoemission

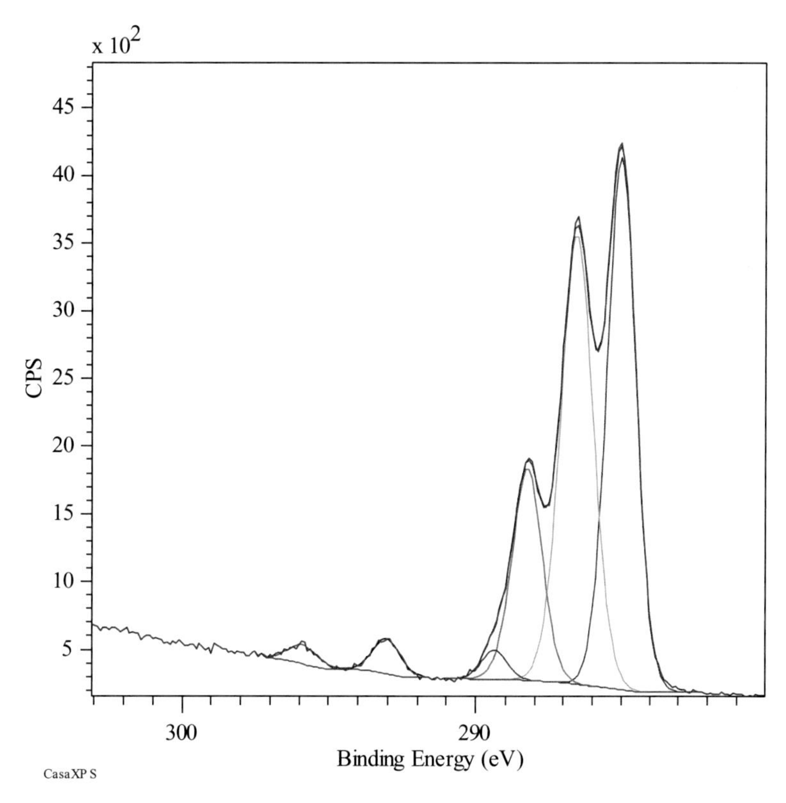

Fig. 2 High-resolution spectrum of C 1s from a sample of *E. coli* with O-antigen. A doublet originating from K 2p electrons is seen at 293 and 296 eV. Fine structure of the C 1s peak can be better observed after peak-fitting: aliphatic C at 285.0 eV, C in C–O or C–N bonds at 285.9–286.5 eV, C in for example aldehydes or peptide bonds 287.9–288.1 eV and C in carboxylate or carboxylic acid groups at 289.0–289.3 eV. Spectrum acquired using Kratos Axis Ultra DLD spectrometer with monochromatic Al Kα radiation ($h\nu$ = 1486.6 eV, 150 W) and liquid nitrogen cooling. An analysis area of 0.3×0.7 mm^2, pass energy of 20 eV, and seven sweeps with a total acquiring time of 422 s were used to obtain this spectrum. Data treatment performed using CasaXPS software

molecules after leaving the surface it will lose kinetic energy and may not reach the detector. To avoid this, the analysis in XPS takes place in ultrahigh vacuum. Furthermore, such energy loss means that only electrons from atoms near the surface will be able to escape the surface with their element and orbital specific binding energies intact. Electrons from deeper layers in the sample will lose some of their energy and become part of the background signal in XPS spectra. This is why the background signal increases at the high BE side of each major peak in the spectra. For further reading about the analysis technique and its application to biological samples, please consult other comprehensive reviews [4–6].

A common method to analyze biological samples in ultrahigh vacuum has been to freeze dry the sample before analysis and obtain the chemical composition of the dehydrated surface. However, this procedure can alter the chemical composition due to surface rearrangements when water is removed [7]. To avoid these changes, a method to analyze bacterial samples using cryo-XPS was developed [8] using gram-negative bacteria (*Escherichia coli*) and was subsequently also applied to study cell wall changes in gram-positive bacteria (*Bacillus subtilis*) [7]. The method allows for prediction of chemical composition of the outermost part of the cell wall, including surface-bound bacterial appendages such as fimbriae or flagella, from the C 1s spectrum of bacteria [8]. It presents the fraction of C atoms in lipids, polysaccharides, and peptides (protein and/or peptidoglycan) in relation to the total amount of C at the surface of the sample (Fig. 3). Proteins and peptidoglycan cannot be distinguished using this method as carbon atoms in these two substances have very similar chemical environment. However, a combination of this approach with protein quantification can often indirectly give the possibility to compare the amount of peptidoglycan between samples.

Furthermore, this method can be applied to other biological systems that are expected to consist of the building blocks polysaccharides, lipids, and peptides (proteins and/or peptidoglycan).

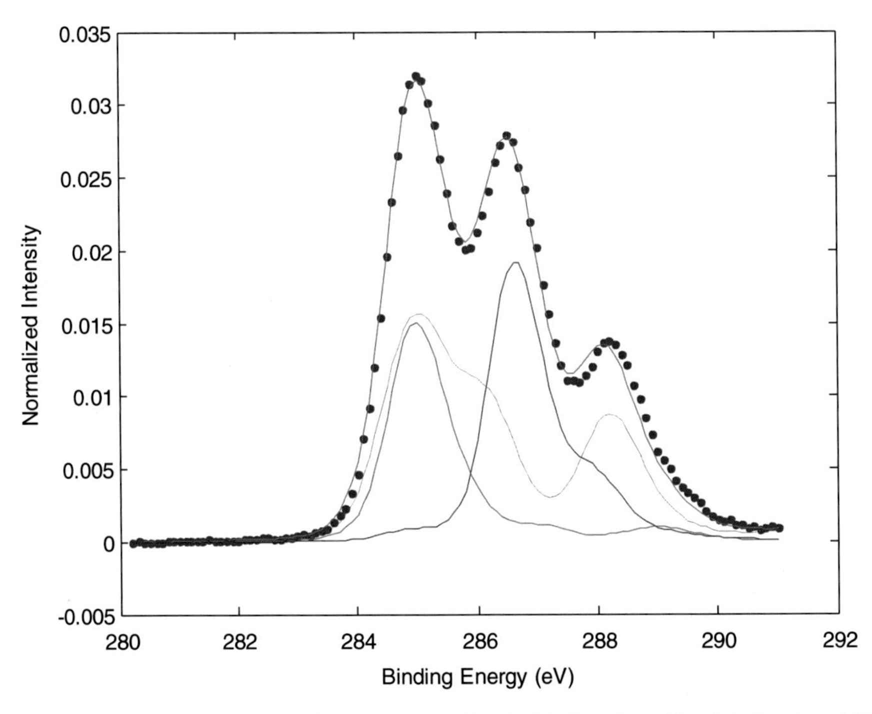

Fig. 3 Matlab-processed C 1s spectrum from a sample of *E. coli* with O-antigen. The data treatment [8] predicts the amount of lipids (*red* 25 %), peptides (*blue* 48 %), and polysaccharides (*green* 27 %) of total C atoms at the surface. (Please note that the output from the algorithm plots the binding energy scale in ascending numbers whereas traditional XPS software plots the binding energy scale in descending binding energies.) *Blue dots* represent measurement data from XPS analysis, *green line* the total prediction

2 Materials

1. X-ray photoelectron spectrometer with facilities for cooling to cryogenic temperature (such as liquid nitrogen temperatures) both in analysis chamber and in load lock.
2. Cooling agent, for example liquid nitrogen.
3. Sample holder enabling rapid cooling of sample.
4. Metal mesh for increasing contact area between sample holder and sample.
5. Pipette with disposable tips for transferring bacterial pellet onto sample holder.
6. Freshly grown bacterial suspension (*see* **Notes 1** and **2**).
7. Carbon free buffer (for example phosphate buffer) or saline with controlled pH.
8. Centrifuge (3000–4500 x*g*).
9. Disinfection for sample holder, e.g., ultrasonic bath, glass beaker with lid, and ethanol (*see* **Note 3**).
10. Container for disposing of biological waste (*see* **Note 4**).
11. Computer with Matlab software with optimization toolbox (or similar).

3 Methods

3.1 Cleaning of Sample Holder

1. The sample holder and metal mesh need to be sterilized and cleaned before usage to avoid any interference from contaminants. This involves immersing the sample holder in ethanol in a small beaker and scrubbing with ethanol. Cleaning can also easily be performed by placing the beaker with ethanol, sample holder and a lid in an ultrasonic bath for a period of approximately 2 min.
2. Allow the sample holder to dry completely.
3. After the analysis this step is repeated to sterilize and decontaminate the sample holder and clean it for the next sample (*see* **Note 3**).

3.2 Sample Preparation

1. Grow the bacterial strain of interest over night in liquid broth or on agar plates (*see* **Notes 1**, **2** and **4**).
2. Collect bacteria from the suspension via centrifugation (3000-4500 ×*g*) or carefully scrape bacteria from a plate. Make sure the quantity is enough to enable adequate sample handling (200 μL cell pellet will suffice).
3. Wash the bacteria using a buffer that do not contain compounds with carbon atoms. For example a phosphate buffer

can be used or a saline with controlled pH. If saline is used, pH should be measured before last centrifugation (below). The surface composition of bacteria is sensitive to pH which is why it is important to control and report pH (*see* **Notes 5** and **6**).

4. Centrifuge (3000–4500 × *g*) and remove the solution from the cell pellet. Keep the cell pellet on ice until analysis.

3.3 Spectrometer Preparation

1. Cool down the sample stage in the spectrometer analysis chamber using liquid nitrogen (or similar cooling agent) (*see* **Note 7**).
2. Cool down the sample stage in the spectrometer load lock using liquid nitrogen (or similar cooling agent) and keep cold (around −170 °C) for at least 20 min before inserting the sample.
3. Directly before sample loading, fill load lock with dry nitrogen gas to obtain a pressure slightly higher than atmospheric pressure.

3.4 Loading of Sample into XPS Instrument

1. Mount the metal mesh into the sample holder.
2. Pipette 10–20 μL of bacterial pellet onto the mesh of the sample holder immediately before insertion onto the precooled sample stage in the spectrometer load lock.
3. Insert the sample holder onto the precooled stage of the spectrometer load lock. Try to do this as quickly as possible and then close the load lock. Despite the overpressure of dry N_2 during venting, the procedure should be done quickly in order to limit the amount of atmospheric water that can enter into the load lock from the laboratory atmosphere.
4. The sample will freeze within seconds. However, wait for 1 min after sample loading before starting evacuating gas from the load lock.
5. If cooling in the load lock was stopped during sample loading, restart the cooling together with the evacuation.
6. When suitable vacuum (~10^{-7} Torr, i.e., ~10^{-5} Pa) is reached, transfer the sample into the sample analysis chamber of the spectrometer and perform the measurement.

3.5 Measurement Procedure

Maintain cooling throughout the entire measurement procedure and carefully monitor the temperature in the sample analysis chamber. Usually a slight increase in temperature is observed, ~5 °C, during the measurement. However, if the temperature rises close to −140 °C ice will start to sublimate. The measurement then has to be stopped and the sample taken out to avoid problems with the measurement and with the vacuum system.

Each brand of XPS equipment has its specific procedures for setting up and executing measurements. Therefore, we here only give a more generalized description of the measurement.

1. Analyze using monochromated radiation (for example Al Kα), to obtain narrow peaks and to avoid unnecessary background from secondary electrons.

2. Use charge-neutralizing equipment in the spectrometer. Bacterial cells are not generally conductive and insulating materials will exhibit a buildup of surface charge during analysis. This will affect data quality and thus charge neutralization is needed.
3. Locate the area for analysis on the sample.
4. Adjust the height of the sample to obtain maximal signal.
5. Acquire a survey spectrum to obtain information on elemental composition of the sample surface.
6. Acquire high resolution spectra of the regions of importance. Spectra from O 1s, N 1s, K 2p, C 1s, P 2p, and S 2p can be obtained from most bacterial samples. Spectra from Na 1s and Cl 2p is also often possible to acquire and generally originates from the buffer or saline used for washing. Please observe that if a phosphate buffer was used the P 2p spectra can also contain phosphate from the buffer (*see* **Notes 8** and **9**).

The total measurement time for one bacterial sample is generally 2–3 h. At least half of that time consists of cooling the spectrometer and the sample loading procedure.

3.6 Data Analysis

Several data-analysis software are available for evaluation of XPS data. These enable peak fitting procedures of the spectra to obtain surface concentrations of elements and functional groups (measured in atomic percentage = percentage of atoms at the surface). A detailed description of this type of data analysis will therefore not be given in this section, but the reader is referred to software manuals for their specific software. Instead a description is given here of how the Matlab code, described in Ramstedt et al. [7, 8], can be used. This code is for predicting surface composition from the C 1s peak of bacterial samples (*see* **Note 10**).

1. Due to sample charging and charge neutralizing procedures, the binding energy scale first needs to be calibrated. This is most conveniently done through curve fitting of the C 1s peak followed by shifting the binding energy scale so that the aliphatic carbon component is set to 285.0 eV (some experimentalists also use 284.7 eV; however, for usage of the Matlab code, described here, the aliphatic carbon should be positioned at 285.0 eV). Thereafter the binding energy scale of all other spectra is calibrated using the same shift.

   After this step has been performed, the curve fitting is no longer needed.
2. Convert your C 1s spectrum so that the binding energy is in the first column (for example column A in an Excel sheet) and the total spectral intensity in the second column (for example column B in an Excel sheet).
3. Import the table into Matlab (software with optimization toolbox is needed).

4. Open the *C.mat* fil in Matlab and run the Matlab file *compfit.m*. Both can be found in supplementary information to Ramstedt et al. [8] (the Matlab version used for creating the files was R2011a).
5. This procedure will predict the fraction of lipids, polysaccharides, and peptides (protein + peptidoglycan) in the C 1s spectra analyzed and present the result both in graphical form and as a number fraction. To obtain percentages of total carbon divide this fraction with the sum of all obtained components and multiply by 100 (*see* **Note 11**).
6. In this procedure only the C 1s spectrum is used to predict the cell wall composition with respect to % of total carbon. However, since the component "peptide" in general should be the only component containing N, the atomic ratio N_{tot}/C_{tot} can be used to cross-check the peptide values. The trends in these two should reflect each other when several samples are compared.

4 Notes

1. Always wear appropriate protection when handling microbiological samples.
2. Ensure that the laboratory is well suited and approved for working at the safety classification level required for your bacterial strain.
3. Sterilize all equipment and tools that have been in contact with bacteria after usage.
4. Ensure appropriate handling of the waste created.
5. The cell wall composition obtained is dependent on pH and thus it is of great importance to carefully control, measure and report pH of the bacterial suspension.
6. The composition of the washing solution may influence the spectra obtained. Therefore, buffers containing substances with C should be completely avoided. Furthermore, the data obtained for elements that exists both in the buffer and in the bacterial cell wall should be interpreted with care.
7. Lab environments with high atmospheric humidity may benefit from lowering air humidity to reduce condensation of water vapor onto the cooling facilities of the spectrometer. Furthermore, lower room temperature generally allows for lower cooling temperatures to be reached.
8. Sublimation of water from the sample inside the spectrometer during measurements can be monitored by monitoring the pressure inside the analysis chamber and also by acquiring a survey spectrum at the beginning and end of a measurement

to ensure that the oxygen content of the sample is not altered. If the sample gets dehydrated inside the spectrometer this can dramatically alter the apparent cell wall composition obtained [7].

9. The cell wall composition is an average value of a larger area of the sample (the size of this area is dependent on the spectrometer and the setting used).
10. The data analysis procedure outlined here uses the mathematical compounds derived and presented in Ramstedt et al. [8]. Constructing new mathematical components, for example predicting the content of other surface components with significantly different C spectra, would require the methodology described in Ramstedt et al. [8] to be repeated. This could be done by analyzing a large number of samples that have different content of these surface components, followed by multivariate analysis of C 1s spectra to define mathematical components that can explain the variation in all analyzed spectra through linear combination.
11. In a 1:1:1 mixture of lipids, polysaccharides, and peptides, the % of C atoms in lipids will be higher than that in peptides and polysaccharides [8] due to differences in chemical composition where lipids simply contain relatively higher amounts of carbon [9].

Acknowledgement

This work was partly funded by the Umeå Center for Microbial Research (UCMR) and the Swedish Research Council.

References

1. Moulder JF, Stickle WF, Sobol PE, Bomben KD (1992) Handbook of X-ray photoelectron spectroscopy. Perkin-Elmer corporation, Physical electronics division. Eden Prairie, Minnesota
2. Beamson G, Briggs D (1992) High resolution XPS of organic polymers: the Scienta ESCA300 database. Wiley, New York
3. Beamson G, Briggs D (2000) The XPS of polymers database, version 1.0. SurfaceSpectra Ltd, Manchester
4. Genet MJ, Dupont-Gillain CC, Rouxhet PG (2008) XPS analysis of biosystems and biomaterials. Springer Science + Business Media, New York
5. Rouxhet P, Genet M (2011) XPS analysis of bio-organic systems. Surf Interface Anal 43: 1453–1470
6. Ramstedt M, Leone L, Shchukarev A (2017) Bacterial surfaces in geochemistry—how can XPS help? In: Veeramani H, Kenney J, Alessi D (eds) Analytical geomicrobiology. Cambridge University Press, Cambridge
7. Ramstedt M, Leone L, Persson P, Shchukarev A (2014) Cell wall composition of Bacillus subtilis changes as a function of pH and Zn^{2+} exposure: insights from cryo-XPS measurements. Langmuir 30:4367–4374
8. Ramstedt M, Nakao R, Wai S, Uhlin B, Boily J (2011) Monitoring surface chemical changes in the bacterial cell wall—multivariate analysis of cryo-X-ray photoelectron spectroscopy data. J Biol Chem 286:12389–12396
9. Dufrêne Y, van der Wal A, Norde W, Rouxhet P (1997) X-ray photoelectron spectroscopy analysis of whole cells and isolated cell walls of gram-positive bacteria: comparison with biochemical analysis. J Bacteriol 179:1023–1028

Part VI

Bioinformatics and Computational Biology Based Approaches

Chapter 17

Biophysical Measurements of Bacterial Cell Shape

Jeffrey P. Nguyen, Benjamin P. Bratton, and Joshua W. Shaevitz

Abstract

A bacteria's shape plays a large role in determining its mechanism of motility, energy requirements, and ability to avoid predation. Although it is a major factor in cell fitness, little is known about how cell shape is determined or maintained. These problems are made worse by a lack of accurate methods to measure cell shape in vivo, as current methods do not account for blurring artifacts introduced by the microscope. Here, we introduce a method using 2D active surfaces and forward convolution with a measured point spread function to measure the 3D shape of different strains of *E. coli* from fluorescent images. Using this technique, we are also able to measure the distribution of fluorescent molecules, such as polymers, on the cell surface. This quantification of the surface geometry and fluorescence distribution allow for a more precise measure of 3D cell shape and is a useful tool for measuring protein localization and the mechanisms of bacterial shape control.

Key words Cell shape, Fluorescent microscopy, Active contours, Bacteria, 3D shape, Deconvolution, Point spread function

1 Introduction

Some of the most basic descriptions of living organisms are size and shape. Without knowing anything about whether a particular cell was eukaryotic or prokaryotic, gram negative or gram positive, early scientists were able to distinguish rod shaped cells from spherical cells or helical cells, and 1 μ m long cells from 20 μ m long ones. From the bacterial fitness perspective, shape is a matter of great importance. Cell radius, for example, has large effects on the energy required for motion. It has been calculated that a 0.1 μ m change in radius can result in up to a 100,000-fold increase in the energy required for chemotaxis [1]. The maintenance of cell length is also vital, as large lengths can prevent predation by larger organisms [2].

Understanding the mechanisms that govern cell shape requires tools to measure shapes of live cells. Bacteria, however, are much smaller and more difficult to analyze than eukaryotes using light microscopy, by far the most amenable modality of study for living cells. With a typical size of approximately 1 μ m, bacteria are just

Hee-Jeon Hong (ed.), *Bacterial Cell Wall Homeostasis: Methods and Protocols*, Methods in Molecular Biology, vol. 1440,
DOI 10.1007/978-1-4939-3676-2_17,

slightly larger than the diffraction limit of light and have a width of about 6–15 pixels on a camera attached to a high powered microscope. This complicates image analysis because cell shapes must have significantly subpixel resolution in order to have an accuracy better than 10%. In addition, the Point Spread Function of the microscope (PSF, the blurred image generated by a single point emitter at the object plane) is relatively large compared to the cell, with a diameter of about 200 nm in x and y and 500 nm in z for the best diffraction-limited imaging modalities. Blurring from the PSF can have a large effect on images and must be accounted for.

Existing bacterial shape analysis algorithms extract 2D contours with subpixel resolution. These techniques, such as those developed by Sliusarenko et al. [3], Locke and Elowitz [4], and Guberman et al. [5], involve a combination of pixel intensity thresholding, edge detection, and active contours to determine the outline of cells from phase contrast, differential interference contrast, or fluorescence microscopy images. All of these methods, however, are strictly 2D and do not account for the effect of PSF blurring from their imaging systems. Furthermore, in contrast-enhanced imaging, such as phase and DIC microscopy, the PSF is poorly defined, making it difficult to account for blurring from light microscopy and thus biasing in the shape estimate.

Fluorescence microscopy addresses these issues. Here, the PSF can be easily measured and 3D images can be taken by moving the focal plane within the sample. With fluorescence microscopy, we can write down the mathematical operation the microscope performs to see the effect of PSF blurring more directly. The image seen in the microscope is the convolution of the microscope PSF with the distribution of fluorescent molecules. An example of this is a circle with a radius of 500 nm convolved with our measured PSF in the $x - z$ plane, shown in Fig. 1. The resulting image is more diamond shape and appears stretched in the z-direction. In addition, the peak intensities are laterally pulled inward because of the blurring, meaning that a measurement of maximum intensity in a

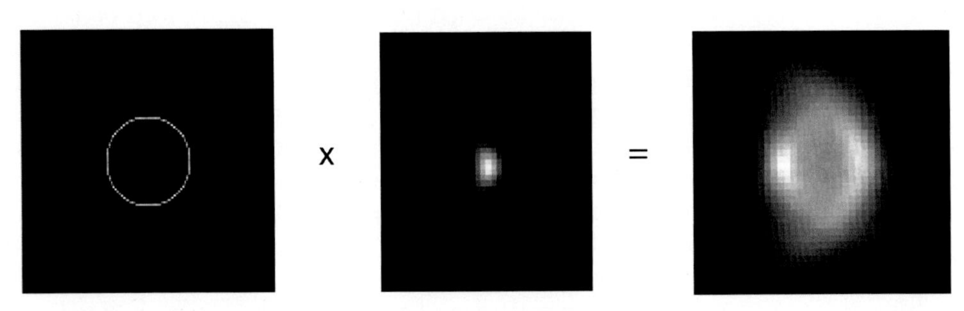

Fig. 1 **An** illustration of the mathematical transformation the microscope performs. The microscope takes the distribution of fluorescent molecules, in this case a circle with a radius of 500 nm (*left*), and convolves it with the PSF (*middle*) in order to produce a blurred image (*right*). The images shown are of the $x - z$ cross section of the PSF

2D image will systematically underestimate the size of the object. A common solution is to use image deconvolution to recover the object. This suffers from two problems. First, deconvolution of a 2D image inherently underestimated the contribution from out of plane objects and thus systematically biases the resultant shape. Second, in the presence of noise, image deconvolution is an inverse problem without a well-defined solution. Noise is amplified and artifacts are produced. Furthermore, the result of the deconvolution is still just an image, which must be analyzed further in order to extract the cell shape. Thus, a better approach is to use a constrained model for the object, and use forward convolution to fit the image. Helmuth and Sbalzarini [6] have implemented one such approach in order to measure shapes from fluorescent images. In their algorithm, the PSF is modeled with a 2D Gaussian and the experimental images are fit with a filled active contour convolved with the PSF. This allowed them to estimate the shape which best describes the fluorescent image observed in the microscope. This work expands on the method of Helmuth and Sbalzarini in order to fit the shapes of cells and polymers using 3D fluorescent images. The method uses active meshes and contours in order to fit 3D cells with fluorescent molecules distributed in different ways.

1.1 Active Contours and Meshes

Active contours (or snakes), first introduced by Kass et al. [7], are a common image analysis tool used to determine boundaries and centers of objects in images. They turn an image analysis problem into a physical problem by modeling an image contour as a relaxing elastic snake under some potential map generated from the image itself. The snake has an internal energy due to stretching and bending, E_{int}, and an external energy based on the image E_{ext}. This potential field is normally either an intensity profile or a gradient of the intensity. The total energy, E, can be expressed as

$$E = E_{\text{int}} + E_{\text{ext}} \tag{1}$$

As the snake moves to minimize its energy, it will align with the image features, but will also maintain some smoothness due to the internal energy constraint.

1.1.1 Active Contours in 2D

In a 2D image, the contour $\mathbf{x} = \{\mathbf{v}_i\} = \{x_i, y_i\}$, like the one shown in Fig. 2, has an internal energy of the form

$$E_{\text{int}}(\mathbf{x}) = \int \frac{\alpha}{2}\left(\frac{\mathrm{d}\mathbf{x}}{ds}\right)^2 + \frac{\beta}{2}\left(\frac{\mathrm{d}^2\mathbf{x}}{ds^2}\right)^2 ds. \tag{2}$$

The first term in the sum is similar to a spring energy between connected points, and the second term models the bending energy in the snake. Here, α and β are parameters that penalize stretching and bending, respectively. The minimization of the total energy from Eqs. 1 and 2 can be found using the Euler–Lagrange equations. The solution $\mathbf{x}$ satisfies

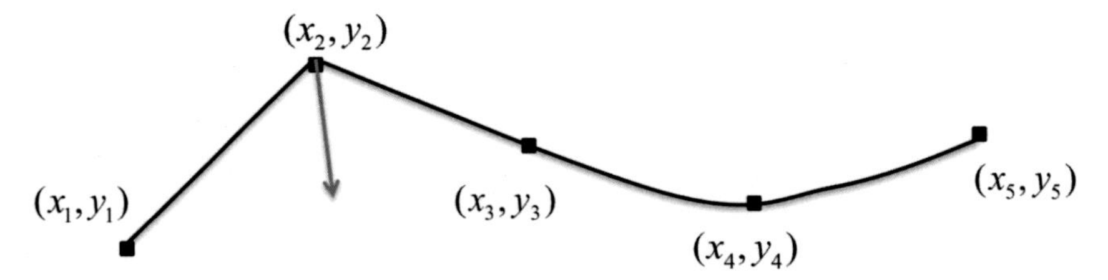

Fig. 2 An example of the parameterization used to describe the active snake. In this example, the point (x_2, y_2) will feel an internal force in the direction of the arrow due to bending and stretching stiffness

$$\alpha \mathbf{x}'' - \beta \mathbf{x}'''' - \frac{\partial E_{\text{ext}}}{\partial \mathbf{v}} = 0 \tag{3}$$

where $\frac{\partial E_{\text{ext}}}{\partial \mathbf{v}} = \left\{ \frac{\partial E_{\text{ext}}}{\partial \mathbf{v}_i} \right\}$. In the discrete case, the derivatives in the internal energy can be written as a multiplication with the matrix **A**, which takes into account all the discrete differences in the derivatives and the values of α and β,

$$\mathbf{A}\mathbf{x} + \frac{\partial E_{\text{ext}}}{\partial \mathbf{v}} = 0, \tag{4}$$

where **A** is a pentadiagonal banded matrix

$$\mathbf{A} = \begin{pmatrix} a_0+b_0 & a_1+b_1 & b_2 & 0 & \cdots & 0 & b_2 & a_1+b_1 \\ a_1+b_1 & a_0+b_0 & a_1+b_1 & b_2 & 0 & \cdots & 0 & b_2 \\ b_2 & a_1+b_1 & a_0+b_0 & a_1+b_1 & b_2 & 0 & \cdots & 0 \\ \vdots & \ddots & \ddots & \ddots & \ddots & \ddots & \ddots & \vdots \\ 0 & \cdots & 0 & b_2 & a_1+b_1 & a_0+b_0 & a_1+b_1 & b_2 \\ b_2 & 0 & \cdots & 0 & b_2 & a_1+b_1 & a_0+b_0 & a_1+b_1 \\ a_1+b_1 & b_2 & 0 & \cdots & 0 & b_2 & a_1+b_1 & a_0+b_0 \end{pmatrix} \tag{5}$$

with $a_0 = 2\alpha$, $a_1 = -\alpha$, $b_0 = 6\beta$, $b_1 = -4\beta$, and $b_2 = \beta$.

Equation 4 can be solved iteratively given an initial contour and a time step size γ via

$$A\mathbf{x}^t + \frac{\partial E_{\text{ext}}}{\partial \mathbf{v}} = -\gamma\left(\mathbf{x}^t - \mathbf{x}^{t-1}\right)$$

$$\mathbf{x}^t = \left(\mathbf{A} + \gamma \mathbf{I}\right)^{-1}\left(\gamma \mathbf{x}^{t-1} - \frac{\partial E_{\text{ext}}}{\partial \mathbf{v}}\right). \tag{6}$$

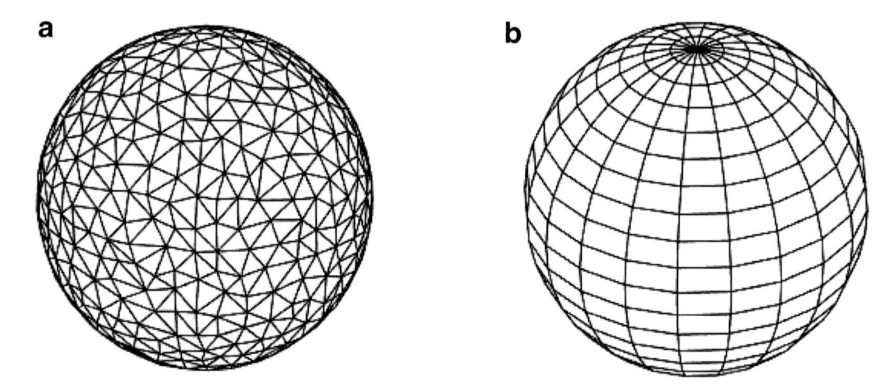

Fig. 3 Examples of spheres represented using (**a**) a triangular framework and (**b**) a rectangular mesh. In the triangular mesh it has no intrinsic coordinate system to describe the points. Vertices have an average of six connections and have more uniform edge lengths. In the rectangular mesh, the surface can be described in cylindrical coordinates as $\mathbf{S}(u,\phi)$, vertices tend to have four connections, but the poles of the sphere have singularities with a large number of connections. In addition, the edge lengths near the poles decrease

Each iteration approaches the minimum energy solution for the active contour. Local minima in the energy landscape can sometimes be overcome with appropriate choices for γ, α, and β. In other circumstances, an annealing step can be used to escape local minima.

1.1.2 3D Active Meshes

In three dimensions, much of the same formulation is used. Let the surface, **S**, of the cell be represented by a set of points $\{\mathbf{v}i\} = \{xi, yi, zi\}$. In practice, any type of connectivity between the points can define the mesh, the most common being either a triangular mesh, such as in Fig. 3a, or a rectangular mesh, shown in Fig. 3b. Regardless of the type of mesh, **S** can be described by the set of vertices and edges. The internal energy penalizes curvature of the surface. The energy of the surface is modeled as the energy of a thin plate under tension, which includes both first and second derivatives of the surface parameterization,

$$E_{\text{int}} = \int \frac{\alpha}{2}(|\nabla \mathbf{S}|^2) + \frac{\beta}{2}|\Delta \mathbf{S}|^2 \, dA. \tag{7}$$

The energy minimization is given by the solution to the Euler–Lagrange equations, and is equivalent to

$$-\alpha \Delta \mathbf{S} + \beta \Delta^2 \mathbf{S} + \frac{\partial E_{\text{ext}}}{\partial \mathbf{v}} = 0. \tag{8}$$

For a discrete set of points on the surface $\{\mathbf{v}_i\}$, the Laplacian at a point $\mathbf{v}k$ is approximated by the umbrella operator,

$$\Delta \mathbf{S}|_{v_k} = \sum_{i \in N(\mathbf{v}_k)} w_{i,k}\left(\mathbf{v}_i - \mathbf{v}_k\right). \tag{9}$$

$N(\mathbf{v}k)$ is the one ring neighborhood around $\mathbf{v}k$.

The weights, w_{ik} can be chosen in several different ways. For simplicity, we chose uniform weights equal to $1/N_k$ where N_k is the number neighbors around point $\mathbf{v}_k$. This is a good approximation when nearest neighbor distances are equal and the points have similar angular spacing between them. The Laplacian operator at all points can then be written as a matrix. If we define $-\alpha\Delta + \beta\Delta^2$ as a matrix $\mathbf{A}$, the minimization problem is identical to Eq. 4 and can be solved by implicit Euler methods using Eq. 6.

1.1.3 Adaptive Meshes

Meshes stretch to fit larger objects, but this is energetically unfavorable as stretched edges increase internal energy. In addition, it is important to keep the connectivity of the mesh as regular as possible in order to use Eq. 9 with uniform weights. Without a method of growing or shrinking to the object size, fitting meshes will be highly dependent on initial conditions. To remove this constraint, we added an adaptive process to facilitate the addition and removal of vertices in triangular meshes during the fitting process. We introduce three mechanisms the mesh can use to relax when fitting objects that require expansion or contraction. The first is vertex switching. Vertex switching loops through all pairs of triangles which share an edge. It then checks which connectivity between the 4 points minimizes the total length. If the current edge is the longer of the two chords in the quadrilateral created by the 4 points (Fig. 4a), it is replaced with the shorter one. This process tries to minimize the total edge length, making edge lengths more similar and triangle faces more equilateral. A corollary of this is that vertices tend to be connected to five, six, or seven other vertices.

The other mechanism which allows for expansion and contraction of the mesh is the addition and removal of points. Edges that are too small are collapsed into a point and the two connected vertices are combined into one (Fig. 4b). The net effect of this is the removal of two triangles. Finally, edge splitting is introduced to add vertices and split long edges in half (Fig. 4c). This adds two triangles and a four connected vertex. Subsequent vertex switching will eventually correct the four-connected vertices, making them 5 or 6 connected. The range of acceptable edge lengths is a user-defined parameter in the algorithm.

1.2 External Energy

Rather than using the image intensity or image edge intensity as a potential energy, we minimize the square difference between the experimentally measured image and the simulated image that arises from convolution of the contour with the PSF. This method is similar to that used by Helmuth and Sbalzarini [6], but extended to 3D. This is similar to modern super resolution techniques, where single fluorescent sources are fit with a Gaussian to determine their location to sub-diffraction accuracy. Here, we have complex distribution of fluorescent sources rather than a point. Therefore, we model the image as a contour convolved with a PSF.

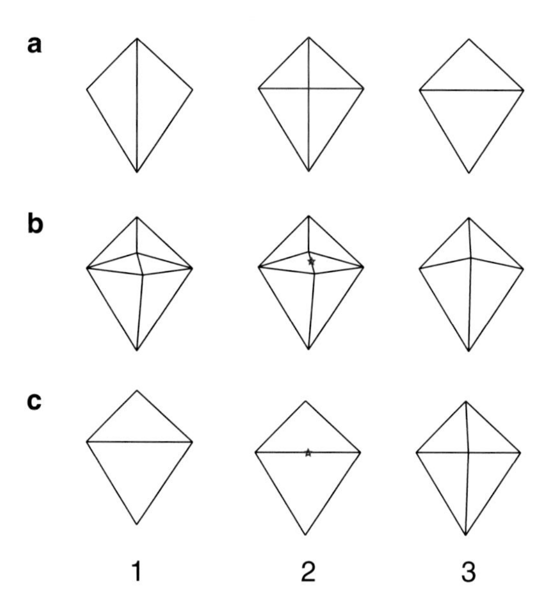

Fig. 4 Examples of two types of mesh refinement during fitting with triangular meshes. (**a**) Vertex switching switches chords in quadrilaterals in order to keep edge lengths more similar. A quadrilateral is formed by two adjacent triangles and both diagonals are measured. Only the shortest of the two chords is kept (**b**) Point merging occurs when edges are too small. The midpoint of the small edge is found and then the two vertices are collapsed into a single point. (**c**) Edge splitting adds a vertex in the middle of edges that are too long, turning two triangles into four

In our model, a surface $\mathbf{S}$ results in the simulated image $Is(\mathbf{x}, \mathbf{S})$, where $\mathbf{x}$ is the coordinate in the image space. The model image is given by the convolution of the experimentally measured PSF, $P(\mathbf{x})$, and the distribution of point sources, given by $O(\mathbf{x}, \mathbf{S})$. $O(\mathbf{x}, \mathbf{S})$ is created by binning the surface points of $\mathbf{S}$ into voxels. The simulated image can then be calculated using Fourier convolution

$$I_s(\mathbf{x}, \mathbf{S}) = O(\mathbf{x}, \mathbf{S}) * P(\mathbf{x}). \quad (10)$$

The external energy is given by the squared difference between our model, $Is(\mathbf{x}, \mathbf{S})$, and the observed image, $I(\mathbf{x})$. This is equivalent to the negative log-likelihood function between the two images

$$E_{\text{ext}} = \int R(\mathbf{x})(\ I(\mathbf{x}) - I_s(\mathbf{x}, \mathbf{S}))^2 \mathrm{d}\mathbf{x}^3, \quad (11)$$

or in the discrete case

$$E_{\text{ext}} = \sum R(\mathbf{x})(\ I(\mathbf{x}) - I_s(\mathbf{x}, \mathbf{S}))^2. \quad (12)$$

Here, $R(\mathbf{x})$ is a weighting function that can account for the noise in the system. For example, in Poisson counting noise, R is proportional to $1 / \sqrt{I(\mathbf{x})}$. For simplicity, we set $R(\mathbf{x}) = 1$. The change in energy from moving one of the vertices $\mathbf{v}k$ in the surface $\mathbf{S}$ is

$$\frac{\partial E_{\text{ext}}}{\partial \mathbf{v}_k} = 2\int\left(I(\mathbf{x}) - I_s(\mathbf{x},\mathbf{S})\right)\frac{\partial I_s(\mathbf{x},\mathbf{S})}{\partial \mathbf{v}_k}\mathrm{d}\mathbf{x}^3. \tag{13}$$

The gradient of the energy depends on the difference between the real and the simulated images, and the change in *Is* with respect to the vertex **v***k*. We can now solve Eq. 6 using Eq. 13. To demonstrate the power of this technique, we calculate the integral in Eq. 13 for two different models of fluorophore distributions. First, when the fluorophores are distributed uniformly on the surface of the cell, and second, when the fluorescent molecules are spread out within the cell interior.

1.2.1 Surface-Labeled Objects

For objects with sources distributed uniformly on the surface, the object intensity image can be modeled by the sum of delta functions, $\delta(\mathbf{x})$, centered at every vertex on the mesh {**v***i*}. This is valid if the vertex density is uniform and the average spacing is much smaller than the width of the PSF. The set of delta functions is added together to form an object intensity image

$$O(\mathbf{x},\mathbf{S}) = c\sum_{v_i \in V}\delta(\mathbf{x} - \mathbf{v}_i).$$

The simulated image, *Is*, is $O(\mathbf{x}, \mathbf{S})$ convolved with the PSF, $P(\mathbf{x})$, multiplied by some scaling constant *c*. It can also be written as

$$I_s(\mathbf{x},\mathbf{S}) = O * P = c\sum_{v_i \in V}P(\mathbf{x} - \mathbf{v}_i).$$

Here, **x** is the coordinate in the image and the **v** coordinate is the position of the membrane in space. The constant *c* corrects for the scaling of the image to the model. The change in the image given a displacement of the *k*th vertex, **v***k*, in the *x* direction is then

$$\frac{\partial I_s(\mathbf{x},\mathbf{S})}{\partial v_{kx}} = c\sum_{v_i \in V}\frac{\partial P(\mathbf{x} - \mathbf{v}_i)}{\partial v_{kx}} = -c\frac{\partial P(\mathbf{x} - \mathbf{v}_k)}{\partial x}.$$

The change in energy from Eq. 13 is

$$\frac{\partial E_{\text{ext}}}{\partial v_{kx}} = -2c\int\left(I(\mathbf{x}) - I_s(\mathbf{x},\mathbf{S})\right)\frac{\partial P(\mathbf{x} - \mathbf{v}_k)}{\partial x}\mathrm{d}\mathbf{x}^3.$$

This integral can be written as a cross correlation ($\star$) between $I(\mathbf{x})$ – $Is(\mathbf{x}, \mathbf{S})$ and $\frac{\partial P(\mathbf{x})}{\partial x}$,

$$\begin{aligned}\frac{\partial E_{\text{ext}}}{\partial v_{kx}} &= -2c(I - I_s)\star\frac{\partial P}{\partial x}\Big|_{\mathbf{v}_k} \\ &= -2c\frac{\partial}{\partial x}\left((I - I_s)\star P\right)\Big|_{\mathbf{v}_k}.\end{aligned}$$

Taking into account all directions yields

$$\frac{\partial E_{\text{ext}}}{\partial \mathbf{v}_k} = -2c\nabla\left(\left(I - I_s\right) \star P\right)\Big|_{\mathbf{v}_k}. \tag{14}$$

1.2.2 Filled Objects

For objects that are filled, the object intensity is modeled differently. Assuming the object is filled with a uniform density of sources, the object intensity can be modeled by

$$O(\mathbf{x},\mathbf{S}) = \begin{cases} c & \text{if for } \mathbf{x} \text{ completely within surface } \mathbf{S} \\ c(d+0.5) & \text{if } d \in [-0.5, 0.5] \\ 0 & \text{if otherwise} \end{cases},$$

where d is the smallest distance between the center of the pixel given by $\mathbf{x}$ and the surface $\mathbf{S}$ and c is a scaling constant. This approximates the proportion of each pixel which is contained within the surface when the pixel is close to the membrane. The value d is positive if the center of the pixel is within the surface and negative if it is outside the surface. This makes O smoother than a binary mask and more sensitive to the position of the surface.

Once again, we start with Eq. 13 with $Is = O * P$. Moving a vertex $\mathbf{v}k$ a distance $d\mathbf{x}$ has the effect of increasing O at $\mathbf{x} = \mathbf{v}k$ if $d\mathbf{x}$ is in the direction of the surface normal $\hat{\mathbf{n}}$, and decreasing O if it is in the opposite direction (Fig. 5). The change in O from moving a point can be approximated by $c d\mathbf{x} \cdot \hat{\mathbf{n}}$. If we consider just motion of the vertex $\mathbf{v}k$ in x direction, the effect on Is is given by

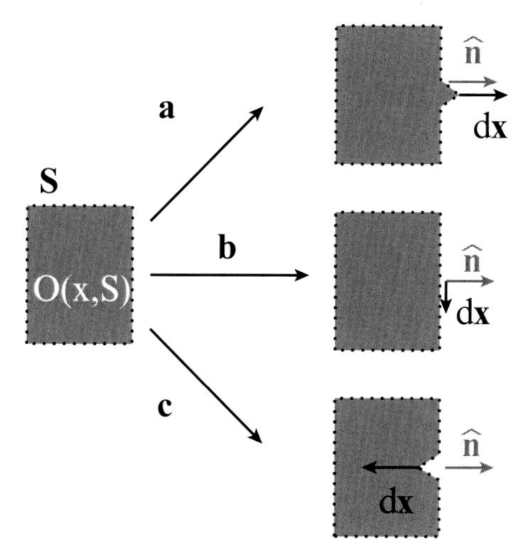

Fig. 5 Schematic of the change in O given motion in one of the points of the surface $\mathbf{S}$ for a filled object. If the change in position $d\mathbf{x}$ is in the direction of the surface normal $\hat{\mathbf{n}}$, then O increases at that point. If $d\mathbf{x}$ and $\hat{\mathbf{n}}$ are perpendicular, there is no change in O, and if they are in opposite directions, O locally decreases. Therefore, the change in O from moving a point can be approximated by $d\mathbf{x} \cdot \hat{\mathbf{n}}$

$$\frac{\partial I_s(\mathbf{x},\mathbf{S})}{\partial v_{kx}} = \frac{\partial O}{\partial v_{kx}} * P$$
$$= c\hat{n}_x \delta(\mathbf{x} - \mathbf{v}_k) * P$$
$$= c\hat{n}_x P(\mathbf{x} - \mathbf{v}_k),$$

where $\hat{n}_x$ is the x component of the surface normal. The corresponding change in E_{ext} is then, by Eq. 13,

$$\frac{\partial E_{\text{ext}}}{\partial v_{kx}} = -2c\int (I(\mathbf{x}) - I_s(\mathbf{x},\mathbf{S}))\hat{n}_x P(\mathbf{x} - \mathbf{v}_k)\mathrm{d}\mathbf{x}^3$$
$$= -2c\hat{n}_x (I(\mathbf{x}) - I_s(\mathbf{x},\mathbf{S})) \star P|_{\mathbf{v}_k} .$$

Taking into account all directions yields

$$\frac{\partial E_{\text{ext}}}{\partial \mathbf{v}_k} = -2c\hat{\mathbf{n}}(I - I_s) \star P|_{\mathbf{v}_k} . \tag{15}$$

1.3 Fitting Polymers on Surfaces

Having a 3D representation of the cell surface allows us to measure the structure of membrane bound fluorescently labelled protein. This is an extension of the model based fitting used to measure cell shapes, but is now applied to features on the cell surface. Many surface associated proteins, such as MreB and FtsZ, form as polymers. We can use model based fitting to find the polymer shape that best describes the output fluorescent image (Fig. 6).

The model for the polymer is an open active contour which is confined to the membrane shape found earlier. The goal is still to find the solution to $\mathbf{x}$ in Eq. 4, using the matrix which accounts for internal forces, $\mathbf{A}$, and the external energy, ∇E_{ext}. For a single polymer, the matrix $\mathbf{A}$ is similar to that of the closed contour, but the open contour has no coupling between the first and last points. The pentadiagonal matrix which represents the internal energies is now

a

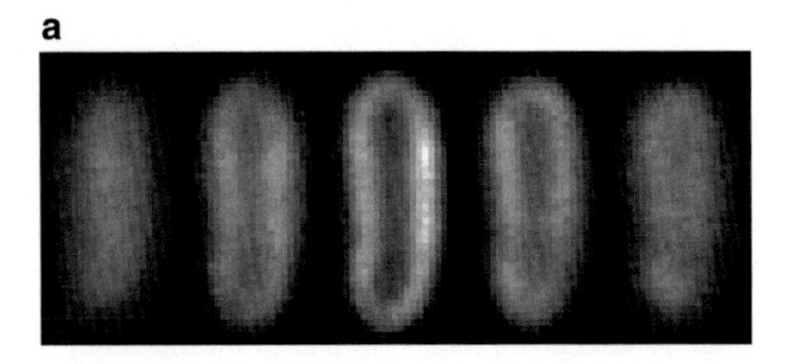

b

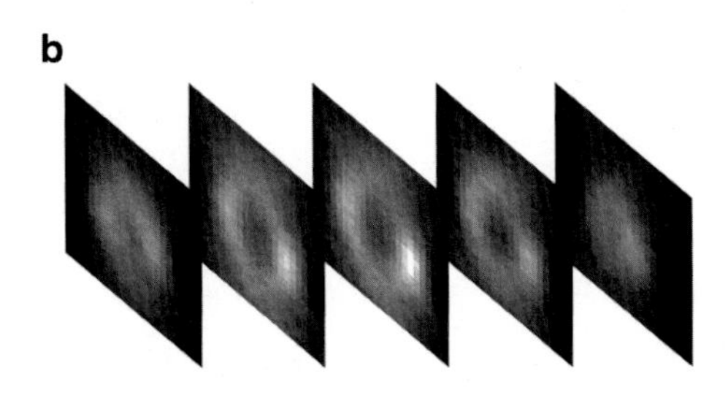

Fig. 6 (a) Several images from an image stack of *E. coli* dyed with the membrane stain FM4-64. The images are spaced by 500 nm of stage motion. (**b**) 2D slices along the cell axes are used to first determine the position of the contour. For each slice, an active contour is fit to the image before they are stitched together to create the initial cell shape estimate. The colormap on these images is fluorescent intensity

$$\mathbf{A}^{*} = \begin{pmatrix} -a_1/2+b_2 & -a_0/2+b_1/2 & b_2 & 0 & \cdots & 0 & 0 & 0 \\ a_1+b_1/2 & a_0+5b_2 & a_1+b_1 & b_2 & 0 & \cdots & 0 & 0 \\ b_2 & a_1+b_1 & a_0+b_0 & a_1+b_1 & b_2 & 0 & \cdots & 0 \\ \vdots & \ddots & \ddots & \ddots & \ddots & \ddots & \ddots & \vdots \\ 0 & \cdots & 0 & b_2 & a_1+b_1 & a_0+b_0 & a_1+b_1 & b_2 \\ 0 & 0 & \cdots & 0 & b_2 & a_1+b_1 & a_0+5b_2 & a_1+b_1/2 \\ 0 & 0 & 0 & \cdots & 0 & -a_1/2+b_2 & -a_0/2+b_1/2 & -a_1/2+b_2 \end{pmatrix}$$

For multiple polymers, the set of points $\mathbf{x}$ is the set of all points, one polymer after another, and the total matrix $\mathbf{A}$ is a block diagonal combination of the $\mathbf{A}^{*}$ from each polymer. The polymers are 3D objects confined to a 2D manifold. Using a rectangular parameterization of the surface, $\mathbf{S}(u,\phi)$, allows us to describe the polymer coordinates as $\{u,\phi\}$, and use the matrix $\mathbf{A}$ on it in the 2D representation. We chose this representation because polymers bound to the membrane in a 3D representation would tend to align with the long axis of the cell because of the straightening force of the internal energy.

For the external energy, we continue to use the difference between the simulated image and the experimental image, as in Eq. 11. The polymer is assumed to be uniformly labelled along its length, so the points that make up the polymer are interpolated to be uniformly distributed. The image intensity map O is created by converting the polymer coordinates in $\{u,\phi\}$ back to 3D cartesian coordinates, $\mathbf{x}$, and then binning of the points. Is is found using convolution as in Eq. 10.

As was the case when measuring cell shapes, the goal is to calculate ∇E_{ext} using Eq. 13. The first step is to calculate the change in the simulated image by moving a vertex. Motion of a point perpendicular to the polymer shifts the fluorescence point sources in the same way it does in membrane stains (Eq. 14). Motion of points parallel to the polymer, however, has no effect on the polymer position or length and the simulated image is unchanged. Thus, the force is as in Eq. 14, but with the directional derivative in the direction normal to the polymer, $\hat{\mathbf{n}}\nabla_{\hat{\mathbf{n}}}$, replacing the normal gradient, ∇,

$$\frac{\partial E_{\text{ext}}}{\partial \mathbf{v}_k} = -2c\hat{\mathbf{n}}\nabla_{\hat{\mathbf{n}}}\left(\left(I - I_s\right) \star P)\right|_{\mathbf{v}_k}. \tag{16}$$

An exception to this occurs at the ends of the polymer, where points moving parallel to the polymer either extend or contract the polymer, increasing the number of fluorophores or decreasing it. This process is similar to points moving in the filled cell in Eq. 15, but with the tangent vector, $\hat{\mathbf{t}}$, as opposed to a surface normal. Combining these results we have

$$\frac{\partial E_{\text{ext}}}{\partial \mathbf{v}_k} = \begin{cases} -2c\hat{\mathbf{n}}\nabla_{\hat{\mathbf{n}}}\left(\left(I - I_s\right) \star P\right)\Big|_{\mathbf{v}_k} - 2c\hat{\mathbf{t}}\left(I - I_s\right) \star P\,\Big|_{\mathbf{v}_k} & \text{if } \mathbf{v}_k \text{ is an end point} \\ -2c\hat{\mathbf{n}}\nabla_{\hat{\mathbf{n}}}\left(\left(I - I_s\right) \star P\right)\Big|_{\mathbf{v}_k} & \text{all other points} \end{cases}$$

The 3D value of $\frac{\partial E_{\text{ext}}}{\partial \mathbf{v}_k}$ is projected onto the manifold to put the force into the same coordinate system as the polymer, and the solution for $\{u, \phi\}$ of each polymer is found iteratively using Eq. 6. Currently, this method works along the body of the cell, but the polar coordinate system causes distortions near the poles. Using a triangular framework for the surface avoids the distortions near the poles, but makes it difficult to parameterize the polymers.

This method's accuracy becomes poorer as polymers get smaller or denser. For polymers much smaller than the diffraction limit of light, changes in the polymer length are indistinguishable from changes in brightness. A lower limit in the measurable size of objects is thus about 300 nm.

1.4 Results

1.4.1 Fitting Accuracy

An example of the actual image I and the simulated image Is is shown in Fig. 6 and Fig. 7. In this plane, there is little noticeable difference between the two images. To better quantify the accuracy of this method, we compared this all optical method to two other methods with higher spatial resolution in different dimensions. To test the XY accuracy of our method, we performed correlated fluorescence-electron microscopy (EM) on dyed cells. We fit the 3D cell shape from fluorescent images using this forward convolution method, and detected the outline of the cell at high spatial

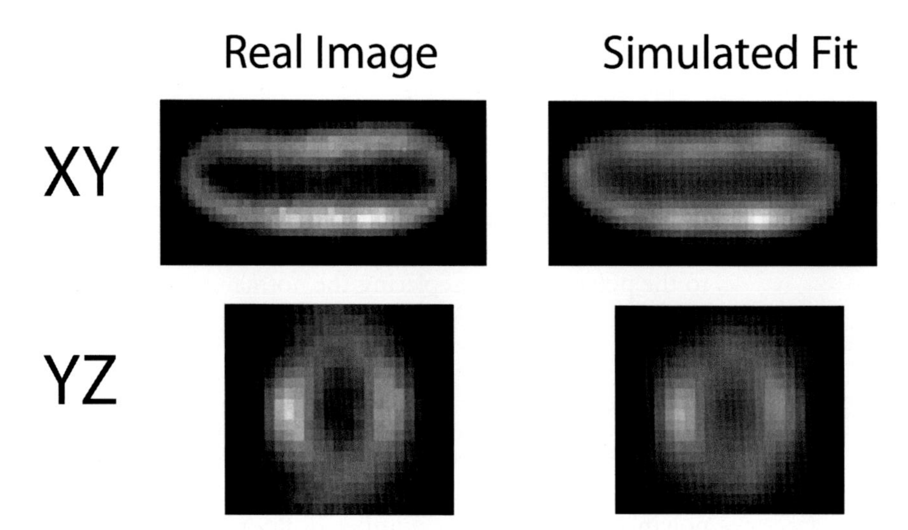

Fig. 7 A comparison between a real image (*left*) and a simulated image (*right*) obtained with our algorithm. An *E. coli* cell is stained with FM4-64 dye and imaged on a microscope to produce an image *I*, shown on the *left*. The output image is fit using our forward convolution method to produce the surface **S** which best describes the image *I*. That surface convolved with the PSF produces the simulated image *Is* shown on the *right*. The image shows the central slice of the cell

resolution using transmission electron microscopy (TEM). An overlay of a fit of a cell with a triangular mesh and the TEM image is shown in Fig. 8. After aligning fluorescent image and the EM image, we measured the outline of the cell from both images. The forward convolution fit method was used to measure the outline of the cell in the fluorescent image and a simple threshold was used to measure the outline of the cell in the electron micrograph. The root mean squared (RMS) difference in these outlines was ~ 30 nm.

To test the spatial resolution of this technique along the optical axis, we used a technique called correlated fluorescence-AFM. We took a 3D fluorescent image of a cell, and then used an AFM to measure the height profile of the same cell. The RMS difference in heights using this technique was also ~ 30 nm. See Materials and Methods (Subheadings 2 and 3, respectively) for details.

We used the polymer fitting technique to measure the polymer properties of MreB in cells expressing $MreB^{msfGFP}$ [8]. The cell shape and protein are measured as before, and the MreB fluorescence is plotted as a function of (u,ϕ). This "unwrapped" image is thresholded to estimate the initial number or polymers. An initial polymer is seeded in each of the regions and is then allowed to relax. An example of this can be seen in Fig. 9. The polymers currently cannot be measured by other techniques, so there is no control for this method.

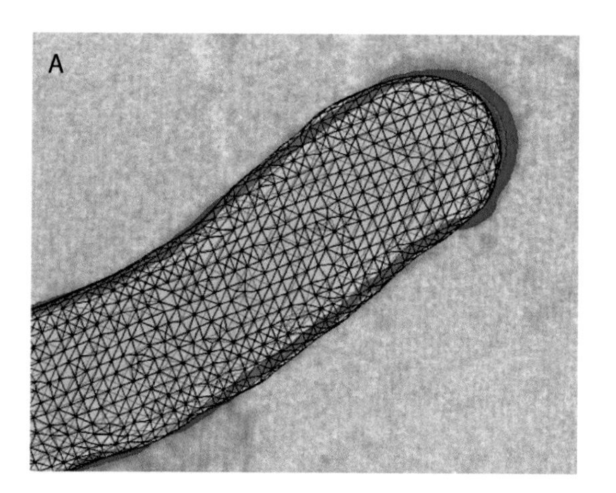

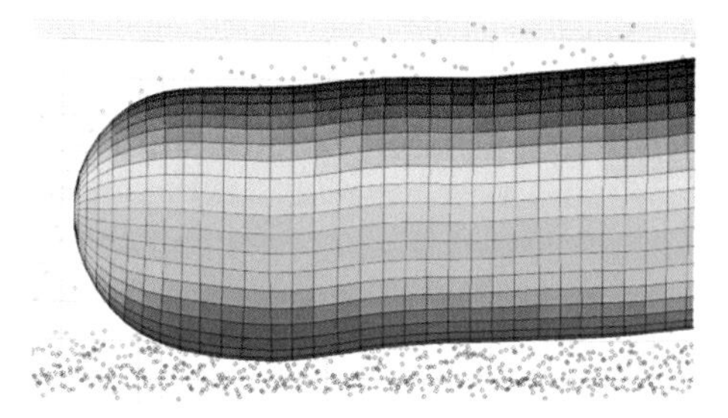

Fig. 8 Comparison of our method with correlated EM and correlated AFM. (**a**) Correlated EM was used to measure the shape of the same cell using two different methods. First the cell was stained with FM4-64, imaged, and fit using our forward convolution method. This outline is shown in the transparent triangular mesh. Below it is an image of the same cell using transmission electron microscopy. The outline of the cell as determined by electron microscopy is shown by the *blue line*. The root mean squared difference between the shapes is ~ 30 nm. (**b**) Correlated AFM applies the same principles as correlated EM, but instead of an EM image, AFM is used to measure the height profile of the cell. The cell outline colormap corresponds to the height of the cell measured using our method. The height of the cell measured by the AFM is shown in the array of points over the cell. The average height difference between the top of the cell as determined by both methods is also ~ 30 nm

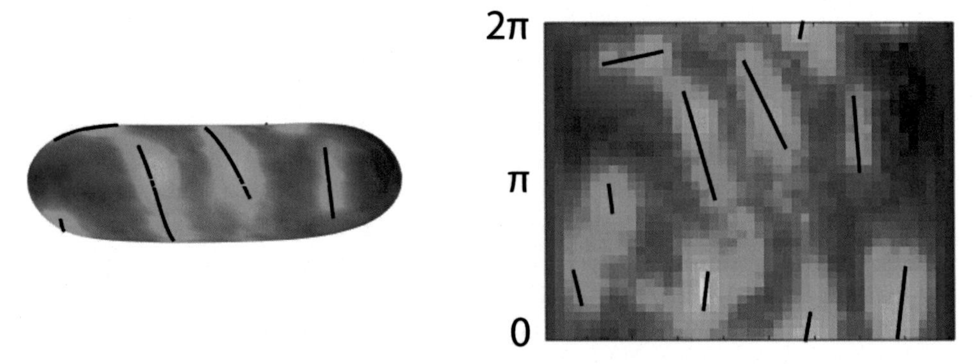

Fig. 9 An example of the reconstruction of a cell with overlaid polymer fits. On the *left* is the 3D representation of the cell and on the *right* is the surface parameterized in (u,ϕ). Note that the 2D representation has large deformations near the poles, and that the pixels in this space have different units on the two axes

1.5 Localization and Curvature

After the contours of the cell are determined, it is possible to look at curvature of the cell at all points. The two principal curvatures, κ_+ and κ_-, are the maximum and minimum curvatures, respectively, at a given point on the surface. Additional curvature metrics include the mean curvature, $H = (\kappa_+ + \kappa_-)/2$, and the Gaussian curvature, $K = \kappa_+ \times \kappa_-$. The Gaussian and mean curvatures can be calculated using the Brioschi formula [9], and the principal curvatures can be derived from K and H. In our rod shaped cells, κ_+ is almost constant. κ_- is the curvature along the axis of the cell and varies with the bumps in the cell or the cell centerline curvature. It is important to note that the Gaussian curvature is independent of the underlying parametrization of the surface and can be calculated continuously everywhere except for the cell poles where the polar parametrization has singularities. Interpolation into a cartesian basis for curvature calculations at the poles remedies this problem and results in a continuous curvature everywhere. In triangular meshes, the curvatures can be estimated at all locations using various numerical techniques [10, 11]

Our method also allows us to map membrane associated proteins on the cell surface. An example of this is shown in Fig. 10. Here, a cell expressing a fluorescently labeled protein, $\text{MreB}^{\text{msfGFP}}$, is stained with an FM4-64 membrane dye. A 3D image is taken for both fluorescent channels. The red membrane channel is first used to extract the cell shape. The coordinates from the surface are then used to interpolate the intensity in green $\text{MreB}^{\text{msfGFP}}$ channel. The result allows us to localize the protein on the 3D cell surface. The left surface shows the fluorescence intensity of a protein $\text{MreB}^{\text{msfGFP}}$, while the right surface shows the Gaussian curvature of the cell.

Analysis of the curvature gives us a way to precisely quantify differences in 3D shape between different cells and mutants. In addition to the standard shape metrics such as cell length, diameter, and volume, the 3D shape allows us to quantify more local features such as curvature. The technique also allows for 3D

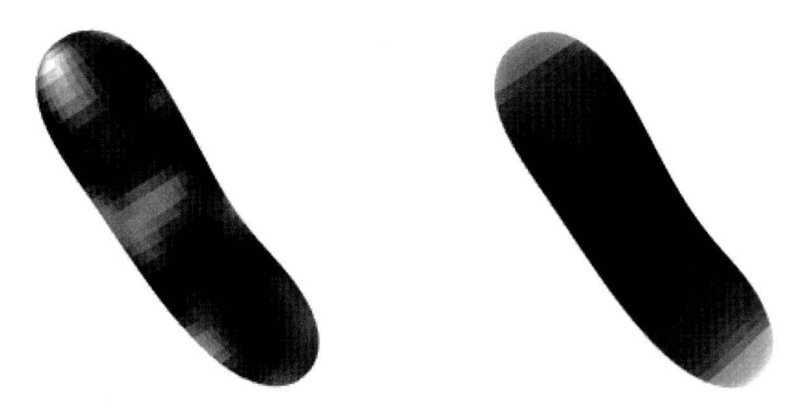

Fig. 10 An example of the reconstruction of a cell with colormap representing Gaussian curvature on the *left* and fluorescence intensity of $MreB^{msfGFP}$ on the *right*. Cells were stained with FM4-64, imaged, and fit using our forward convolution method. In addition to an FM4-64 image, $MreB^{msfGFP}$ was also imaged. The *left image* shows the MreB fluorescence intensity mapped onto the surface. The *right image* shows the calculated Gaussian curvature at each point on the surface

mapping of fluorescent proteins onto the cell surface and characterization of those proteins. This has already been demonstrated by Ursell et al. [8], where it was shown that $MreB^{msfGFP}$ preferentially localizes at areas of negative Gaussian curvature or low mean curvature in *E. coli*.

2 Materials

1. Cells were grown in M63 media with casamino acids and glucose, which is 15 mM $(NH^4)^2SO^4$, 100 mM KH^2PO^4, 1.7 μM $FeSO^4$ EDTA, 1 mM $MgSO^4$, adjusted to pH 7 with KOH, 0.2 % casamino acids, 0.5 % glucose (*see* **Note** 1).
2. Agarose pads were made by melting 1 % ultra pure agarose in M63. When necessary, FM46-4 membrane dye was added to the pad at 50 ng/mL (*see* **Note** 2). Pads were made by pressing 200 μL of media between glass slides. Spacers were used to make the pads approximately 1 mm thick.
3. Coverslips were cleaned prior to use in imaging in order to remove any impurities that may pollute fluorescent signals. Coverslips were placed in a teflon rack, submersed 15 % KOH in ethanol, and sonicated for 1 h. After sonicating, coverslips were rinsed with water and ethanol, and stored at room temperature in ethanol.
4. Samples were sealed with VALAP, which is a 1:1:1 mixture of vaseline, lanolin, and paraffin wax.
5. PSF images were taken from 100 nm Tetraspeck™ microspheres (Invitrogen).

3 Methods

3.1 PSF Measurements

A sample of microspheres was added to an agarose pad and imaged using a custom built microscope (the same microscope used to image cells). Image stacks were taken by moving the stage vertically in 50 nm steps over a 6 μm range. Since imaging noise has a large effect on the image of a single particle, over 200 single particles were cropped, aligned, and averaged to create the PSF used for image convolution. A $3 \times 3 \times 6\,\mu m^3$ cropping volume around each particle is used. Axial spacing for image slices was adjusted to account for the mismatch in refractive index (*see* **Note** 3).

3.2 Image Acquisition

Cell shape analysis was done on *E. coli* expressing $MreB^{msfGFP}$. Overnight cultures were diluted 100× in M63 media with glucose and casamino acids. Cells were grown at 37°C to OD 0.3 and 1 μL of cell culture was seeded on agarose pads with FM4-64 dye. The droplet is allowed to absorb into the pad for 2 min. The pads were covered with cleaned coverslips and sealed using VALAP to prevent drying. Image stacks were taken on a custom built fluorescent microscope taking a series of images at 100 nm intervals in z over a 6 μm range. The microscope was controlled by custom-made software in LabVIEW. FM4-64 images were taken with a 561 nm wavelength laser. For $MreB^{msfGFP}$ localization, an additional stack was taken using a 488 nm laser. Again, axial spacing of image planes was adjusted using the focal shift (*see* **Note** 3). Cells are then individually cropped so that each image stack contains a single cell (*see* **Note** 4).

3.3 Shape Initialization

The 3D images were analyzed using custom MATLAB code. An initial surface was created depending on the type of mesh being used to fit the shape. For rod-shaped *E. coli* fit with a rectangular mesh, we first find an approximate centerline by thresholding and skeletonizing the 3D image stack (*see* **Note** 5). The centerline is then used to produce a curvilinear coordinate system about the cell. Slices perpendicular to the centerline yield cross-sectional images of the cell, as shown in Fig. 6. In each slice of the cell, an active contour is used to find the membrane outline. These slices are then combined to generate the cell body. Cell poles cannot be fit this way since resolution is lost when image slices are taken tangent to the membrane. The poles were instead parameterized with spherical coordinates and slices were taken at different angles much like a sliced orange. The final membrane surface parametrization is $\mathbf{S}(u,\phi)$, where u runs along the length of the cell and ϕ is the rotation angle (*see* Fig. 3b). The contour is then relaxed using our method to find the final shape of the cell.

Cells that are amorphous or do not have a defined long axis are better fit with a triangular mesh. For triangular meshes, cells were initialized with a sphere in the middle of the image (*see* **Note** 6). With the adaptive mesh, the sphere is able to expand and fit the shape of the cell.

3.4 Shape Fitting

The initial shapes are relaxed using the algorithm described in Fig. 11 (*see* **Note** 7). Parameters for fitting, namely the stiffnesses, α and β, the step size, γ, and the convergence threshold can be estimated by fitting a small number of cells and inspecting the results. The fitting output should be robust over an order of magnitude for α and β. They should be chosen so that the surface fits the image without over fitting noise. For increased throughput, images can be analyzed in parallel using computing clusters.

3.5 Polymer Fitting

When fitting the fluorescent membrane bound polymers, an initialized set of contours is made by segmenting the unwrapped image (Fig. 9) and seeding an initial polymer in each region. The initial polymer has the length and orientation of the segmented region. Although the polymers are initially found in the 2D unwrapped image, they are returned to 3D for the image fitting process. The algorithm in Fig. 11 is then applied with the polymers confined to the cell surface. The polymer length and orientation can then be measured from the final position of the polymers (*see* **Note** 8).

3.6 Benchmarks

Correlated EM was performed using the same custom built fluorescent microscope. Cells were attached to an electron microscopy grids (Electron Microscopy Sciences™) using polyethylenimine

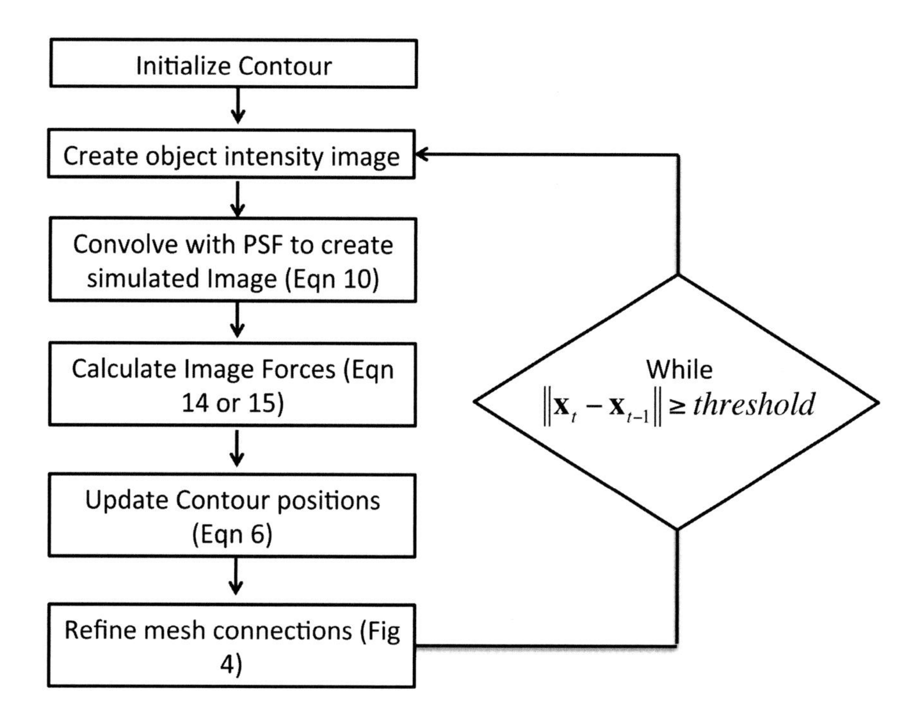

Fig. 11 A flow chart outlining the steps of the algorithm for fitting 3D fluorescent images using an active mesh with convolution with the microscope PSF. The algorithm continues until the change in the surface points {**x**} is below some threshold

(PEI) (Sigma™). Cells were imaged on the fluorescent microscope before being imaged on the electron microscope (CARL ZEISS LEO OMEGA 912 EF-TEM). The outline of the cell in the EM micrograph was found by applying a threshold and taking the outline of the binary image. The outline from the EM image was aligned with the fit from the fluorescent image by minimizing the RMS difference between the two outlines. The RMS difference between the two outlines was then measured.

Correlated AFM was performed using a custom built combined fluorescent microscope—AFM system. Cells were adhered to cover glass using PEI and dyed with FM4-64 dye in phosphate buffered saline solution. Fluorescent image stacks were taken and the AFM was used to immediately measure the height profile of the cell. A spline sheet was fit to the array of heights from the AFM, and the height difference between the sheet and the cell surface near the top of the cell was measured.

4 Notes

1. It is important to use an optically clear media during imaging. Media such as LB have signifiant autofluorescence which will disrupt the image fitting process.
2. FM4-64 in large amounts can be toxic to cells and can stain impurities in the sample.
3. Focal shift, an aberration due to the mismatch between the refractive index of the sample and the glass, has a large effect on the size of objects imaged using optical sectioning. When the objective or stage moves 100 nm in the z-direction, the focal plane moves between 60 and 80 nm depending on the numerical aperture (NA) of the objective lens, the mismatch of the refractive indices, and the depth of the imaging plane in the sample. We measured the focal shift for our microscope experimentally and found that 100 nm of stage motion in z corresponded to a 60 nm change in the position of the focal plane.
4. Cells that appeared to be forming a division septum were removed from the analysis during this step. Because the PSF may not be circularly symmetric, it is important that images are not rotated during image cropping. The image orientation must match that orientation used to measure the PSF.
5. Fitting with a rectangular mesh also works well for filamented and helical cells.
6. The initial sphere in the triangular mesh must be in contact with the cell in order to fit it. The triangular mesh works best for cells that are filled rather than just membrane stained because

filled images can interact with the surface throughout the entire volume of the cell as opposed to just at its boundaries.

7. The object intensity image is created at a higher resolution than the image from the camera. This makes the image forces more dependent on the precise position of the surface. To calculate the simulated image and the image forces, the PSF must also be up-sampled to the same resolution as the object intensity image. In practice, there is a tradeoff between the amount of up-sampling and the computational cost of the convolution. We often use a resolution twice that of the original image.
8. In order to meaningfully detect polymers, the surface fluorescence must appear as elongated structures like in Fig. 9. The algorithm will attempt to fit a single polymer to any bright connected structure, so large foci and patches will not be described properly.

References

1. Mitchell J (2002) The energetics and scaling of search strategies in bacteria. Am Nat 160(6): 727–740
2. Justice S, Hung C, Theriot J, Fletcher D, Anderson G, Footer M, Hultgren S (2004) Differentiation and developmental pathways of uropathogenic *Escherichia coli* in urinary tract pathogenesis. Proc Natl Acad Sci USA 101(5), 1333–1338
3. Sliusarenko O, Heinritz J, Emonet T, Jacobs-Wagner C (2011) High-throughput, subpixel precision analysis of bacterial morphogenesis and intracellular spatio-temporal dynamics. Mol Microbiol 80(3): 612–627
4. Locke JC, Elowitz MB (2009) Using movies to analyse gene circuit dynamics in single cells. Nat Rev Microbiol 7(5):383–392
5. Guberman JM, Fay A, Dworkin J, Wingreen NS, Gitai Z (2008) Psicic: noise and asymmetry in bacterial division revealed by computational image analysis at sub-pixel resolution. PLoS Comput Biol 4(11):e1000233–e1000233
6. Helmuth JA, Sbalzarini IF (2009) Deconvolving active contours for fluorescence microscopy images. In: Advances in visual computing. Springer, Berlin, pp 544–553
7. Kass M, Witkin A, Terzopoulos D (1987) Snakes - active contour models. Int J Comput Vis 1(4):321–331
8. Ursell TS, Nguyen J, Monds RD, Colavin A, Billings G, Ouzounov N, Gitai Z, Shaevitz JW, Huang KC (2014) Rod-like bacterial shape is maintained by feedback between cell curvature and cytoskeletal localization. Proc Natl Acad Sci USA 111(11):E1025–E1034
9. Gray A, Abbena E, Salamon S (2006) Modern differential geometry of curves and surfaces with Mathematica. Textbooks in mathematics, vol 10, 3rd edn. Chapman and Hall/CRC, p. 12 (1016 pp)
10. Rusinkiewicz S (2004) Estimating curvatures and their derivatives on triangle meshes. In: Proceedings of 3D Data processing, visualization and transmission, 2004, 2nd International symposium on 3DPVT 2004. IEEE, Washington, DC, pp 486–493
11. Dong C-S, Wang G-Z (2005) Curvatures estimation on triangular mesh. J Zhejiang Univ Sci 6(1):128–136

Chapter 18

Coarse-Grained Molecular Dynamics Simulations of the Bacterial Cell Wall

Lam T. Nguyen, James C. Gumbart, and Grant J. Jensen

Abstract

Understanding mechanisms of bacterial sacculus growth is challenging due to the time and length scales involved. Enzymes three orders of magnitude smaller than the sacculus somehow coordinate and regulate their processes to double the length of the sacculus while preserving its shape and integrity, all over a period of tens of minutes to hours. Decades of effort using techniques ranging from biochemical analysis to microscopy have produced vast amounts of data on the structural and chemical properties of the cell wall, remodeling enzymes and regulatory proteins. The overall mechanism of cell wall synthesis, however, remains elusive. To approach this problem differently, we have developed a coarse-grained simulation method in which, for the first time to our knowledge, the activities of individual enzymes involved are modeled explicitly. We have already used this method to explore many potential molecular mechanisms governing cell wall synthesis, and anticipate applying the same method to other, related questions of bacterial morphogenesis. In this chapter, we present the details of our method, from coarse-graining the cell wall and modeling enzymatic activities to characterizing shape and visualizing sacculus growth.

Key words Coarse-grained modeling, Molecular dynamics simulations, Cell wall synthesis, Bacterial morphogenesis, Rod shape maintenance

1 Introduction

Most bacterial cells are surrounded by a sacculus that prevents lysis from turgor pressure and determines the cell's shape (e.g., a rod, in the case of *E. coli*) [1]. How the cell coordinates sacculus growth so that breaks introduced to allow insertion of new material do not cause lysis remains an open question.

Considerable work has revealed the structure of the sacculus. Paper chromatography and high-performance liquid chromatography (HPLC) revealed that the *E. coli* sacculus is made of peptidoglycan (PG) [2–4]. The glycan strand is polymerized from disaccharides of an *N*-acetylglucosamine (NAG) and an *N*-acetylmuramic (NAM) acid, each attached to a stem L-Ala-D-iGlu-m-A_2pm-D-Ala-D-Ala penta-peptide. Peptides on adjacent strands form crosslinks, most at

Hee-Jeon Hong (ed.), *Bacterial Cell Wall Homeostasis: Methods and Protocols*, Methods in Molecular Biology, vol. 1440, DOI 10.1007/978-1-4939-3676-2_18, © Springer Science+Business Media New York 2016

the fourth (D-Ala) residues of the donors and the third (m-A_2pm) residues of the acceptors, resulting in a mesh-like PG network. Early electron microscopy studies revealed that purified sacculi retain the cell's rod shape [5]. Later, electron cryo-tomography was used to show that glycan strands run circumferentially around the rod [6]. This is consistent with a classical model of sacculus architecture which posits that long and stiff glycan strands run circumferentially, bearing the greatest stress, while short and flexible peptide crosslinks run parallel to the rod's long axis, bearing half as much stress [1].

Other work revealed the enzymatic details of the PG synthesis machinery. PG precursors are synthesized in the cytoplasm and then transferred to the periplasm [7] where they are polymerized and crosslinked into the sacculus by transglycosylases and transpeptidases, also known as penicillin-binding proteins (PBPs) [8]. Also essential to the process are endopeptidases that cleave covalent bonds to open space for the new material [9]. X-ray crystallography has revealed the structures of many synthases and hydrolases from multiple species [8, 10], and the enzymatic activities of *E. coli* synthases have been characterized in vitro [11, 12]. Affinity chromatography and bacterial two-hybrid studies showed that many synthases and hydrolases interact with one another [12–16] and with the outer membrane proteins LpoA/B [17, 18], suggesting that the enzymes exist in a complex spanning the periplasm. While affinity chromatography showed that both cytoplasmic and periplasmic enzymes interact with the morphogenetic proteins MreB/C/D and Rod A/Z [19–21], fluorescence microscopy has yielded conflicting results as to whether these proteins co-localize/move in the same complex [21–25].

Despite decades of experiments, how the activities of PG synthesis enzymes are coordinated at the molecular and cellular levels remains unclear. Several models have been proposed. For instance, a "make-before-break" strategy was proposed in which autolysins would cleave crosslinks along the template strand to liberate it only after new strands are fully crosslinked to the sacculus underneath the existing strand, thus preventing lysis [26]. Whether the enzymes could actually be coordinated to execute such temporally and spatially separated operations is unclear, however. Similarly, it was proposed that the cytoskeletal protein MreB forms an extended filament that guides PG insertion to maintain rod shape [27]. Disagreement on the oligomeric form and driving force of movement of MreB [23, 24, 27–41], however, obscures its role in PG synthesis.

We realized that another approach is needed to shed light on the coordination of PG remodeling enzymes. Coarse-grained simulation of cell wall remodeling, pioneered by Huang and colleagues [42], has proven to be a valuable method to test different models suggested by experiments. The Huang model, however, has mainly focused on different mechanisms by which MreB might guide insertion sites of new PG. To do that, the incorporation of each new PG strand into the sacculus has been modeled as a single event in which an entire glycan strand is introduced, and all necessary peptide crosslinks cleaved and re-formed, all in one step before any relaxation of the sacculus can occur [28, 41, 43, 44]. This reduces the computational cost but has prevented an exploration of the properties and coordination of PG remodeling enzymes.

In order to explore different molecular mechanistic models of sacculus growth, we have developed a simulation method that allows us to vary properties of PG-remodeling enzymes and their coordination [45]. To make our model as realistic as possible, PG is represented by a coarse-grained model whose mechanical properties were derived from all-atom molecular dynamics (MD) simulations of isolated glycan strands and peptides. For the first time to our knowledge, individual enzymes, including transglycosylases, transpeptidases, and endopeptidases, are explicitly represented.

The rich literature of biochemical data and hypothetical models made it challenging to build our initial model. One approach would have been to implement all the models proposed in the literature. Many models, however, are contradictory, e.g., multi-enzyme complex [46] vs. diffusive transpeptidase [25], extended helical MreB filament [27, 29–35] vs. circumferentially moving MreB spots [23, 24, 28], or single-strand insertion [47, 48] vs. strands inserted in pairs [49–52]. And even if the combined models worked, it would be impossible to dissect which models are required and which redundant. We instead decided to pursue the simplest model that works. We started with a very simple model (named Remodeler 1.0, as explained below) and implemented additional hypotheses only when necessary. A schematic of this process is presented in Fig. 1, and readers are referred to [45] for details. For each model that failed to maintain rod shape, we

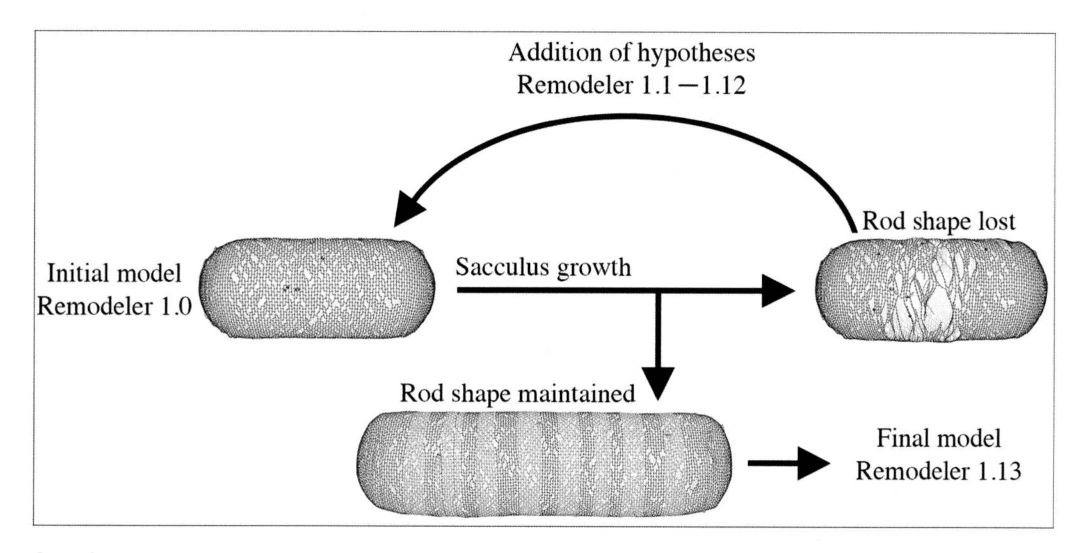

Fig. 1 Schematic of the process of iteratively building a complex model from a simple initial model

analyzed the most obvious cause and added a molecular hypothesis based on evidence from the literature and/or biophysical plausibility to fix the problem. This process was iterated until rod shape was maintained (Remodeler 1.1–1.12). Finally, we removed hypotheses one by one from Remodeler 1.12 to check if any were rendered redundant by the other hypotheses. The final model (Remodeler 1.13) thus comprised one simple set of hypotheses capable of maintaining rod shape during sacculus growth.

We have used our model to study how rod shape might be maintained in Gram-negative bacteria (*see* **Note 1**). We distributed the simulation codes for each stage of this model, named Remodeler 1.0–1.13 [45]. In the future, we anticipate further developing our model to study other PG-synthesis related topics, from rod shape maintenance in Gram-positive bacteria, to shape recovery of perturbed cells, to cell division and even sporulation (*see* **Note 2**). In each case, we will name our models Remodeler 2.x, 3.x, and so on to help readers who wish to use our codes, starting with any stage of the model.

In the following sections, the details of our model are described and a brief discussion of results is presented. Readers are advised to watch https://www.youtube.com/watch?v=_5Ov3vp6Qyg&feature=youtu.be since video represents the model building process better than static figures. For further details of results and discussion, the readers are referred to [45].

2 Materials

All-atom MD simulations were conducted using the software NAMD [53]. The coarse-grained simulation software was written using Fortran language. Visualization of simulation data was done using Visual Molecular Dynamics (VMD) [54]. Images were processed using Photoshop. Movies were made using QuickTime Pro and concatenated using Final Cut Pro.

3 Method

The general procedure for sacculus growth simulations is illustrated in Fig. 2. First (Setup), we built the initial system, composed of a sacculus and a set of enzymes including transglycosylases, transpeptidases, and endopeptidases. Next (PG remodeling), in each time step, each enzyme performed its function with a certain probability. Forces exerted on the sacculus and enzymes were then calculated and the coordinates of the system updated (PG relaxation). The process of PG remodeling and relaxation was repeated until the sacculus reached the desired mass. We discuss the details of the procedure below.

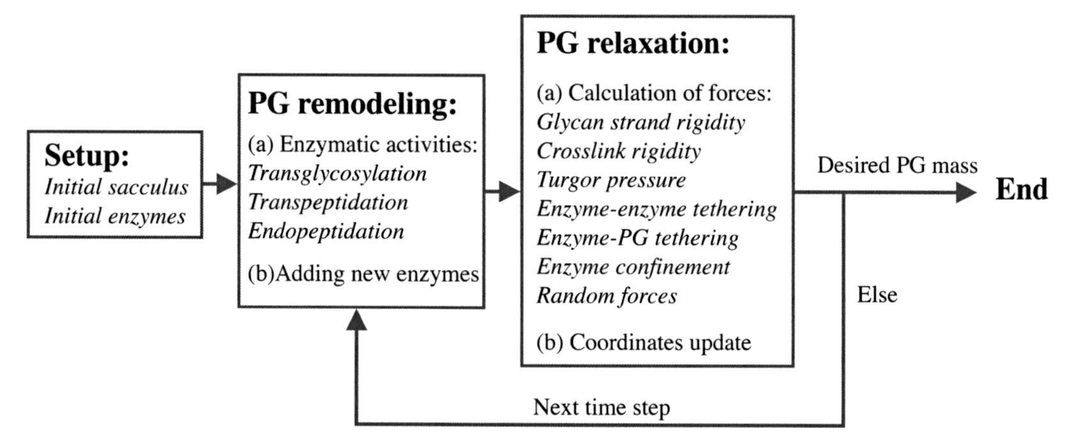

Fig. 2 Computational procedure for sacculus growth. For clarity, the names of code sections are italicized in all flowcharts

3.1 Coarse-Grained Peptidoglycan Model

Most MD simulation software provides atomic-level insights into processes that occur on the nanosecond timescale; for example, NAMD was used to simulate an HIV-1 capsid of ~10^7 atoms over a period of ~500 ns [55]. By contrast, the sacculus is a giant molecule (on the order of 10^8 atoms) that doubles its size over a period of minutes to hours. Coarse-graining the sacculus helps reduce the computational cost, allowing observation of phenomena occurring at both the molecular and cellular levels. To do this, we represented each pair of disaccharides as one bead and connected the beads with springs to form chain-like glycan strands. As adjacent disaccharides are rotated 90° with respect to each other [56–58], half of the peptides presumably protrude perpendicular to the sacculus surface and do not participate in crosslinking. We therefore ignored these out-of-plane peptides. Thus, each bead in our coarse-grained model was attached to one in-plane peptide (Fig. 3a).

3.1.1 Glycan Mechanical Properties

We previously developed all-atom force fields for PG [59, 60] that allowed us to set the mechanical properties of the coarse-grained model to match the behavior of all-atom MD simulations. To calculate the stiffness of glycan, a fully solvated system of an 80-tetrasaccharide strand without stem peptides was equilibrated for 6.6 ns using the software NAMD [53] (Fig. 3b). For scale, this system contained nearly one million atoms (mostly solvent) and was over 150 nm in length, yet represents only a miniscule fraction of the entire sacculus. During the simulation, the strand shrank slightly, by about 2%, but maintained an extended conformation overall. We then extracted histograms of distances and bending angles between adjacent tetrasaccharides. MD simulations were next run on an equivalently sized coarse-grained (CG) strand where adjacent beads, each representing one tetrasaccharide, were connected by springs of constant k_g and relaxed length l_g, and a bending angle θi at bead i was penalized with an energy of

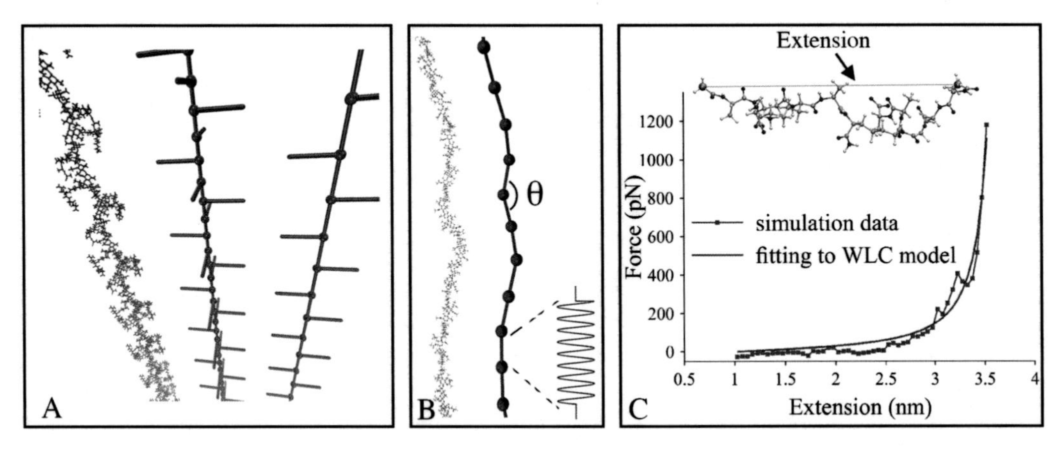

Fig. 3 Parameterization of the coarse-grained model, adapted from [45]. (**a**) The glycan strand—disaccharides in *blue* and peptides in *red*—in atomic representation (*left*) was coarse-grained as a chain of beads, each representing first a disaccharide attached to a peptide (*middle*) and finally a tetrasaccharide attached to an in-plane peptide (*right*). (**b**) Snapshots of a glycan strand in all-atom MD (*left*) and coarse-grained simulations (*right*). In the latter, the strand was modeled as a chain of beads connected by springs. (**c**) Extension dependence of force on a peptide crosslink extracted from all-atom MD simulations (*blue*), and after fitting to a worm-like chain model (*red*)

$$E_{\mathrm{b}}^{(i)} = \frac{1}{2} k_{\mathrm{b}} \left(\theta_i - \theta_0\right)^2 \quad (1)$$

where k_b is the bending stiffness and θ_0 is the relaxed angle. A Langevin damping term of $\gamma = 2\mathrm{ps}^{-1}$ was added to mimic water viscosity; values ranging from 1 to 5 ps^{-1} were tested and found to have no effect on the resulting sampled bond and angle lengths. A time step of 100 fs was used for the CG simulations and they were run for 500 ns (note that times are not directly comparable between the simulations due to the significantly simplified potential of CG simulations). The parameters for the CG model were iteratively sampled until the histograms extracted from CG simulations matched those of the all-atom simulations by visual inspection. The CG parameters that produced the best match were $k_g = 5570$ pN/nm, $l_g = 2.0$ nm, $k_b = 8.36 \times 10^{-20}$ J, and (as expected) $\theta_0 = 3.14$ rad.

3.1.2 Peptide Crosslink Mechanical Properties

Initially, we tried to fit the peptide-crosslink bond strength using a similar histogram matching procedure. However, it quickly became apparent that no match could be obtained; the bond-distance histogram from the atomistic simulations of three crosslinked glycan strands was not symmetric as would be expected for a harmonic bond. To better understand the length distribution of peptide crosslinks, we simulated a single peptide crosslink, i.e., an Ala(1)-isoGlu(2)-A_2pm(3)-Ala(4)-Ala(5) pentapeptide linked to an Ala(1)-isoGlu(2)-A_2pm(3)-Ala(4) tetrapeptide through an A_2pm(3)-Ala(4) peptide bond. Specifically, we determined the potential of mean force (PMF) as a function of end-to-end extension. We used all-atom MD adaptive biasing force (ABF) simulations so the full energy landscape could be assessed quickly. This is a quasi-equilibrium method in which the biasing forces exerted on the two terminal (reaction) atoms are iteratively calculated as the *positive* gradient of the PMF, thus making the two atoms diffuse freely [61, 62]. The reaction coordinate (extension) was divided into four 10-Å windows to accelerate convergence; each was run for between 5 and 8 ns. Based on the resulting PMF and associated mean-force profile, we determined that the peptide crosslink is better modeled as a worm-like chain (WLC) than a spring, i.e., the force is almost zero at small extension, increases only moderately for extensions less than the contour length, but then increases dramatically at large extension (Fig. 3c). We therefore fit the mean force vs. extension curve to the following formula:

$$F\left(x^*\right) = k_{\mathrm{WLC}}\left[\frac{L_c^*}{4\left(1 - x^* / L_c^*\right)^2} - \frac{L_c^*}{4} + x^*\right] \tag{2}$$

where $L_c^* = L_c - x_0$ is the effective contour length, L_c is the contour length, x_0 is the extension (end-to-end distance) x at which the force is zero, $x^* = x - x_0$ is the effective extension, and k_{WLC} is a force constant. We then determined the parameters that produced the best fit as $k_{\mathrm{WLC}} = 15.0$ pN/nm, $L_c = 4.8$ nm, and $x_0 = 1.0$ nm (*see* **Note 3**). Consistently, Braun et al. used space-filling models to show that the peptide crosslink is ~4.2 nm long when fully extended and ~1.0 nm long when maximally collapsed [63], which agrees well with the L_c and x_0 derived from our simulations.

3.1.3 Initial Sacculus

The initial sacculus model was built by placing glycan strands along circumferential hoops (Fig. 4a) and connecting opposing peptides on adjacent hoops to form crosslinks (Fig. 4b). We used hoops of the same diameter to form a cylindrical waist, and those of gradually decreasing diameter to form two polar caps. Initially, the lengths of glycan strands were chosen uniformly randomly within a range from 10 to 20 tetrasaccharides (later, during sacculus growth, the length was determined by the enzyme processivity). To reduce the computational cost, we used an initial sacculus of circumference 100 tetrasaccharides, ~10 times smaller than typical wild-type *E. coli* cells. We did, however, test the effect of size by running simulations on sacculi of diameters twice and four times as large (about the size of a small rod-shaped Gram-negative cell such as *Acetonema longum*), and obtained similar results [45] (*see* **Note 4**).

Note that even though the initial sacculus was built with ordered glycan strands, its shape after relaxation in the presence of turgor pressure is a function of the mechanical properties of glycan strands and peptide crosslinks.

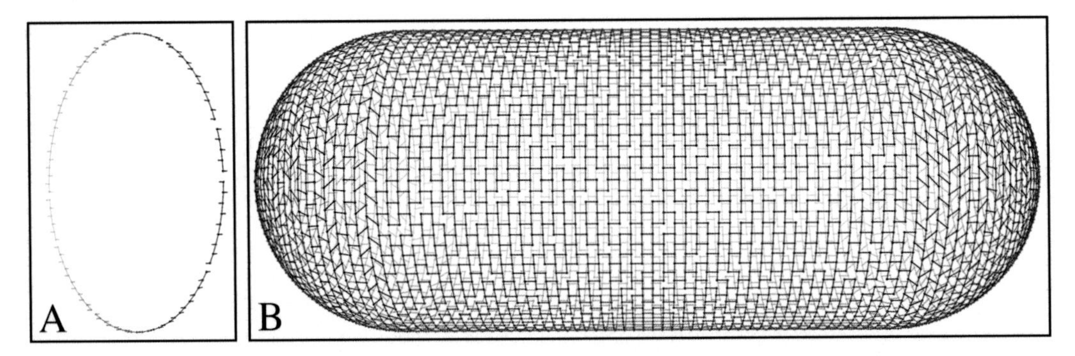

Fig. 4 Building a coarse-grained sacculus. (**a**) Coarse-grained glycan strands were arranged in hoops. (**b**) A sacculus was formed by connecting peptides on adjacent hoops

3.2 Turgor Pressure

Turgor pressure plays an important role in sacculus growth as it inflates the sacculus, allowing new material to be incorporated. We therefore added to the total energy of the system the work done by turgor pressure P to inflate the sacculus to volume V:

$$E_{\mathrm{vol}} = -PV \tag{3}$$

To calculate V, the volume enclosed by the sacculus, the mesh-like surface was divided into a series of polygons (Fig. 5). The polygons were then further divided into triangles from which tetrahedrons were built using the sacculus center as the fourth vertex. V was then calculated as the sum volume of the tetrahedrons:

$$V = \sum_i V_i = \sum_i A_i h_i / 3 \tag{4}$$

where Ai is the area of triangle i and hi is the distance from the sacculus center to the plane of triangle i. The force on the sacculus due to turgor pressure was then calculated as $F = -\nabla E_{\mathrm{vol}}$. As most measurements of turgor pressure within Gram-negative bacteria have been reported to be between 2 and 4 atm [64–66], we used a turgor pressure of 3.0 atm in most of our simulations.

3.3 Coarse-Grained Model of Enzymes

While current MD simulation software is limited to the study of "closed" systems, our coarse-grained model allows exploration of "open" systems by implementing enzymatic activities that could add or remove beads and bonds. To explore possible molecular mechanisms of PG synthesis, generic transglycosylases, transpeptidases, and endopeptidases were modeled explicitly as individual coarse-grained beads (Fig. 6). They were modeled to diffuse within the confines of the periplasm and interact with each other and with the sacculus while performing their functions. By modeling enzymatic activities step-by-step, we could investigate different molecular mechanisms for spatial and temporal coordination of the enzymes.

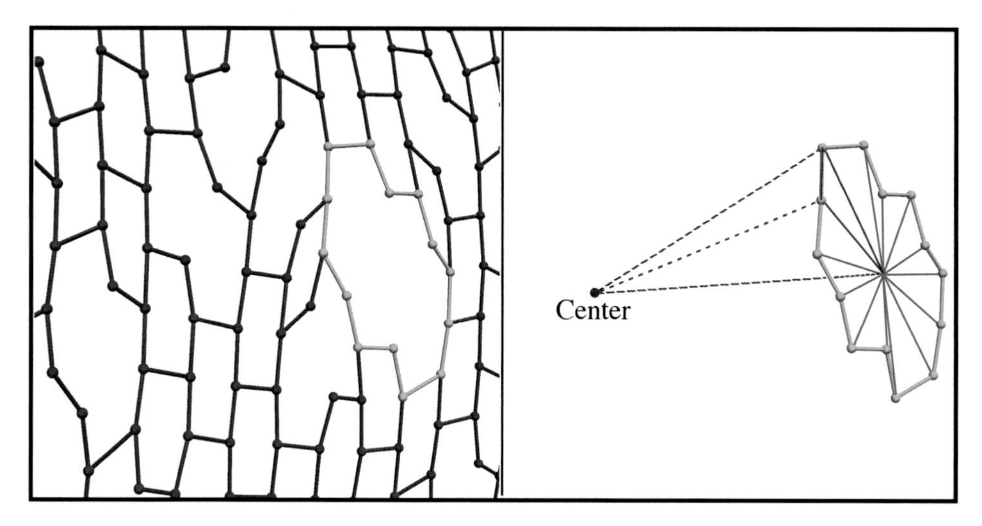

Fig. 5 Volume determination. A polygon (*green*) on the sacculus surface (*left*) was divided into triangles, each having an edge on the polygon and sharing the polygon's center as the third vertex (*right*). Tetrahedrons were then built from the triangle using the sacculus center as the fourth vertex

Note that in our later simulations, glycosidic bond hydrolysis and carboxypeptidation were also implemented without explicitly modeling the corresponding enzymes [45].

We implemented stepwise enzymatic activities using flags, which we capitalize for clarity in the following descriptions. Each enzymatic activity was modeled to occur with a probability that was arbitrarily chosen since we were unaware of any biochemical data on the rates of PG synthesis enzymes in vivo.

3.3.1 Transglycosylation

For *transglycosylation*, the enzymes were modeled as INACTIVE (not synthesizing but diffusing around) or ACTIVE (ready to synthesize a new strand). ACTIVE transglycosylases were modeled as STRAND-FREE (not holding a new strand), STRAND-BOUND (donor domain holding a new strand), PRECURSOR-FREE (not loaded with a precursor in the acceptor domain), or PRECURSOR-LOADED (loaded with a precursor in the acceptor domain). STRAND-BOUND enzymes could be PRE-TRANSLOCATED (immediately after initiating a new strand or adding a bead to the growing strand), or TRANSLOCATED (translocated to the strand tip after initiating a new strand or adding a new bead to the growing strand, ready to be loaded with a precursor). A flowchart of the *transglycosylation* loop in our simulation code is presented in Fig. 7.

An INACTIVE transglycosylase, upon interaction with a lipoprotein was "activated" with a probability of once every 10^4 steps. ACTIVE but STRAND-FREE transglycosylases became STRAND-BOUND once they were "loaded" with a PG precursor bead in the active site (Fig. 6). Precursor loading was modeled to occur with a probability of once every 10^3 time steps. Precursor reloading on a

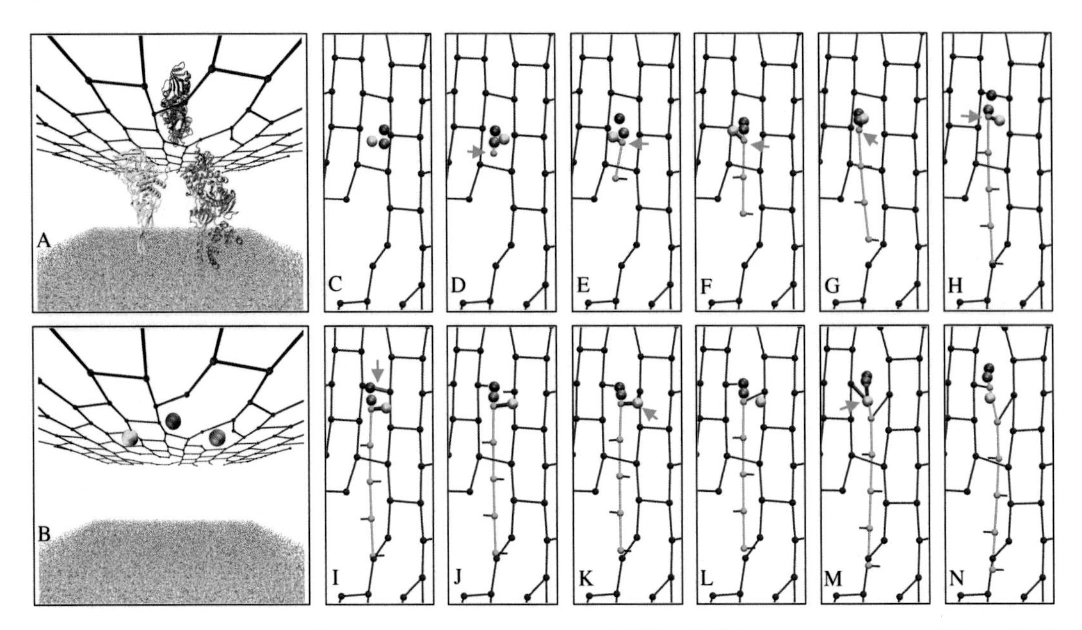

Fig. 6 Coarse-grained model of enzymes. A transglycosylase (3FWM, [10]) in *orange*, a transpeptidase (3EQV, [85]) in *yellow*, and an endopeptidase (2EX2, [86]) in *gray* are shown in crystal structures (**a**), and modeled as beads (**b**). Inner membrane is shown for context. (**c–n**) Visual depiction of enzymatic activities (noted by *blue arrows*). A transglycosylase initiates a new strand (shown in *green*) (**d**), and elongates it (**e–h**). An endopeptidase cleaves a peptide crosslink (**i**). A transpeptidase crosslinks the new strand to the sacculus (**k, m**)

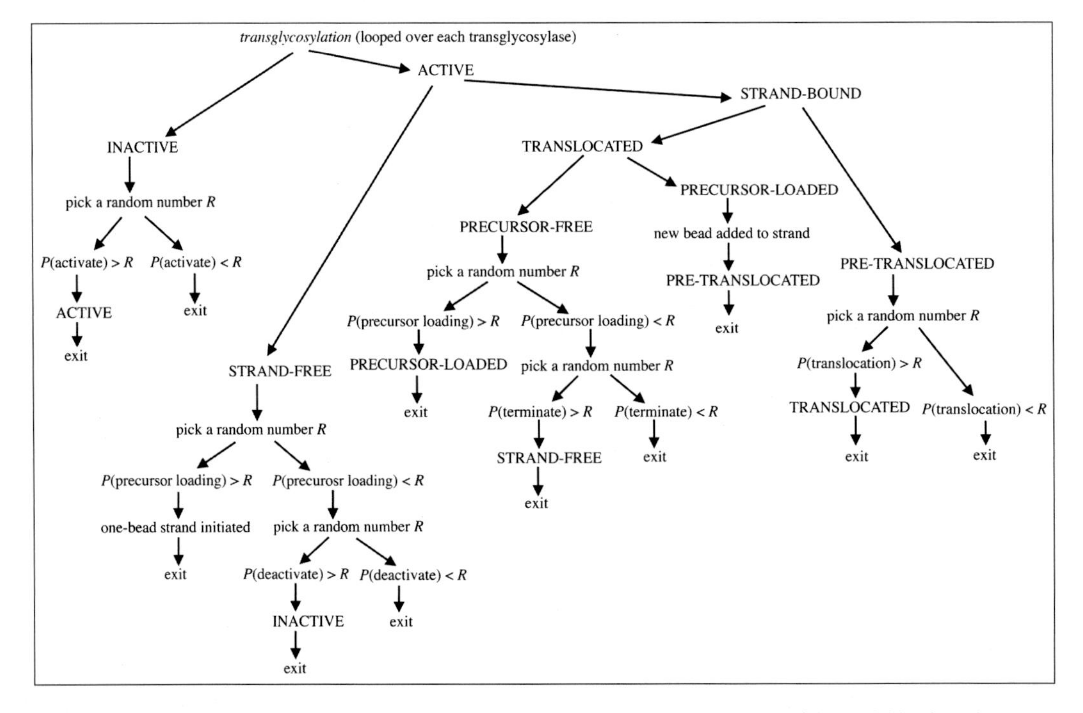

Fig. 7 Flowchart of *transglycosylation* in the simulation code. P denotes probabilities of the activities in a time step

STRAND-BOUND enzyme was prohibited until the enzyme TRANSLOCATED to the strand tip [67–69], which occurred with a probability of once every 2×10^4 time steps. A TRANSLOCATED transglycosylase could either be reloaded with another precursor, leading to further strand elongation (Fig. 6), or termination could occur, with a probability of once every 10^6 steps, leaving the transglycosylase once again in an ACTIVE but STRAND-FREE state. While an ACTIVE, STRAND-FREE transglycosylase could initiate a new strand, it could also be "inactivated," with a probability of once every 5×10^4 steps.

3.3.2 Transpeptidation

It is widely accepted that transpeptidation occurs in an ordered fashion in which the enzyme first binds to a donor peptide, forming an intermediary complex which later catalyzes crosslink formation when an acceptor peptide is captured [70–72]. Transpeptidases were therefore modeled to be either DONOR-FREE (not loaded with a peptide in the donor domain) or DONOR-LOADED (loaded with a peptide in the donor domain). DONOR-LOADED enzymes could exist as either ACCEPTOR-FREE (not loaded with a peptide in the acceptor domain) or ACCEPTOR-LOADED (loaded with a peptide in the acceptor domain).

A flowchart of the *transpeptidation* loop in the simulation code is presented in Fig. 8. A DONOR-FREE transpeptidase within a reaction distance, $d_0 = 2.0$ nm, of a bead bearing an uncrosslinked

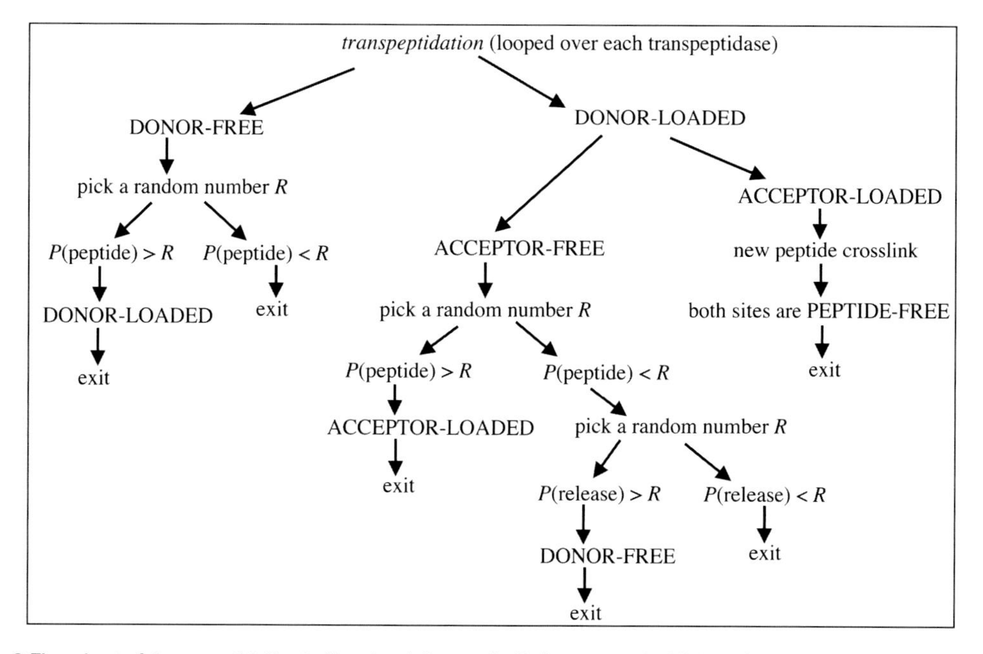

Fig. 8 Flowchart of *transpeptidation* in the simulation code. P denotes probabilities of the activities in a time step

peptide was "loaded" with (became bound to) that peptide with a probability that was a function of the distance d, $P = (1 - d / d_0)^2$. The enzyme was now DONOR-LOADED. Beyond the reaction distance, the peptide-loading probability was zero. A DONOR-LOADED transpeptidase could release the peptide (becoming DONOR-FREE again) with a smaller probability, once in 10^4 steps. If the enzyme instead loaded, with the same distance-dependent probability, an acceptor peptide as well (becoming ACCEPTOR-LOADED), a new crosslink between the corresponding beads was added to the model and the enzyme was released (becoming DONOR-FREE again) (Fig. 6). Because the fifth residues of peptides are quickly removed [73], preventing them from acting as donors, only peptides on a growing strand can be donors [1], and this restriction was implemented in our model.

3.3.3 Endopeptidation

Endopeptidases were modeled as PEPTIDE-FREE (not bound to any peptide), or PEPTIDE-BOUND (bound to one or both peptides released from crosslink cleavage). If during a time step a PEPTIDE-FREE endopeptidase diffused across a peptide crosslink, the crosslink was cleaved with a probability of 0.1 (Fig. 6), and the two peptides remained bound to the enzyme (enzyme became PEPTIDE-BOUND). In early models, the two peptides were released from the endopeptidase immediately after cleavage (enzyme became PEPTIDE-FREE). Later a "cleaved crosslink capture" hypothesis was added to the model specifying that, until competed off by transpeptidases, endopeptidases bind tightly to cleaved crosslinks (remaining PEPTIDE-BOUND), only releasing peptides (to become PEPTIDE-FREE) with a low probability, on average once every 10^7 time steps [45]. A flowchart of *endopeptidation* is presented in Fig. 9.

3.3.4 Enzyme Diffusion

Enzyme diffusion was modeled by exerting a random force on each enzyme in each time step. To generate random forces, a set of Gaussian distributed random numbers was first generated using the Box-Muller transformation [74]:

$$r_1 = \cos(2\pi u_2)\sqrt{-2\ln(u_1)} \tag{5}$$

$$r_2 = \sin(2\pi u_2)\sqrt{-2\ln(u_1)} \tag{6}$$

where u_1 and u_2 are two random numbers from a uniform 0–1 distribution. Each Cartesian component of the random force was then obtained by scaling a Gaussian random number by a force constant of 500 pN. We assumed that random forces on the small PG beads were negligible, and thus could be ignored.

3.3.5 Periplasmic Confinement

In cells, PG remodeling enzymes are confined within the thin periplasmic space. To model this confinement, the enzymes in our

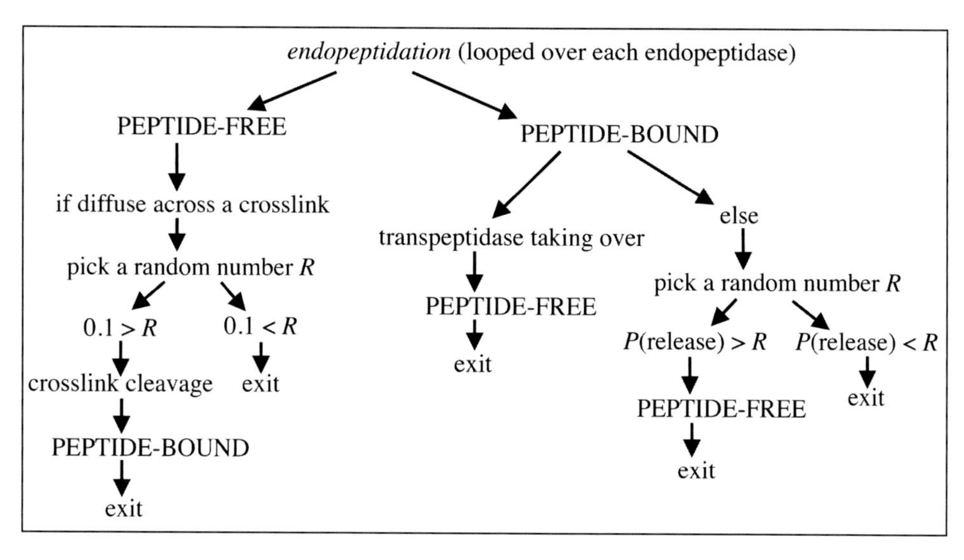

Fig. 9 Flowchart of *endopeptidation* in the simulation code. P denotes probabilities of the activities in a time step

model were constrained to the sacculus surface (Fig. 10a). As an enzyme moved a distance d_s away from the surface, a Hookean spring-like force normal to the surface was exerted on the enzyme:

$$F_{\text{surf}} = -k_{\text{surf}} d_{\text{s}} \tag{7}$$

where k_{surf} is a spring constant chosen as 500 pN/nm.

3.3.6 Interaction with LpoA and LpoB

In *E. coli*, the outer-membrane lipoproteins LpoA and LpoB interact with and activate the bifunctional transglycosylases PBP1A and PBP1B, which are partially embedded in the inner membrane [17, 18]. Thus active transglycosylase-lipoprotein complexes, spanning the periplasm from the outer membrane through the sacculus to the inner membrane, presumably cannot cross through strands or crosslinks. To model this constraint, as an active transglycosylase approached the edge of a hole in the network, a repulsive force was applied on the enzyme and on the two PG beads at either end of the edge (Fig. 10b). For simplicity, in calculating the repulsive force we assumed that each transglycosylase-lipoprotein complex was rigid and extended perpendicular to the sacculus surface. When the distance Δd from the enzyme's projection on the surface to the edge was less than $\Delta D = 0.5$ nm, the repulsive force was calculated as:

$$F_r = k_r \left(\frac{\Delta D}{\Delta d} - 1 \right)^2 \tag{8}$$

where $k_{\text{r}} = 100\text{pN}$ is a force constant.

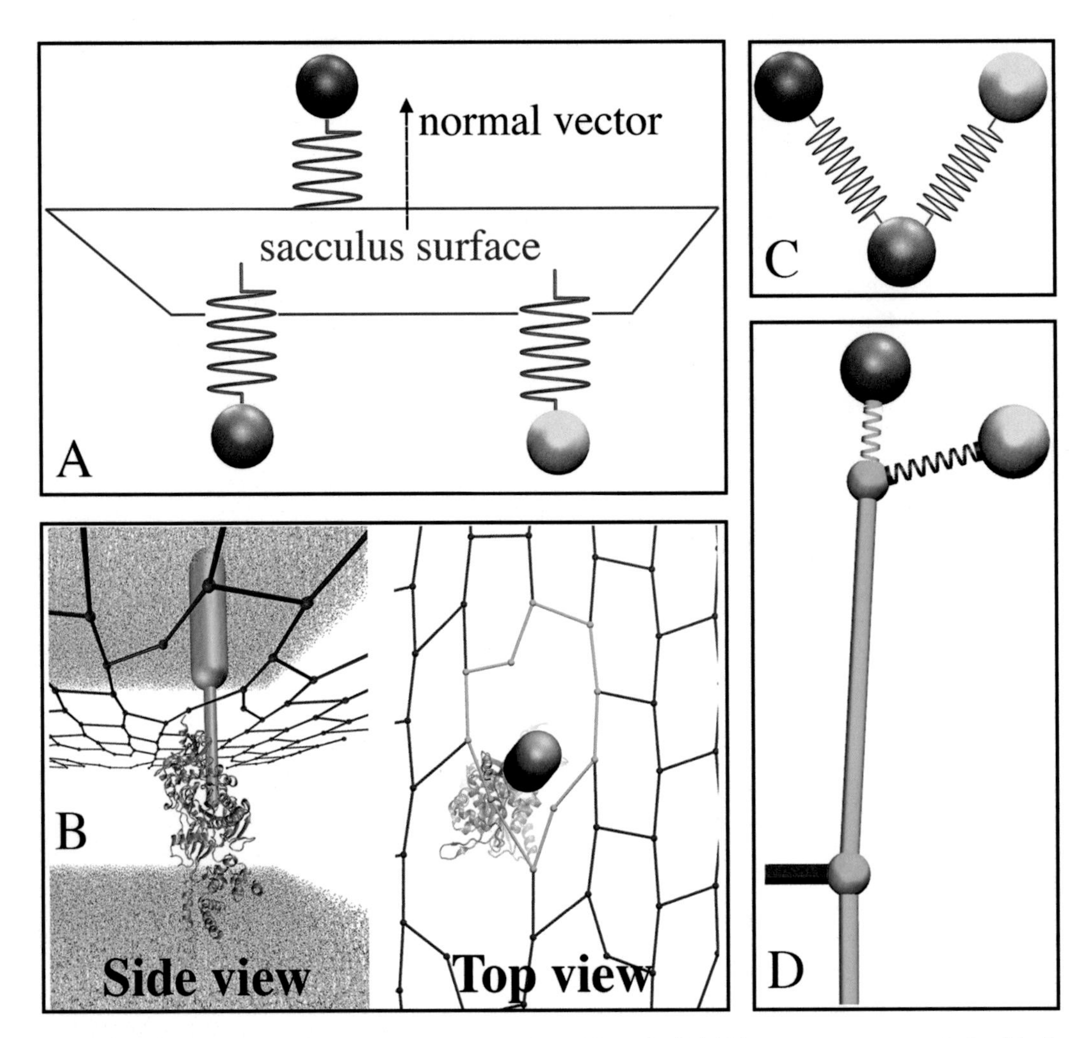

Fig. 10 Schematic of constraints on the enzymes, adapted from [45]. (**a**) The enzymes are constrained to the sacculus surface by Hookean spring-like forces. (**b**) In complex with outer membrane lipoproteins (cyan cylinder), an active transglycosylase (*orange*) is constrained within holes (*green*) formed by surrounding strands and peptide crosslinks. (**c**) In the multi-enzyme complex model, a transglycosylase is tethered to a transpeptidase (*yellow*) and an endopeptidase (*gray*) by spring-like forces. (**d**) A transglycosylase (*orange*) is linked to the tip of the growing strand by a spring-like force. An enzyme, either endopeptidase or transpeptidase (*yellow*), once bound to a peptide (*red*), is linked to the associated PG bead by a spring-like force

3.3.7 Enzyme-Enzyme Tethering

As the bifunctional transglycosylase/transpeptidases PBP1A and PBP1B are the major synthases in *E. coli* [11], in our model transglycosylases and transpeptidases were modeled as complexes in which they were linked together via a spring-like force (Fig. 10c). In the initial model, endopeptidase was not part of the complex. After a "multi-enzyme complex" hypothesis was added, transglycosylases were tethered to both transpeptidases and endopeptidases. To model enzyme tethering, if the distance d_{ez} between two tethered enzymes became larger than $D_0 = 1.0\text{nm}$, a spring-like force was applied to draw them closer together:

$$F_{ez} = -k_{ez}\left(d_{ez} - D_0\right) \quad (9)$$

where $k_{ez} = 10\mathrm{pN}/\mathrm{nm}$ is a force constant.

3.3.8 Enzyme-PG Interaction

We also modeled binding of enzymes to PG while they remodeled the sacculus (Fig. 10d). As a transglycosylase was elongating a new strand, the enzyme was linked to the PG bead at the strand tip via a spring (called a G-spring) of constant $k_{gt} = 50\mathrm{pN}/nm$ and relaxed length $d_{gt} = 0.5\mathrm{nm}$. The bending stiffness of glycan was taken into account at the tip using Eq. 1 with θ now representing the angle between the G-spring and the strand. Enzymes also transiently bound peptides. For instance, transpeptidases bound to peptides before crosslinking them, and endopeptidases might remain bound to peptides released from cleaved crosslinks. To model the binding of peptides to an enzyme, a restoring force $F_{pt} = -k_{pt}\left(d - d_{pt}\right)$, where $k_{pt} = 50\mathrm{pN}/\mathrm{nm}$, was applied to both the enzyme and the peptide-associated PG bead if the distance d between them was more than $d_{pt} = 1.0\mathrm{nm}$. Within distance d_{pt}, the force was zero.

3.4 Relaxation

To relax sacculi after initial generation and during growth we used a simple MD simulation of the coarse-grained model. Specifically, coordinate $X(t)$ of each bead was evolved following the Langevin equation:

$$M\frac{d^2X}{dt^2} = -\nabla U(X) - \gamma\frac{dX}{dt} + R(t) \quad (10)$$

where M is the mass of the bead, U the interaction potential, γ the damping constant and R the random force on the bead. Assuming inertia of the bead was negligible, and thus $M = 0$, displacement was therefore simply a linear function of force:

$$dX = \frac{1}{\gamma}\left[-\nabla U(X) + R\right]dt \quad (11)$$

In principle, one might be able to estimate viscous drag coefficients from the masses and sizes of the PG beads and enzymes. However, since sacculi are linked to the outer membrane through lipoproteins [75, 76], and since PBPs might exist in complexes with other proteins such as MreBCD, RodA, or RodZ [77], their response to viscosity might differ. For simplicity, the effective viscous drag coefficients of the enzymes were estimated to be four times that of the PG beads.

Using a fixed time step could make the system unstable because a large force might move a bead too far. To prevent this instability, we constrained the maximal displacement of the PG beads, corresponding to the maximal force F_{max}, in every time step, to $D_{max} = 0.005$ nm. Displacement D of each bead was then calculated as

$$D = \frac{D_{max}}{\gamma F_{max}} F \tag{12}$$

where F is the force on the bead, and $\gamma = 1.0$ for PG beads and 4.0 for enzymes.

3.5 Shape Characterization

We developed measures to quantify preservation of sacculus integrity and maintenance of rod shape. First, we calculated hole size since large holes in the sacculus could threaten cell integrity. We then quantified bulges, straightness, and roughness of the sacculus surface to analyze rod shape maintenance.

3.5.1 Hole Size

We quantified hole size by calculating the surface area covering the hole. A hole on the sacculus surface is a polygon whose edges connect neighboring beads into a closed loop that cannot be further divided by a glycan or peptide bond (Fig. 5). This polygon was divided into triangles sharing the polygon's center as their third vertex. The hole size was then calculated as the sum of the triangles' areas.

3.5.2 Central Line

To characterize maintenance of rod shape, we first had to define a central line through the sacculus between the polar caps. To do this, we constructed a central "axis" chain of beads extending the length of the cylinder, connected by unstretched springs of uniform spring constant. This axis chain was then connected to the sacculus by dividing the PG cylinder into segments, each corresponding to one axis bead, and connecting each axis bead to the PG beads in its corresponding segment with identical springs. The axis chain was relaxed by minimizing the energy

$$E = \sum \frac{1}{2} k_b \left(d_{ij} - d_0\right)^2 + \sum \frac{1}{2} k_\theta \left(\theta_i - \theta_0\right)^2 + \sum \frac{1}{2} k_{pg} l_{ij}^{\,2} \tag{13}$$

The first term represents the axis springs, where dij is the distance between axis bead i and axis bead j, $k_b = 10^3\,\mathrm{pN/nm}$, and $d_0 = 2.0$ nm. The second term represents the bending stiffness of the axis chain, where θi is the angle at axis bead i, $k_\theta = 10^{-20}$ J, and $\theta_0 = 3.14$ rad. The third term represents the springs connecting the axis beads to the sacculus, where lij is the distance between axis bead i and PG bead j, and $k_{pg} = 10^{-2}\,\mathrm{pN/nm}$. The central line was then defined as this relaxed axis chain.

3.5.3 Bulges

To quantify bulges we assessed fluctuations in local radii. The PG cylinder was divided into short segments, and each local radius was calculated as the average distance from the PG beads of that segment to the central line.

3.5.4 Straightness

Sacculus straightness was defined as the ratio of the shortest distance between the end points of the central line to its contour length.

3.5.5 Roughness

Surface roughness was defined as the ratio of standard deviation to mean of the local radii.

3.6 Visualization of Sacculus Growth

Simulation codes were written in Fortran. Visualization of sacculus growth using Visual Molecular Dynamics (VMD) [54], however, was a challenge. Sacculus growth involved addition of new beads, cleavage of old bonds, and formation of new ones, but VMD could only visualize systems with constant numbers of beads and identical topology. To overcome this problem, the following two strategies were applied.

First, to ensure a constant number of beads, the number of "future" PG beads was predicted and their coordinates were initially set to be at the center of the sacculus, forming a reservoir of available beads. Once a bead was "added," its coordinates were simply changed to match the location of the corresponding enzyme, and then evolved as part of the dynamic system using Eq. 12.

The second problem was maintaining topology. Due to the mesh-like nature of the PG network, many bonds, either glycan or peptide, were formed on common beads, so that bond cleavage and formation violated the topological constraint. To overcome this problem, instead of using one visualization bead (V-bead) to visualize one PG bead, a bonded pair of V-beads (blue = existing glycan, green = new glycan, red = peptide) was used to visualize each glycan/peptide bond. Thus, a PG bead at the junction of N bonds was visualized with N V-beads overlapping one another, ensuring that each bond could be added/removed independently from the others. So, for example, when a new peptide bond was added, two bonded V-peptide-beads were moved from the central reservoir to the location of the corresponding PG beads of the bond. When a new glycan bead was added, two bonded V-glycan-beads were moved from the central reservoir to the location of the new bond, one overlapping the existing bead at the strand tip and the other forming the new strand tip. To visualize removal of a glycan bead or peptide bond, the corresponding V-glycan/peptide-beads were moved back to the central reservoir. A schematic of visualization is presented in Fig. 11.

To show a dynamic process like sacculus growth, moving images obviously work better than static ones. We therefore created movies to document simulated sacculus growth events, analysis of causes of shape loss, and hypothetical mechanisms to fix problems [45]. To generate movies, we first captured individual snapshots of sacculus remodeling using VMD. Text and graphical schematics were then embedded using Photoshop. Frames were imported into QuickTime Pro to generate individual movies, and Final Cut Pro used to concatenate movies.

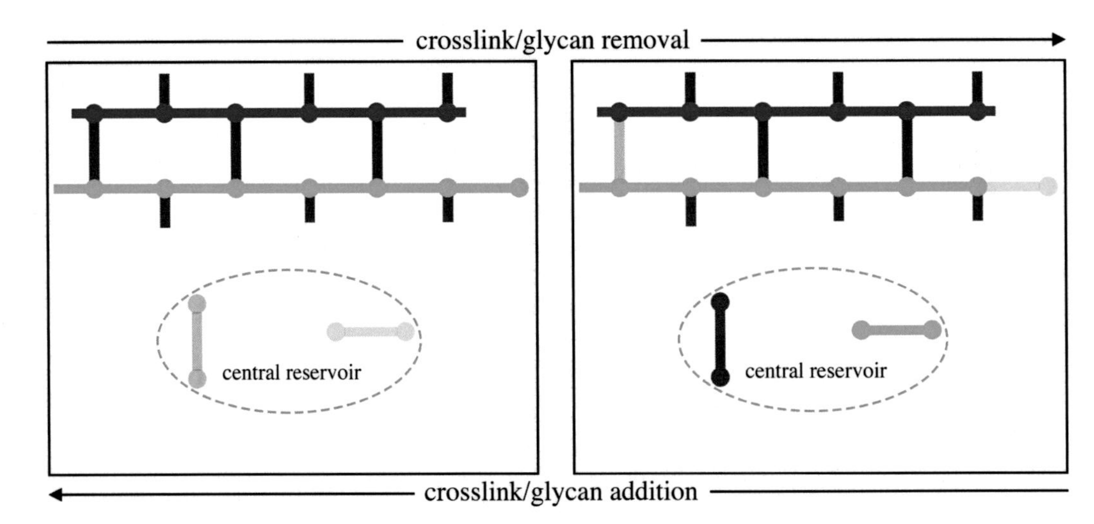

Fig. 11 Schematic showing the use of visualization beads (shown in central reservoir) to show the addition/removal of PG beads or peptide crosslinks. Existing PG is visualized in *blue*, new PG in *green*, and peptides in *red*

4 Notes

1. In the first round of our work, we used our model to reveal many challenges the cell might face while remodeling its wall, as well as possible molecular mechanisms the cell might use to preserve its integrity and characteristic rod shape during cell elongation [45]. We highlight some of the results of our simulations here.

   First and foremost, as hydrolases must cleave peptide bonds in order for new PG material to be incorporated, their activities must be regulated to preserve sacculus integrity in the presence of large internal turgor pressure. We have shown that not only do synthases and hydrolases likely form a complex, but their activities are likely temporally coordinated in such a way that a peptide released by bond cleavage would be captured quickly by new crosslink formation. Second, activities of synthases are also likely regulated spatially and temporally to prevent aggregation of new material. To ensure processivity, not only might the orientation of transglycosylases be fixed, but their translocation along the new glycan strand might also be facilitated by transpeptidation. Further, termination of transglycosylase is likely not purely stochastic but rather regulated, for instance by crosslinkage of the growing strand and/or by hole size. Interestingly, we found that the presence of a housekeeping glycosidase that removes uncrosslinked glycan tails could help prevent aggregation. While our manuscript was under revision, such a glycosidase was identified in cells [78], proving the usefulness of our approach in generating testable biological hypotheses.

Maintenance of rod shape requires maintenance of regular order of glycan strands, since disordered PG leads to bulges and shape distortion. We have shown that the presence of multiple synthases in the complex could help preserve the regular order and shape of the sacculus. While bifunctional transpeptidases likely form crosslinks only on one side, the presence of a monofunctional transpeptidase would ensure crosslink formation on the other side of new strands, perhaps explaining why the monofunctional PBP2 of *E. coli* is essential for shape maintenance [79]. Finally, in a single-strand insertion mode, new peptides do not line up with old peptides, causing circumferential stress and, gradually, distortion. By contrast, we show that the presence of two transglycosylases incorporating two strands into the sacculus concomitantly brings the peptides into register, thus preventing defects.

Our results show that rod shape maintenance can occur with only local coordination of the enzymes within individual, randomly diffusing complexes, and that coordination of PG insertion sites over long distances by cytoskeletal filament scaffolds, a role previously suggested for MreB, is not required.

2. In the future, it will be interesting to expand our model to include cytoplasmic proteins that regulate PG synthesis, notably MreB during cell elongation and FtsZ during division [77]. Several roles have been proposed for MreB including serving as a cytoskeletal scaffold to direct PG insertion sites [27, 44] and/or simply tagging along [23, 24, 28], bridging cytoplasmic and periplasmic enzymes [45, 80], and organizing and/or orienting the PG remodeling enzyme complex [45]. To test whether MreB directs PG insertion sites, as in the Huang model, insertion of new strands could be constrained to sites that implicitly represent the location of MreB [28, 41, 43, 44]. To test whether MreB helps form the PG remodeling complex by channeling PG precursors and/or organizing the enzymes, the presence/absence of MreB could be represented by a high/low probability of loading precursors onto transglycosylases and/or a long/short lifetime of the complex.

   It has been proposed that FtsZ may serve as a scaffold to recruit divisome proteins and/or exert a constricting force on the membrane during division [81]. To model the former, the localization of PG synthesis enzymes could simply be biased to the midcell. To model the latter, assuming that forces exerted on the membrane would be transferred to the stress-bearing sacculus, forces perpendicular to the sacculus surface could be applied to PG beads at sites representing the location of FtsZ.

   We plan to use the same method to study many related topics, including, for example, shape maintenance of Gram-positive bacteria, lemon-to-rod transition and rod shape recovery, cell division, and even sporulation, where PG synthesis/hydrolysis is thought to drive prespore engulfment [82–84].

3. We originally adopted the spring model for peptide crosslinks from the work of Huang et al. [44]. This model, however, failed to stabilize the system if the diameter of the sacculus was large or there were big holes on the surface. As the sacculus diameter increased, the cross-sectional area of the sacculus and therefore the stress on peptide crosslinks from turgor pressure increased quadratically, while the number of crosslinks along the circumference increased only linearly. As the sacculus reached a certain diameter, the system therefore became unstable, preventing realistic representation of the sacculus' mechanical properties and exploration of the effect of size.
4. In simulating a sacculus four times larger, to reduce the computational cost only the cylindrical part of the sacculus was modeled, without including the two caps. In the other simulations, PG synthesis was not modeled to occur at the caps due to experimental evidence that the caps are inert, so this should not affect the conclusions in any way.

Acknowledgements

The authors wish to thank Catherine Oikonomou for revising the manuscript for clarity.

References

1. Höltje JV (1998) Growth of the stress-bearing and shape-maintaining murein sacculus of Escherichia coli. Microbiol Mol Biol Rev 62:181–203
2. Primosigh J, Pelzer H, Maass D, Weidel W (1961) Chemical characterization of mucopeptides released from the E. coli B cell wall by enzymic action. Biochim Biophys Acta 46:68–80
3. Glauner B (1988) Separation and quantification of muropeptides with high-performance liquid chromatography. Anal Biochem 172:451–464. doi:10.1016/0003-2697(88)90468-X
4. Harz H, Burgdorf K, Höltje J-V (1990) Isolation and separation of the glycan strands from murein of Escherichia coli by reversed-phase high-performance liquid chromatography. Anal Biochem 190:120–128. doi:10.1016/0003-2697(90)90144-X
5. Weidel W, Pelzer H (1964) Bagshaped macromolecules—a new outlook on bacterial cell walls. Adv Enzymol Relat Areas Mol Biol 26:193–232
6. Gan L, Chen S, Jensen GJ (2008) Molecular organization of Gram-negative peptidoglycan. Proc Natl Acad Sci 105:18953–18957. doi:10.1073/pnas.0808035105
7. Barreteau H, Kovac A, Boniface A et al (2008) Cytoplasmic steps of peptidoglycan biosynthesis. FEMS Microbiol Rev 32:168–207. doi:10.1111/j.1574-6976.2008.00104.x
8. Sauvage E, Kerff F, Terrak M et al (2008) The penicillin-binding proteins: structure and role in peptidoglycan biosynthesis. FEMS Microbiol Rev 32:234–258. doi:10.1111/j.1574-6976.2008.00105.x
9. Singh SK, SaiSree L, Amrutha RN, Reddy M (2012) Three redundant murein endopeptidases catalyse an essential cleavage step in peptidoglycan synthesis of Escherichia coliK12. Mol Microbiol 86:1036–1051. doi:10.1111/mmi.12058
10. Sung M-T, Lai Y-T, Huang C-Y et al (2009) Crystal structure of the membrane-bound bifunctional transglycosylase PBP1b from

Escherichia coli. Proc Natl Acad Sci U S A 106: 8824–8829. doi:10.1073/pnas.0904030106

11. Vollmer W, Bertsche U (2008) Murein (peptidoglycan) structure, architecture and biosynthesis in Escherichia coli. Biochim Biophys Acta 1778:1714–1734, doi: 16/j.bbamem.2007.06.007
12. Banzhaf M, van den Berg van Saparoea B, Terrak M et al (2012) Cooperativity of peptidoglycan synthases active in bacterial cell elongation. Mol Microbiol 85:179–194. doi:10.1111/j.1365-2958.2012.08103.x
13. Romeis T, Höltje JV (1994) Specific interaction of penicillin-binding proteins 3 and 7/8 with soluble lytic transglycosylase in Escherichia coli. J Biol Chem 269:21603–21607
14. von Rechenberg M, Ursinus A, Höltje JV (1996) Affinity chromatography as a means to study multienzyme complexes involved in murein synthesis. Microb Drug Resist 2:155–157
15. Vollmer W, von Rechenberg M, Holtje J-V (1999) Demonstration of molecular interactions between the murein polymerase PBP1B, the lytic transglycosylase MltA, and the scaffolding protein MipA of Escherichia coli. J Biol Chem 274:6726–6734. doi:10.1074/jbc.274.10.6726
16. Bertsche U, Kast T, Wolf B et al (2006) Interaction between two murein (peptidoglycan) synthases, PBP3 and PBP1B, in Escherichia coli. Mol Microbiol 61:675–690. doi:10.1111/j.1365-2958.2006.05280.x
17. Paradis-Bleau C, Markovski M, Uehara T et al (2010) Lipoprotein cofactors located in the outer membrane activate bacterial cell wall polymerases. Cell 143:1110–1120. doi:10.1016/j.cell.2010.11.037
18. Typas A, Banzhaf M, van den Berg van Saparoea B et al (2010) Regulation of peptidoglycan synthesis by outer-membrane proteins. Cell 143:1097–1109. doi:10.1016/j.cell.2010.11.038
19. Divakaruni AV, Loo RRO, Xie Y et al (2005) The cell-shape protein MreC interacts with extracytoplasmic proteins including cell wall assembly complexes in Caulobacter crescentus. Proc Natl Acad Sci U S A 102:18602–18607. doi:10.1073/pnas.0507937102
20. van den Ent F, Leaver M, Bendezu F et al (2006) Dimeric structure of the cell shape protein MreC and its functional implications. Mol Microbiol 62:1631–1642. doi:10.1111/j.1365-2958.2006.05485.x
21. White CL, Kitich A, Gober JW (2010) Positioning cell wall synthetic complexes by the bacterial morphogenetic proteins MreB and MreD. Mol Microbiol 76:616–633. doi:10.1111/j.1365-2958.2010.07108.x
22. Dye NA, Pincus Z, Theriot JA et al (2005) Two independent spiral structures control cell shape in Caulobacter. Proc Natl Acad Sci U S A 102:18608–18613. doi:10.1073/pnas.0507708102
23. Domínguez-Escobar J, Chastanet A, Crevenna AH et al (2011) Processive movement of MreB-associated cell wall biosynthetic complexes in bacteria. Science 333:225–228. doi:10.1126/science.1203466
24. Garner EC, Bernard R, Wang W et al (2011) Coupled, circumferential motions of the cell wall synthesis machinery and MreB filaments in B. subtilis. Science 333:222–225. doi:10.1126/science.1203285
25. Lee TK, Tropini C, Hsin J et al (2014) A dynamically assembled cell wall synthesis machinery buffers cell growth. Proc Natl Acad Sci 201313826. doi: 10.1073/pnas.1313826111
26. Koch AL (1990) Additional arguments for the key role of "smart" autolysins in the enlargement of the wall of gram-negative bacteria. Res Microbiol 141:529–541
27. Jones LJF, Carballido-López R, Errington J (2001) Control of cell shape in bacteria: helical, actin-like filaments in Bacillus subtilis. Cell 104:913–922, doi: 16/S0092-8674(01)00287-2
28. van Teeffelen S, Wang S, Furchtgott L et al (2011) The bacterial actin MreB rotates, and rotation depends on cell-wall assembly. Proc Natl Acad Sci 108:15822–15827. doi:10.1073/pnas.1108999108
29. Daniel RA, Errington J (2003) Control of cell morphogenesis in bacteria: two distinct ways to make a rod-shaped cell. Cell 113:767–776, doi: 16/S0092-8674(03)00421-5
30. Kruse T, Møller-Jensen J, Løbner-Olesen A, Gerdes K (2003) Dysfunctional MreB inhibits chromosome segregation in Escherichia coli. EMBO J 22:5283–5292. doi:10.1093/emboj/cdg504
31. Shih Y-L, Le T, Rothfield L (2003) Division site selection in Escherichia coli involves dynamic redistribution of Min proteins within coiled structures that extend between the two cell poles. Proc Natl Acad Sci 100:7865–7870. doi:10.1073/pnas.1232225100
32. Soufo HJD, Graumann PL (2003) Actin-like proteins MreB and Mbl from Bacillus subtilis are required for bipolar positioning of replication origins. Curr Biol 13:1916–1920. doi:10.1016/j.cub.2003.10.024
33. Soufo HJD, Graumann PL (2004) Dynamic movement of actin-like proteins within bacterial cells. EMBO Rep 5:789–794. doi:10.1038/sj.embor.7400209

34. Figge RM, Divakaruni AV, Gober JW (2004) MreB, the cell shape-determining bacterial actin homologue, co-ordinates cell wall morphogenesis in Caulobacter crescentus. Mol Microbiol 51:1321–1332. doi:10.1111/j.1365-2958.2003.03936.x
35. Gitai Z, Dye N, Shapiro L (2004) An actin-like gene can determine cell polarity in bacteria. Proc Natl Acad Sci U S A 101:8643–8648. doi:10.1073/pnas.0402638101
36. Vats P, Rothfield L (2007) Duplication and segregation of the actin (MreB) cytoskeleton during the prokaryotic cell cycle. Proc Natl Acad Sci 104:17795–17800. doi:10.1073/pnas.0708739104
37. Swulius MT, Chen S, Jane Ding H et al (2011) Long helical filaments are not seen encircling cells in electron cryotomograms of rod-shaped bacteria. Biochem Biophys Res Commun 407: 650–655. doi:10.1016/j.bbrc.2011.03.062
38. Swulius MT, Jensen GJ (2012) The helical MreB cytoskeleton in Escherichia coli MC1000/pLE7 is an artifact of the N-terminal yellow fluorescent protein tag. J Bacteriol 194:6382–6386. doi:10.1128/JB.00505-12
39. Reimold C, Defeu Soufo HJ, Dempwolff F, Graumann PL (2013) Motion of variable-length MreB filaments at the bacterial cell membrane influences cell morphology. Mol Biol Cell 24:2340–2349. doi:10.1091/mbc.E12-10-0728
40. Olshausen PV, Defeu Soufo HJ, Wicker K et al (2013) Superresolution imaging of dynamic MreB filaments in B. subtilis—a multiple-motor-driven transport? Biophys J 105:1171–1181. doi:10.1016/j.bpj.2013.07.038
41. Ursell TS, Nguyen J, Monds RD et al (2014) Rod-like bacterial shape is maintained by feedback between cell curvature and cytoskeletal localization. Proc Natl Acad Sci 111:E1025–E1034. doi:10.1073/pnas.1317174111
42. Huang KC, Mukhopadhyay R, Wen B et al (2008) Cell shape and cell-wall organization in Gram-negative bacteria. Proc Natl Acad Sci 105:19282–19287. doi:10.1073/pnas.0805309105
43. Wang S, Furchtgott L, Huang KC, Shaevitz JW (2012) Helical insertion of peptidoglycan produces chiral ordering of the bacterial cell wall. Proc Natl Acad Sci 109:E595–E604. doi:10.1073/pnas.1117132109
44. Furchtgott L, Wingreen NS, Huang KC (2011) Mechanisms for maintaining cell shape in rod-shaped Gram-negative bacteria. Mol Microbiol 81:340–353. doi:10.1111/j.1365-2958.2011.07616.x
45. Nguyen LT, Gumbart JC, Beeby M, Jensen GJ (2015) Coarse-grained simulations of bacterial cell wall growth reveal that local coordination alone can be sufficient to maintain rod shape. Proc Natl Acad Sci 112:E3689–E3698. doi:10.1073/pnas.1504281112
46. Höltje J-V (1996) A hypothetical holoenzyme involved in the replication of the murein sacculus of Escherichia Coli. Microbiology 142:1911–1918. doi:10.1099/13500872-142-8-1911
47. Cooper S, Hsieh ML, Guenther B (1988) Mode of peptidoglycan synthesis in Salmonella typhimurium: single-strand insertion. J Bacteriol 170:3509–3512
48. de Jonge BL, Wientjes FB, Jurida I et al (1989) Peptidoglycan synthesis during the cell cycle of Escherichia coli: composition and mode of insertion. J Bacteriol 171:5783–5794
49. Burman LG, Park JT (1984) Molecular model for elongation of the murein sacculus of Escherichia coli. Proc Natl Acad Sci U S A 81: 1844–1848
50. Zijderveld CA, Aarsman ME, den Blaauwen T, Nanninga N (1991) Penicillin-binding protein 1B of Escherichia coli exists in dimeric forms. J Bacteriol 173:5740–5746
51. Charpentier X, Chalut C, Rémy M-H, Masson J-M (2002) Penicillin-binding proteins 1a and 1b form independent dimers in Escherichia coli. J Bacteriol 184:3749–3752. doi:10.1128/JB.184.13.3749-3752.2002
52. Bertsche U, Breukink E, Kast T, Vollmer W (2005) In vitro murein peptidoglycan synthesis by dimers of the bifunctional transglycosylase-transpeptidase PBP1B from Escherichia coli. J Biol Chem 280:38096–38101. doi:10.1074/jbc.M508646200
53. Zhao G, Perilla JR, Yufenyuy EL et al (2013) Mature HIV-1 capsid structure by cryo-electron microscopy and all-atom molecular dynamics. Nature 497:643–646. doi:10.1038/nature12162
54. Burge RE, Fowler AG, Reaveley DA (1977) Structure of the peptidoglycan of bacterial cell walls. J Mol Biol 117:927–953. doi:10.1016/S0022-2836(77)80006-5
55. Labischinski H, Barnickel G, Bradaczek H, Giesbrecht P (1979) On the secondary and tertiary structure of murein. Eur J Biochem 95:147–155.doi:10.1111/j.1432-1033.1979.tb12949.x
56. Kim SJ, Singh M, Preobrazhenskaya M, Schaefer J (2013) Staphylococcus aureus peptidoglycan stem packing by rotational-echo double resonance NMR spectroscopy. Biochemistry (Mosc) 52:3651–3659. doi:10.1021/bi4005039
57. Beeby M, Gumbart JC, Roux B, Jensen GJ (2013) Architecture and assembly of the Gram-positive cell wall. Mol Microbiol 88:664–672. doi:10.1111/mmi.12203
58. Gumbart JC, Beeby M, Jensen GJ, Roux B (2014) Escherichia coli peptidoglycan structure

and mechanics as predicted by atomic-scale simulations. PLoS Comput Biol 10, e1003475. doi:10.1371/journal.pcbi.1003475

59. Phillips JC, Braun R, Wang W et al (2005) Scalable molecular dynamics with NAMD. J Comput Chem 26:1781–1802. doi:10.1002/jcc.20289
60. Darve E, Pohorille A (2001) Calculating free energies using average force. J Chem Phys 115:9169–9183. doi:10.1063/1.1410978
61. Hénin J, Chipot C (2004) Overcoming free energy barriers using unconstrained molecular dynamics simulations. J Chem Phys 121:2904–2914. doi:10.1063/1.1773132
62. Braun V, Gnirke H, Henning U, Rehn K (1973) Model for the structure of the shape-maintaining layer of the Escherichia coli cell envelope. J Bacteriol 114:1264–1270
63. Reed RH, Walsby AE (1985) Changes in turgor pressure in response to increases in external NaCl concentration in the gas-vacuolate cyanobacterium Microcystis sp. Arch Microbiol 143:290–296. doi:10.1007/BF00411252
64. Koch AL, Pinette MF (1987) Nephelometric determination of turgor pressure in growing gram-negative bacteria. J Bacteriol 169:3654
65. Cayley DS, Guttman HJ, Record MT (2000) Biophysical characterization of changes in amounts and activity of Escherichia coli cell and compartment water and turgor pressure in response to osmotic stress. Biophys J 78:1748–1764
66. Lovering AL, De Castro LH, Lim D, Strynadka NCJ (2007) Structural insight into the transglycosylation step of bacterial cell-wall biosynthesis. Science 315:1402–1405. doi:10.1126/science.1136611
67. Yuan Y, Barrett D, Zhang Y et al (2007) Crystal structure of a peptidoglycan glycosyltransferase suggests a model for processive glycan chain synthesis. Proc Natl Acad Sci 104:5348–5353. doi:10.1073/pnas.0701160104
68. Perlstein DL, Wang T-SA, Doud EH et al (2010) The role of the substrate lipid in processive glycan polymerization by the peptidoglycan glycosyltransferases. J Am Chem Soc 132:48–49. doi:10.1021/ja909325m
69. Wise EM, Park JT (1965) Penicillin: its basic site of action as an inhibitor of a peptide cross-linking reaction in cell wall mucopeptide synthesis. Proc Natl Acad Sci U S A 54:75–81
70. Tipper DJ, Strominger JL (1965) Mechanism of action of penicillins: a proposal based on their structural similarity to acyl-D-alanyl-D-alanine. Proc Natl Acad Sci U S A 54:1133–1141
71. Ghuysen JM (1997) Penicillin-binding proteins. Wall peptidoglycan assembly and resistance to penicillin: facts, doubts and hopes. Int J Antimicrob Agents 8:45–60
72. de Pedro MA, Schwarz U (1981) Heterogeneity of newly inserted and preexisting murein in the sacculus of Escherichia coli. Proc Natl Acad Sci U S A 78:5856–5860
73. Box GEP, Muller ME (1958) A note on the generation of random normal deviates. Ann Math Stat 29:610–611. doi:10.1214/aoms/1177706645
74. Braun V (1975) Covalent lipoprotein from the outer membrane of Escherichia coli. Biochim Biophys Acta 415:335–377
75. Silhavy TJ, Kahne D, Walker S (2010) The bacterial cell envelope. Cold Spring Harb Perspect Biol. doi:10.1101/cshperspect.a000414
76. Typas A, Banzhaf M, Gross CA, Vollmer W (2012) From the regulation of peptidoglycan synthesis to bacterial growth and morphology. Nat Rev Micro 10:123–136. doi:10.1038/nrmicro2677
77. Humphrey W, Dalke A, Schulten K (1996) VMD: visual molecular dynamics. J Mol Graph 14:33–38. doi:10.1016/0263-7855(96)00018-5
78. Cho H, Uehara T, Bernhardt TG (2014) Beta-lactam antibiotics induce a lethal malfunctioning of the bacterial cell wall synthesis machinery. Cell 159:1300–1311. doi:10.1016/j.cell.2014.11.017
79. Spratt BG (1975) Distinct penicillin binding proteins involved in the division, elongation, and shape of Escherichia coli K12. Proc Natl Acad Sci U S A 72:2999–3003
80. Rueff A-S, Chastanet A, Domínguez-Escobar J et al (2014) An early cytoplasmic step of peptidoglycan synthesis is associated to MreB in Bacillus subtilis. Mol Microbiol 91:348–362. doi:10.1111/mmi.12467
81. Erickson HP, Anderson DE, Osawa M (2010) FtsZ in bacterial cytokinesis: cytoskeleton and force generator all in one. Microbiol Mol Biol Rev 74:504–528. doi:10.1128/MMBR.00021-10
82. Meyer P, Gutierrez J, Pogliano K, Dworkin J (2010) Cell wall synthesis is necessary for membrane dynamics during sporulation of Bacillus subtilis. Mol Microbiol 76:956–970. doi:10.1111/j.1365-2958.2010.07155.x
83. Tocheva EI, Matson EG, Morris DM et al (2011) Peptidoglycan remodeling and conversion of an inner membrane into an outer membrane during sporulation. Cell 146:799–812. doi:10.1016/j.cell.2011.07.029
84. Tocheva EI, López-Garrido J, Hughes HV et al (2013) Peptidoglycan transformations during Bacillus subtilis sporulation. Mol Microbiol 88:673–686. doi:10.1111/mmi.12201
85. Powell AJ, Tomberg J, Deacon AM et al (2009) Crystal structures of penicillin-binding

protein 2 from penicillin-susceptible and -resistant strains of Neisseria gonorrhoeae reveal an unexpectedly subtle mechanism for antibiotic resistance. J Biol Chem 284:1202–1212. doi:10.1074/jbc.M805761200

86. Kishida H, Unzai S, Roper DI et al (2006) Crystal structure of penicillin binding protein 4 (dacB) from Escherichia coli, both in the native form and covalently linked to various antibiotics. Biochemistry (Mosc) 45:783–792. doi:10.1021/bi051533t

Chapter 19

Structural Comparison and Simulation of Pneumococcal Peptidoglycan Hydrolase LytB

Xiao-Hui Bai*, Qiong Li*, Yong-Liang Jiang, Jing-Ren Zhang, Yuxing Chen, and Cong-Zhao Zhou

Abstract

Three-dimensional structural determination combined with comprehensive comparisons with the homologs is a straightforward strategy to decipher the molecular function of an enzyme. However, in many cases it's difficult to obtain the complex structure with the substrate/ligand. Structure-based molecular simulation provides an alternative solution to predict the binding pattern of a substrate/ligand to the enzyme. The *Streptococcus pneumoniae* LytB is a peptidoglycan hydrolase that cleaves the glycosidic bond and therefore involves the cell division; however, the details of catalytic mechanism remain unknown. Based on the crystal structure of the catalytic domain of LytB (termed $LytB_{CAT}$), we describe here how to assign the molecular functions of three $LytB_{CAT}$ modules: SH3b, WW, and GH73, using structural comparisons. Moreover, we dock a putative tetrasaccharide-pentapeptide substrate of peptidoglycan onto $LytB_{CAT}$ to provide the details of substrate binding pattern. The tetrasaccharide-pentapeptide is well accommodated in a T-shaped substrate binding pocket formed by the three modules. The conclusions deduced from structural comparison and simulation are further proved by the hydrolytic activity assays in combination with site-directed mutagenesis.

Key words *Streptococcus pneumoniae*, Peptidoglycan, Peptidoglycan hydrolase, LytB, Structural comparison, Simulation, Hydrolytic activity assay

1 Introduction

Peptidoglycan (PG), also known as murein, is the major and specific component of bacterial cell wall. It withstands cell turgor in order to maintain cell shape and preserve cell integrity [1]. PG comprises alternating β(1,4)-linked *N*-acetylglucosamine (NAG) and *N*-acetylmuramic acid (NAM) residues, attached by cross-linked short peptides to form a three-dimensional structure [1, 2]. Subtle "destruction" or remodeling of PG is crucial for bacterial cell growth and division [3]. It requires highly diverse group of hydrolases to cleave different covalent bonds of PG [4]. In the past decades, several PG hydrolases had been identified in human

*Author contributed equally with all other contributors

Hee-Jeon Hong (ed.), *Bacterial Cell Wall Homeostasis: Methods and Protocols*, Methods in Molecular Biology, vol. 1440, DOI 10.1007/978-1-4939-3676-2_19, © Springer Science+Business Media New York 2016

pathogen *Streptococcus pneumoniae*, such as autolysin LytA [5], lysozyme LytC [6], and so on. In 1999, LytB was initially characterized as a PG hydrolase for the reason that the *lytB* knockout pneumococci were deficient in cell separation and formed long-chains [7]. Subsequently, García et al. found that the purified recombinant LytB is capable of dispersing the long-chains of *lytB* knockout pneumococci, indicating that LytB possesses a glucosaminidase activity to cleave the β(1,4)-linked glycosidic bond between NAG and NAM [8]. Thus, LytB may play an indispensable role in cell division. Recently, we reported the crystal structure of the catalytic domain of LytB (residues Lys375-Asp658, termed $LytB_{CAT}$) [9].

As we know, the similarity analysis of protein structure is a vital step in understanding protein's function. Here, we divide $LytB_{CAT}$ into three distinct modules: a C-terminal α-helix module and two all-β modules, and then identify their function by comparing the structure of each module with the known structures, respectively. According to primary sequence analysis, the C-terminal α-helix module (residues Gly494-Asp658) is classified into the glycoside hydrolase family 73 (GH73) [10]. Then we superimpose this module onto the only two known structures of GH73: the surface associated autolysin Auto from *Listeria monocytogenes* (PDB code 3fi7) [11] and the flagellar protein FlgJ from *Sphingomonas* sp. (PDB code 2zyc) [12], using SUPERPOSE [13] as a part of the CCP4i [14] on the basis of secondary structure matching (SSM) algorithm. The results suggest that this α-helix module possesses a GH73 fold and functions as a catalytic module, with Glu564 as the catalytic residue. Concerning the first all-β module (residues Asn385-Ser450), we use Dali server [15] to search homologous structures, which are in turn applied to structural superpositions against the input structure. The results indicate that the first all-β module may resemble SH3b domain and contribute to PG recognition. However, the Dali search against the second all-β module (residues Lys451-Asp493) yields no significant homologs. Instead, after searching against the Structural Classification of Proteins (SCOP) database, it is identified as a WW domain-like fold which probably binds to the carbohydrate moiety of PG, and can be well superimposed onto the chitin binding domain (ChBD) of *Serratia marcescens* chitinase ChiB (PDB code 1e15) [16]. Hence, $LytB_{CAT}$ is divided into three structurally independent modules: $LytB_{SH3b}$, $LytB_{WW}$, and $LytB_{GH73}$.

Though LytB has been proved to cleave the NAG-(β-1,4)-NAM glycosidic bond of PG at the septum to separate two daughter cells [8], its bona fide physiological substrate remains undefined. Due to the commercial unavailability of the complex fragments of PG, we choose to simulate a PG fragment that mimics the physiological substrate, to provide the details of substrate binding pattern of $LytB_{CAT}$. Molecular simulation is a computational procedure that attempts to predict noncovalent binding of a macromolecule (receptor) and a small molecule (substrate/ligand). Among various tools of simulation, AutoDock has been proved to be able to effectively and accurately predict the conformations and binding affinity of a substrate/ligand towards the target macromolecule [17]. AutoDock Vina automatically calculates the grid maps and clusters the results in a transparent way [18]. It speeds up the gradient optimization by using a simpler scoring function and therefore significantly improves the accuracy of the binding mode predictions. A T-shaped substrate binding pocket can be found from the electrostatistic potential diagram of $LytB_{CAT}$, which is reminiscent of a PG fragment: *t*etra*s*accharide-*p*enta*p*eptide NAM-NAG-NAM (-L-Ala-D-iGln-L-Lys-D-Ala-D-Ala)-NAG (TSPP) as the putative substrate. Then we generate the atomic coordinates of TSPP using PRODRG Server [19] and dock it onto $LytB_{CAT}$ using AutoDock Vina [18]. The final simulated model suggests that the tetrasaccharide moiety of TSPP is accommodated in the groove of $LytB_{GH73}$, whereas the pentapeptide moiety stretches into the cleft between $LytB_{SH3b}$ and $LytB_{WW}$.

In order to prove the above results of structural comparisons and simulation, we test the contribution of each module of $LytB_{CAT}$ to the hydrolytic activity of $LytB_{CAT}$. LytB hydrolyzes the wild-type PG at a much lower velocity compared to the PG purified from the *lytB* knockout strain (Δ*lytB* PG) [9], in agreement with that LytB probably prefers immature PG [8]. Thus Δ*lytB* PG is applied to all hydrolytic activity assays. We label Δ*lytB* PG with Remazol Brilliant Blue (RBB), and then incubate it with different versions of recombinant $LytB_{CAT}$ protein ($LytB_{CAT}$, $LytB_{WW\text{-}GH73}$, $LytB_{GH73}$, $LytB^{E564Q}$) at 37 °C for 10 h. After terminating the reaction, the activity of each protein sample is calculated by detecting the amount of RBB-labeled Δ*lytB* PG released to the supernatant upon hydrolysis. The results show that Glu564 plays a crucial role in hydrolysis, and none of the three modules is dispensable for the activity of $LytB_{CAT}$. The results indicate the reliability of structural comparisons and simulation.

2 Materials

2.1 The Atomic Coordinates

The atomic coordinates used in structural comparisons are listed in Table 1 (*see* **Note 1**).

2.2 Websites and Programs

1. UniProt: Universal Protein Resource, http://www.uniprot.org/. It provides the scientific community with a comprehensive, high-quality, and freely accessible resource of protein sequences and functional information.

Table 1
The atomic coordinates used in structural comparisons

	PDB code	Bacterial species	Description
$LytB_{CAT}$.pdb	4q2w	*S. pneumoniae*	The catalytic domain of LytB
3fi7.pdb	3fi7	*L. monocytogenes*	The GH73 domain of the surface associated autolysin Auto
2zyc.pdb	2zyc	*Sphingomonas* sp.	The GH73 domain of the flagellar protein FlgJ
2hbw.pdb	2hbw	*A. variabilis*	The SH3b domain of the γ-D-glutamyl-L-diamino acid endopeptidase AvPCP
1r77.pdb	1r77	*S. capitis*	The SH3b domain of peptidoglycan hydrolase ALE-1
1e15.pdb	1e15	*S. marcescens*	The chitin binding domain of chitinase ChiB

2. PyMOL: http://www.pymol.org/. A user-sponsored molecular visualization system on an open-source foundation.
3. CCP4: A world-leading, integrated suite of programs that allows researchers to determine macromolecular structures by X-ray crystallography, and other biophysical techniques [14] (*see* **Note 2**).
4. Dali server: http://ekhidna.biocenter.helsinki.fi/dali_server/. A network service for comparing protein structures in 3D, comparing the submitted coordinates of a query protein structure against those in the PDB [15].
5. SCOP: Structural Classification of Proteins, http://scop.mrc-lmb.cam.ac.uk/scop/. It aims to provide a detailed and comprehensive description of the structural and evolutionary relationships between all proteins whose structure is known.
6. PRODRG server: http://davapc1.bioch.dundee.ac.uk/cgi-bin/prodrg. It takes a description of a small molecule and from it generates a variety of topologies for use with GROMACS, Autodock, and other programs, as well as energy-minimized coordinates in a variety of formats [19].
7. Autodock: A suite of automated docking tools. It is designed to predict how small molecules, such as substrates or drug candidates, bind to a receptor of known 3D structure. AutoDock Tools (ADT) 1.5.4 [18] and AutoDock Vina software (version 1.0) [20] are used.
8. GraphPad: A powerful combination of biostatistics, curve fitting (nonlinear regression) and scientific graphing.

2.3 Hydrolytic Activity Assays

1. Recombinant proteins: The wild-type $LytB_{CAT}$ protein and different mutated versions of $LytB_{CAT}$ protein ($LytB_{WW\text{-}GH73}$, $LytB_{GH73}$, $LytB^{E564Q}$) are constructed and purified according to a previous report [9].
2. Δ*lytB* PG is purified from the *lytB* knockout TIGR4 strain as previously reported [21]. The chromosomal *lytB* knockout strain from *S. pneumoniae* wild-type TIGR4 strain is generated by allelic replacement according to Bricker and Camilli [22].
3. 20 mM Remazol Brilliant Blue (RBB; Sigma): The RBB powder is dissolved in 0.25 M NaOH (*see* **Note 3**).
4. 0.25 M HCl: diluted from the 11 M HCl with double-distilled water (ddH_2O) to neutralize the reaction.
5. Reaction buffer: 50 mM Na_2HPO_4/NaH_2PO_4, pH 7.0
6. Centrifuge (HITACHI, Japan).
7. DU800 spectrophotometer (Beckman Coulter, Fullerton, CA).

3 Methods

After careful structural analyses, the overall structure of $LytB_{CAT}$ is divided into three distinct modules packing against each other: two all-β modules (residues Asn385-Ser450 and Lys451-Asp493, respectively) followed by a C-terminal α-helix module (residues Gly494-Asp658).

3.1 Structural Comparison of the C-Terminal α-Helix Module

1. Search "LytB in *S. pneumoniae*" in the UniProt website to collect the related information of LytB.
2. Based on the primary sequence analysis of Pfam database showed in UniProt, the C-terminal α-helix module of $LytB_{CAT}$ is defined as a glucosaminidase (PF01832), belonging to the glycoside hydrolase family 73 (GH73). GH73 is a family of glycoside hydrolases that include peptidoglycan hydrolases of endo-β-N-acetylglucosaminidase specificity. Therefore, the C-terminal α-helix module is assigned to the catalytic module of $LytB_{CAT}$. To date, only the structures of two members in this family: the surface associated autolysin Auto from *L. monocytogenes* [11] and the flagellar protein FlgJ from *Sphingomonas* sp. [12], had been solved according to the summary of the Pfam database [10] (*see* **Note 4**).
3. Open the atomic coordinates of $LytB_{CAT}$ (PDB code 4q2w, $LytB_{CAT}$.pdb) by PyMOL, and show the protein sequence. Select residues Gly120-Asp284 (corresponding to Gly494 to Asp658 in the full-length protein sequence) and then save it as $LytB_{GH73}$.pdb.

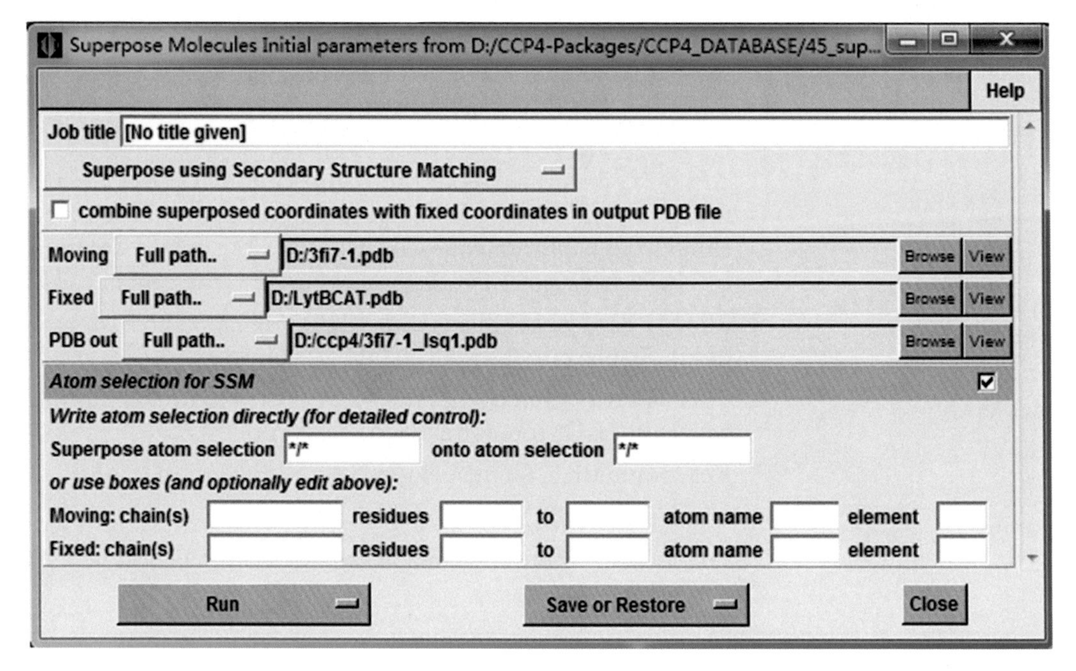

Fig. 1 The Superpose interface in CCP4i program suite 6.4.0 of superimposing LytB$_{CAT}$ against the GH73 domain of Auto (3fi7.pdb)

4. Superimpose the LytB$_{GH73}$.pdb with the GH73 domains of Auto (PDB code 3fi7) and FlgJ (PDB code 2zyc) by SUPERPOSE [13] as a part of the CCP4i [14] program suite, respectively. The atomic coordinates of the GH73 domains of Auto and FlgJ are termed 3fi7.pdb and 2zyc.pdb, respectively (*see* **Note 5**).
5. Open Superpose interface in CCP4i program suite 6.4.0, select "Superpose using Secondary Structure Matching". Fixed LytB$_{GH73}$.pdb, moving 3fi7.pdb or 2zyc.pdb, and then run the program (Fig. 1) (*see* **Note 6**).
6. View and analyze the output PDBs. Open the output 3fi7_lsq1.pdb or 2zyc_lsq1.pdb using PyMOL and then open LytB$_{GH73}$.pdb in the same window. The LytB$_{GH73}$.pdb will automatically superimposed onto the 3fi7_lsq1.pdb or 2zyc_lsq1.pdb. The results show that LytB$_{GH73}$ can be well superimposed with the GH73 domains of Auto and FlgJ, with a root mean square deviation (RMSD) of 2.12 and 1.96 Å over 94 and 86 Cα atoms, respectively. Furthermore, LytB$_{GH73}$ possesses a similar active site compared with the two GH73 domains, especially the catalytic residue. Altogether, it indicates that the C-terminal α-helix module of LytB$_{CAT}$ possesses a GH73 fold and functions as a catalytic module, with Glu564 as the catalytic residue.

3.2 Homology Search of the Two All-β Modules

1. In PyMOL, open $LytB_{CAT}$.pdb and show its protein sequence. The two all-β modules: residues Asn385-Ser450 and Lys451-Asp493 numbering in the full-length LytB, correspond to Asn11-Ser76 and Lys77-Asp119 in the $LytB_{CAT}$.pdb file. Select the residues of each all-β module, then save as β1.pdb and β2.pdb, respectively (*see* **Note 7**).
2. The Pfam database cannot classify the two all-β modules into any known family on the basis of primary sequence. Thus Dali server is chosen as an alternative tool for comparisons with structures deposited in the PDB database, to identify to which family the two all-β modules may belong and their probable function.
3. In the Dali server website, upload the atomic coordinate file (β1.pdb or β2.pdb), enter your own email address and then press "submit" (*see* **Note 8**).
4. The output is normally received in an hour or several hours later. Carefully check all hits and summarize (*see* **Note 9**).
5. In the output of the first all-β module, most proteins with a Z-score of ≥5.1 contain SH3b domain, which were predicted or hypothetical bacterial cell wall hydrolases. The first all-β module may resemble SH3b domain, thus termed $LytB_{SH3b}$. The only two well-characterized hits are the SH3b domain of the γ-D-glutamyl-L-diamino acid endopeptidase AvPCP from *A. variabilis* [23] and that of *S. capitis* peptidoglycan hydrolase ALE-1 [24], both of which appear to contribute to substrate binding.
6. Superimpose $LytB_{SH3b}$ against the SH3b domain of AvPCP (PDB code 2hbw) or ALE-1 (PDB code 1r77). The atomic coordinates of the SH3b domains of AvPCP and ALE-1 are termed 2hbw.pdb and 1r77.pdb, respectively. Run the superposition as **step 5** and **step 6** in Subheading 3.1. $LytB_{SH3b}$ shares a fold quite similar to the SH3b domains of AvPCP and ALE-1, with an RMSD of 2.5 and 2.1 Å over 60 and 58 Cα atoms, respectively. It suggests that $LytB_{SH3b}$ might contribute to substrate binding.
7. However, concerning the second all-β module, no significant results have been found (*see* **Note 10**).
8. Alternatively, process a homology search for the second all-β module in SCOP. Choose the ASTRAL database (SCOP domain sequences and pdb-style coordinate files) in the "Access methods" item, and analyze all structures in the "all beta proteins" class. The fold No. 70, called WW domain-like, is the only fold that consists of a 3-stranded meander beta-sheet similar to the second all-β module, which is in consequence termed $LytB_{WW}$. Superimpose each known structure of WW domain-like fold against $LytB_{WW}$ as **step 5** and **step 6** in

Subheading 3.1. Only the chitin binding domain (ChBD) of *S. marcescens* chitinase ChiB (PDB code 1e15, 1e15.pdb) [16] can be well superimposed onto $LytB_{WW}$, with an RMSD of 1.6 Å over 26 Cα atoms. The ChBD belongs to the carbohydrate binding domain superfamily in WW domain-like fold, indicating that $LytB_{WW}$ may also contribute to binding carbohydrate substrates (*see* **Notes 11** and **12**).

3.3 Simulation of $LytB_{CAT}$ Against the Putative Substrate

1. Analyze $LytB_{CAT}$.pdb to check whether it exists a possible substrate binding pocket on the surface of $LytB_{CAT}$. Open $LytB_{CAT}$.pdb with PyMOL and generate its "protein contact potential (local)" in vacuum electrostatistic item. A T-shaped pocket can be clearly seen from the electrostatistic potential diagram, which is most likely the putative substrate binding pocket. The T-shaped pocket comprises a groove through the catalytic module $LytB_{GH73}$, in addition to a cleft between $LytB_{SH3b}$ and $LytB_{WW}$.
2. Considering the reported structure of PG [1], this T-shaped substrate binding pocket is clearly reminiscent of an extended repetitive unit of PG, namely the *t*etra*s*accharide-*p*enta*p*eptide NAM-NAG-NAM(-L-Ala-D-iGln-L-Lys-D-Ala-D-Ala)-NAG (termed TSPP). Simulating TSPP, which mimics the physiological substrate, onto $LytB_{CAT}$ may provide the details of substrate binding pattern.
3. Generate the atomic coordinates of TSPP by the GlycoBioChem PRODRG2 Server [19]. Click "Get started…" in the bottom of the PRODRG website to open the compound submission window. Firstly, submit your email address to the server to get a valid token before using (*see* **Note 13**).
4. Secondly, paste the obtained token and click "Draw the molecule with JME". JME is a molecular editor tool for structure input and editing. In the new opened window, draw the chemical formula of TSPP. However, do not close the original window. After finish drawing, click "transfer to PRODRG window", and the automatically generated coordinate data will be displayed in the compound submission window (*see* **Note 14**).
5. Finally, run PRODRG. In the result page, download the generated coordinate file in pdb format. Thus the coordinate file of TSPP is termed TSPP.pdb (*see* **Note 15**).
6. Open $LytB_{CAT}$.pdb using AutoDock Tools (ADT) 1.5.4 [20]. Then edit it to add polar hydrogen atoms and save as a PDBQT format in the Grid item. Select a grid box with dimensions of $40 \times 45 \times 50$ points around the active site to accommodate TSPP. Write down the number of points in *x,y,z* dimensions and numerical values of Center Grid Box in *x,y,z* (*see* **Note 16**).

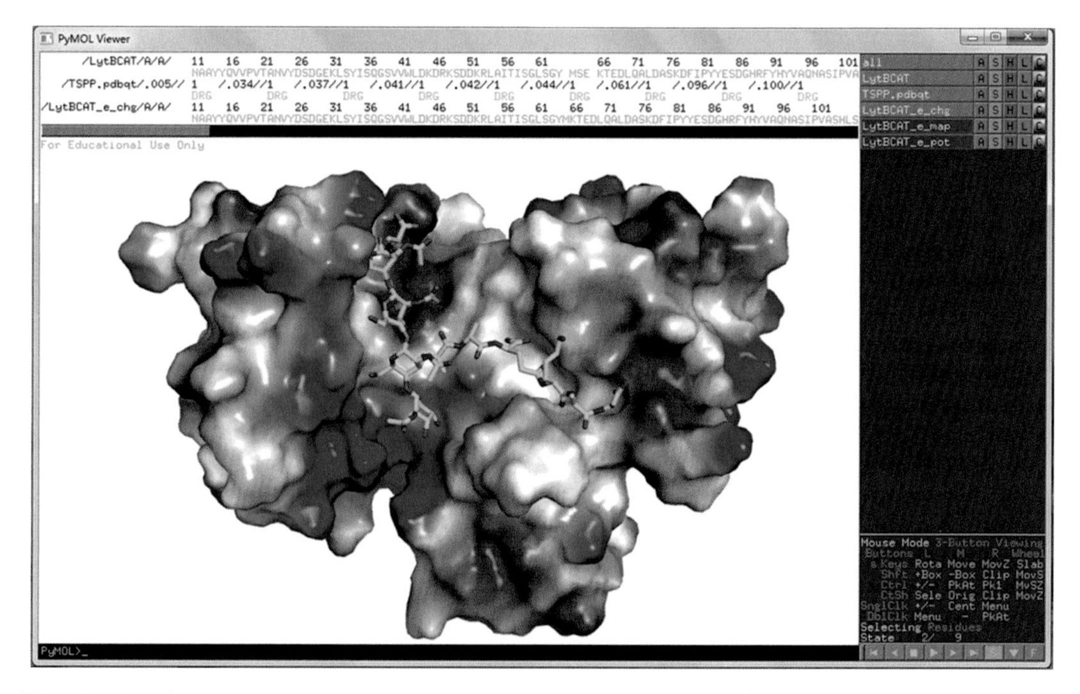

Fig. 2 The PyMOL interface that shows TSPP onto $LytB_{CAT}$

7. Delete the opened $LytB_{CAT}$.pdb and input the TSPP.pdb. In the torsion tree item, all single-bonds within the TSPP are set to allow rotation. Afterwards, convert the TSPP.pdb from a PDB format to a PDBQT format (*see* **Note 17**).
8. Build a new txt file that includes the names of receptor ($LytB_{CAT}$.pdbqt) and ligand (TSPP.pdbqt), the six parameters of the grid box and the exhaustiveness parameter (*see* **Note 18**).
9. Invoke the AutoDock Vina.exe and the above built txt file. Now the Vina will run to simulate the TSPP onto $LytB_{CAT}$ (*see* **Note 19**).
10. Open the output PDBQT file and the original $LytB_{CAT}$.pdb file using PyMOL. Analyze the given conformations and orientations of TSPP at the active site of $LytB_{CAT}$ one by one, and select the most rational one as the final model (Fig. 2). The simulated model showed that the tetrasaccharide moiety is accommodated in the groove of $LytB_{GH73}$, whereas the pentapeptide moiety stretches into the cleft between $LytB_{SH3b}$ and $LytB_{WW}$ (*see* **Note 20**).

3.4 Hydrolytic Activity Assays

1. Label the purified Δ*lytB* PG with RBB: Incubate Δ*lytB* PG with 20 mM RBB in 0.25 M NaOH at 37 °C overnight, and subsequently neutralize the reaction system with 0.25 M HCl. Then centrifuge the mixture at 21,000 ×*g* for 20 min at 20 °C

to collect the precipitate. Wash the RBB-labeled Δ*lytB* PG six times with ddH_2O to remove the free RBB, and then weigh it after lyophilizing (*see* **Note 21**).

2. Dissolve the lyophilized RBB-labeled Δ*lytB* PG and dilute different versions of protein ($LytB_{CAT}$, $LytB_{WW\text{-}GH73}$, $LytB_{GH73}$, $LytB^{E564Q}$) with the reaction buffer.
3. Mix 10 μM purified protein and 1 mg/mL RBB-labeled Δ*lytB* PG in a 150 μL system and react at 37 °C for 10 h (*see* **Note 22**).
4. Heat the mixture at 95 °C for 5 min to terminate the reaction.
5. Afterwards, centrifuge the mixture at 130,000 ×*g* for 20 min at 20 °C to remove the remaining insoluble PG that has not been hydrolyzed (*see* **Note 23**).
6. Apply the soluble RBB-labeled PG fragments, which are released to the supernatant upon hydrolysis, to a DU800 spectrophotometer to measure the optical density at 595 nm (*see* **Note 24**).
7. Perform each reaction for three times. Analyze the results using GraphPad software. The results further prove that Glu564 plays a crucial role in the hydrolysis and none of the three modules is dispensable for the activity of $LytB_{CAT}$.

4 Notes

1. All pdb files mentioned here are download from the RCSB protein data bank (http://www.rcsb.org/pdb/), unless otherwise specified.
2. The used version of CCP4 software needs to be compatible with the computer system. Otherwise, the running may fail.
3. The RBB powder should be dissolved in 0.25 M NaOH, but not in water, for NaOH supplies an alkaline buffer system for the labeling of PG.
4. The primary sequence analysis of LytB or the structural information of GH73 family can also be obtained in the Pfam Homepage (http://pfam.xfam.org/) by sequence search or key word search, respectively.
5. The pdb file used for superposition should contain only the residues of protein itself, but not other molecules, such as water molecules, glycerol molecules and so on.
6. The job title can be blank, and there is no need to change other default options. Better not to check the "combine superposed coordinates with fixed coordinates in output PDB file" option. If you check this option, the two superimposed structures will be combined in the output pdb file, which is not convenient for graphing.

7. This step can be performed simultaneously with **step 3** in Subheading 3.1.
8. Run the server once for only one structure. When the search has finished, you will receive an email notification. It is better to give each running a job title when doing more than one structural comparisons successively.
9. Many superimposed structures with different Z scores will be given, in which many are redundant. A higher Z score means a structure more similar to the input structure. Summarize the hits and consider the functional relationship with the input structure.
10. The output of the second all-β module with the Dali sever includes several functionally unrelated proteins with a Z-score of ≤ 2.5. It is hard to classify the second all-β module to any family of structure-known proteins.
11. The ASTRAL database has now been integrated into the new SCOPe website (http://scop.berkeley.edu/). Go to the new website to get the new versions of both SCOPe and ASTRAL.
12. There could be many structures in every class, so it is necessary to analyze them carefully and patiently. With regard to the "all beta proteins" class, consider the number of strands in each fold first.
13. Receive the valid token immediately or several minutes later. A valid token could be used for five PRODRG runs.
14. If the Java version of the browser is outdated, the JME window may display with error. The JME help is in the bottom of the website to help draw the molecule. Pay attention to the chirality of the molecule.
15. Download the PDB file in four formats: all H's, polar/aromatic H's, polar H's only and no H's, which differ from each other in the number of H atoms in the coordinate file. It is better to choose the all H's format.
16. The size of the grid box must cover the entire active site and allow the ligand to move freely.
17. Delete the atomic coordinates of the receptor before inputting the atomic coordinates of the ligand. Or reopen the AutoDock Tools software and then input the ligand.pdb. Choosing torsion depends on your request.
18. The exhaustiveness parameter sets the number of runs, telling the program how hard to search. It is an optional setting with a default value of 8.
19. When invoking, the $LytB_{CAT}$.pdbqt, the TSPP.pdbqt, and the built txt file must be saved in the same folder.

20. The AutoDock Vina may give a set of docked poses. The pose with the highest affinity may not be the most rational one. Compare different metrics, such as the interaction between receptor and ligand, free energy of binding, RMSD, van der Waals, and so on, in a general consideration when choosing the final simulated model.
21. Discard the supernatant carefully without touching the precipitate.
22. The concentration of protein and the RBB-labeled Δ*lytB* PG reminded here means the final concentration in the reaction mix. The volume of the reaction mix can be enlarged to 200 μL. Keep the protein and the RBB-labeled Δ*lytB* PG on ice before starting the reaction. It is better to add the RBB-labeled Δ*lytB* PG to the reaction system in prior of adding protein.
23. Avoid disturbing the precipitate when pipetting the supernatant.
24. Use the same volume reaction buffer without protein as the blank control.

Acknowledgements

This work was supported by the Ministry of Science and Technology of China (Grants No. 2013CB835300 and 2014CB910100), the National Natural Science Foundation of China (Grants No. 31270781 and U1332114), the Grand Challenges Exploration of the Bill and Melinda Gates Foundation (Grant No. OPP1021992), and the Fundamental Research Funds for the Central Universities.

References

1. Vollmer W, Blanot D, de Pedro MA (2008) Peptidoglycan structure and architecture. FEMS Microbiol Rev 32(2):149–167
2. Meroueh SO, Bencze KZ, Hesek D, Lee M, Fisher JF, Stemmler TL, Mobashery S (2006) Three-dimensional structure of the bacterial cell wall peptidoglycan. Proc Natl Acad Sci U S A 103(12):4404–4409
3. Neuhaus FC, Baddiley J (2003) A continuum of anionic charge: structures and functions of D-alanyl-teichoic acids in Gram-positive bacteria. Microbiol Mol Biol Rev 67(4):686–723
4. Vollmer W, Joris B, Charlier P, Foster S (2008) Bacterial peptidoglycan (murein) hydrolases. FEMS Microbiol Rev 32(2):259–286
5. Mosser JL, Tomasz A (1970) Choline-containing teichoic acid as a structural component of pneumococcal cell wall and its role in sensitivity to lysis by an autolytic enzyme. J Biol Chem 245(2):287–298
6. Garcia P, Paz Gonzalez M, Garcia E, Garcia JL, Lopez R (1999) The molecular characterization of the first autolytic lysozyme of *Streptococcus pneumoniae* reveals evolutionary mobile domains. Mol Microbiol 33(1): 128–138
7. Garcia P, Gonzalez MP, Garcia E, Lopez R, Garcia JL (1999) LytB, a novel pneumococcal murein hydrolase essential for cell separation. Mol Microbiol 31(4):1275–1281
8. De Las Rivas B, Garcia JL, Lopez R, Garcia P (2002) Purification and polar localization of pneumococcal LytB, a putative endo-N-acetylglucosaminidase: the chain-dispersing

murein hydrolase. J Bacteriol 184(18): 4988–5000

9. Bai XH, Chen HJ, Jiang YL, Wen Z, Huang Y, Cheng W, Li Q, Qi L, Zhang JR, Chen Y, Zhou CZ (2014) Structure of pneumococcal peptidoglycan hydrolase LytB reveals insights into the bacterial cell wall remodeling and pathogenesis. J Biol Chem 289(34): 23403–23416
10. Cantarel BL, Coutinho PM, Rancurel C, Bernard T, Lombard V, Henrissat B (2009) The Carbohydrate-Active EnZymes database (CAZy): an expert resource for glycogenomics. Nucleic Acids Res 37(Database issue): D233–D238
11. Bublitz M, Polle L, Holland C, Heinz DW, Nimtz M, Schubert WD (2009) Structural basis for autoinhibition and activation of Auto, a virulence-associated peptidoglycan hydrolase of *Listeria monocytogenes*. Mol Microbiol 71(6):1509–1522
12. Hashimoto W, Ochiai A, Momma K, Itoh T, Mikami B, Maruyama Y, Murata K (2009) Crystal structure of the glycosidase family 73 peptidoglycan hydrolase FlgJ. Biochem Biophys Res Commun 381(1):16–21
13. Krissinel E, Henrick K (2004) Secondary-structure matching (SSM), a new tool for fast protein structure alignment in three dimensions. Acta Crystallogr D Biol Crystallogr 60(Pt 12 Pt 1):2256–2268
14. Collaborative Computational Project N (1994) The CCP4 suite: programs for protein crystallography. Acta Crystallogr D Biol Crystallogr 50(Pt 5):760–763
15. Holm L, Rosenstrom P (2010) Dali server: conservation mapping in 3D. Nucleic Acids Res 38(Web Server issue):W545–W549.
16. van Aalten DM, Synstad B, Brurberg MB, Hough E, Riise BW, Eijsink VG, Wierenga RK (2000) Structure of a two-domain chitotriosidase from *Serratia marcescens* at 1.9-A resolution. Proc Natl Acad Sci U S A 97(11):5842–5847
17. Goodsell DS, Olson AJ (1990) Automated docking of substrates to proteins by simulated annealing. Proteins 8(3):195–202
18. Trott O, Olson AJ (2010) AutoDock Vina: improving the speed and accuracy of docking with a new scoring function, efficient optimization, and multithreading. J Comput Chem 31(2):455–461
19. Schuttelkopf AW, van Aalten DM (2004) PRODRG: a tool for high-throughput crystallography of protein-ligand complexes. Acta Crystallogr D Biol Crystallogr 60(Pt 8): 1355–1363
20. Morris GM, Huey R, Lindstrom W, Sanner MF, Belew RK, Goodsell DS, Olson AJ (2009) AutoDock4 and AutoDockTools4: automated docking with selective receptor flexibility. J Comput Chem 30(16):2785–2791
21. Morlot C, Uehara T, Marquis KA, Bernhardt TG, Rudner DZ (2010) A highly coordinated cell wall degradation machine governs spore morphogenesis in *Bacillus subtilis*. Genes Dev 24(4):411–422
22. Bricker AL, Camilli A (1999) Transformation of a type 4 encapsulated strain of *Streptococcus pneumoniae*. FEMS Microbiol Lett 172(2): 131–135
23. Xu Q, Sudek S, McMullan D, Miller MD, Geierstanger B, Jones DH, Krishna SS, Spraggon G, Bursalay B, Abdubek P, Acosta C, Ambing E, Astakhova T, Axelrod HL, Carlton D, Caruthers J, Chiu HJ, Clayton T, Deller MC, Duan L, Elias Y, Elsliger MA, Feuerhelm J, Grzechnik SK, Hale J, Han GW, Haugen J, Jaroszewski L, Jin KK, Klock HE, Knuth MW, Kozbial P, Kumar A, Marciano D, Morse AT, Nigoghossian E, Okach L, Oommachen S, Paulsen J, Reyes R, Rife CL, Trout CV, van den Bedem H, Weekes D, White A, Wolf G, Zubieta C, Hodgson KO, Wooley J, Deacon AM, Godzik A, Lesley SA, Wilson IA (2009) Structural basis of murein peptide specificity of a gamma-D-glutamyl-L-diamino acid endopeptidase. Structure 17(2):303–313
24. Lu JZ, Fujiwara T, Komatsuzawa H, Sugai M, Sakon J (2006) Cell wall-targeting domain of glycylglycine endopeptidase distinguishes among peptidoglycan cross-bridges. J Biol Chem 281(1):549–558

Index

Hee-Jeon Hong (ed.), *Bacterial Cell Wall Homeostasis: Methods and Protocols*, Methods in Molecular Biology, vol. 1440, DOI 10.1007/978-1-4939-3676-2, © Springer Science+Business Media New York 2016

Printed in the United States
By Bookmasters